Cell Junctions

Edited by
Susan E. LaFlamme and
Andrew Kowalczyk

Related Titles

Unsicker, K., Krieglstein, K. (Eds.)

Cell Signaling and Growth Factors in Development

From Molecules to Organogenesis

2006
ISBN: 978-3-527-31034-0

Meyers, R. A. (Ed.)

Encyclopedia of Molecular Cell Biology and Molecular Medicine

16 Volume Set

2005
ISBN: 978-3-527-30542-1

Kropshofer, H., Vogt, A. B. (Eds.)

Antigen Presenting Cells

From Mechanisms to Drug Development

2005
ISBN: 978-3-527-31108-8

Wedlich, D. (Ed.)

Cell Migration in Development and Disease

2005
ISBN: 978-3-527-30587-2

Yawata, Y.

Cell Membrane

The Red Blood Cell as a Model

2003
ISBN: 978-3-527-30463-9

von Bohlen and Halbach, O., Dermietzel, R.

Neurotransmitters and Neuromodulators

Handbook of Receptors and Biological Effects

2006
ISBN: 978-3-527-31307-5

Cell Junctions

Adhesion, Development, and Disease

Edited by
Susan E. LaFlamme and Andrew Kowalczyk

WILEY-VCH Verlag GmbH & Co. KGaA

The Editors

Prof. Dr. Susan E. LaFlamme
Albany Medical College
Center for Cell Biology and Cancer Research
47 New Scotland Avenue
Albany, NY 12208
USA

Prof. Dr. Andrew Kowalczyk
Emory University
Departments of Cell Biology
  and Dermatology
615 Michael Street
Atlanta, GA 30322
USA

Cover
Cell layers of the epidermis: The epidermis connects to the underlying dermis via a specialized basement membrane (orange layer). Desmosomes (red dots) and hemidesmosomes (blue dots) serve as anchorage points for cytokeratin intermediate filaments (green lines) to the plasma membrane. (Courtesy of Peter J. Koch and Ansgar Schmidt, see Chapter 12)

■ All books published by Wiley-VCH are carefully produced. Nevertheless, authors, editors, and publisher do not warrant the information contained in these books, including this book, to be free of errors. Readers are advised to keep in mind that statements, data, illustrations, procedural details or other items may inadvertently be inaccurate.

Library of Congress Card No.:
applied for

British Library Cataloguing-in-Publication Data
A catalogue record for this book is available from the British Library.

Bibliographic information published by the Deutsche Nationalbibliothek
Die Deutsche Nationalbibliothek lists this publication in the Deutsche Nationalbibliografie; detailed bibliographic data are available on the Internet at <http://dnb.d-nb.de>.

© 2008 WILEY-VCH Verlag GmbH & Co. KGaA, Weinheim

All rights reserved (including those of translation into other languages). No part of this book may be reproduced in any form – by photoprinting, microfilm, or any other means – nor transmitted or translated into a machine language without written permission from the publishers. Registered names, trademarks, etc. used in this book, even when not specifically marked as such, are not to be considered unprotected by law.

Printed in the Federal Republic of Germany
Printed on acid-free paper

Composition   SNP Best-set Typesetter Ltd., Hong Kong
Printing   Betz-Druck GmbH, Darmstadt
Bookbinding   Litges & Dopf GmbH, Heppenheim

ISBN: 978-3-527-31882-7

Preface

The evolutionary appearance of multicellular organisms is coincident with the emergence of molecular mechanisms that allow cells to contact their neighbors and to receive positional cues essential for the development of complex tissue architecture and function. Cells interact with one another and with the extracellular matrix (ECM) through a variety of mechanisms. Often, large macromolecular complexes assemble at sites of contact in order to perform specific functions, including adhesion, communication, and to form barriers between tissue compartments. These complex molecular assemblies – or "cell junctions" – are now thought of as cellular organelles that not only mediate physical interactions, but also couple cell contact to signaling pathways that influence cell shape, cell-cycle progression, and gene expression. The manifestation of these various functions becomes apparent in model organisms in which genes encoding cell junction proteins have been ablated, and in the growing list of human disorders that are now known to result from gene mutations in junction components. Indeed, the phenotypes and clinical presentations of these genetic and acquired disorders are impressive in both scope and variety.

The chapters of this book cover a wide range of cell–cell and cell–matrix interactions. The first section of the book focuses on cell–matrix interactions, including focal adhesions and hemidesmosomes. Embedded within these chapters are discussions of integrin activation and cytoskeletal interactions, as well as signaling mediated by integrins and their cytoplasmic binding partners. The importance of the physical three-dimensional structure of the ECM in regulating cell behavior is underscored. The second section of the book concentrates on cell–cell interactions, including anchoring junctions such as adherens junctions and desmosomes, which utilize cadherins as the major adhesive molecules. Tight junctions and gap junctions are also highlighted, with unique insights provided into the roles of these proteins in complex epithelia and disease pathogenesis. Several chapters in the volume converge on the regulation of cell junction dynamics, including the turnover of membrane proteins, the exchange of components into and out of junctions, and the interface between membrane-trafficking pathways and adhesion receptors. The book concludes with a chapter on how cell–cell and cell–matrix interactions are coordinated during complex morphogenic processes.

Cell Junctions. Adhesion, Development, and Disease.
Edited by Susan E. LaFlamme and Andrew Kowalczyk
Copyright © 2008 WILEY-VCH Verlag GmbH & Co. KGaA, Weinheim
ISBN: 978-3-527-31882-7

We hope that the assembly of these chapters – all in one bound volume and authored by some of the world's leading investigators – provides a resource that is valuable to the new student of cell junctions, to experienced scientists in the field, and to those educators who require a comprehensive assembly of overviews in the discipline of cell contact and adhesion. We are enormously grateful to the authors of each chapter for their outstanding contributions. It is our hope that this book provides not only a resource and compilation of existing knowledge and insight, but also a foundation for future studies designed to understand how cell contact and adhesion influence development and human disease.

Susan E. LaFlamme
Albany Medical College
Albany, NY
USA
January 2008

Andrew Kowalczyk
Emory University
Atlanta, GA
USA

Contents

Preface *V*
List of Contributors *XIII*

Part One Cell–Matrix Junctions

1   **The Ins and Outs of Integrin Signaling** *3*
           Asoka Banno and Mark H. Ginsberg
1.1   General Overview *3*
1.1.1   Integrin Receptors *3*
1.1.2   Functions *4*
1.2   Integrin Activation *5*
1.2.1   Definition *5*
1.2.1.1   Platelets *6*
1.2.1.2   Leukocytes *6*
1.2.2   Structural Basis of Activation *6*
1.2.2.1   Extracellular Rearrangements *7*
1.2.2.2   Transmembrane Propagation *7*
1.2.2.3   Intracellular Rearrangements *8*
1.2.2.4   Interactions at the Integrin Cytoplasmic Domains *9*
1.2.3   Regulation of Integrin Activation *13*
1.2.4   Future Research Directions *16*
           References *18*

2   **Integrin Signaling Through Focal Adhesion Kinase** *25*
           Youngdong Yoo and Jun-Lin Guan
2.1   Introduction *25*
2.2   Structure of FAK *26*
2.3   Regulation of FAK Activity *28*
2.3.1   Activation of FAK *28*
2.3.2   Inhibition of FAK Activity *30*
2.4   Regulation of Cellular Functions by FAK Signaling Pathways *30*
2.4.1   Cell Adhesion and Spreading *31*
2.4.2   Cell Migration and Invasion *32*

Cell Junctions. Adhesion, Development, and Disease.
Edited by Susan E. LaFlamme and Andrew Kowalczyk
Copyright © 2008 WILEY-VCH Verlag GmbH & Co. KGaA, Weinheim
ISBN: 978-3-527-31882-7

2.4.3	Cell Survival and Proliferation	36
2.5	Recent Analysis of FAK Functions *In Vivo*	37
2.6	Conclusions	39
	Acknowledgment	39
	References	39

3	**The Paxillin Family and Tissue Remodeling**	**47**
	David A. Tumbarello and Christopher E. Turner	
3.1	Focal Adhesions and the Paxillin Superfamily	47
3.1.1	Focal Adhesions	47
3.1.2	The Paxillin Superfamily	49
3.1.3	Structure of Paxillin	50
3.1.4	Structure of Hic-5	51
3.2	The Paxillin Superfamily and Tissue Remodeling	52
3.2.1	Epithelial–Mesenchymal Transformation	52
3.2.2	Integrin-Mediated Signaling and EMT	53
3.2.3	Paxillins α and δ in EMT	56
3.2.4	Hic-5 and EMT	58
	References	60

4	**Adhesion Dynamics in Motile Cells**	**71**
	Christa L. Cortesio, Keefe T. Chan, and Anna Huttenlocher	
4.1	Introduction	71
4.2	Focal Adhesion Dynamics	72
4.2.1	Focal Adhesion Composition	73
4.2.2	Mechanisms of Focal Adhesion Assembly	74
4.2.2.1	Rho GTPases	75
4.2.2.2	Talin and Phosphoinositides	75
4.2.3	Mechanisms of Focal Adhesion Disassembly	75
4.2.3.1	Calpain	76
4.2.3.2	Tyrosine Phosphorylation and Contractile Machinery	76
4.3	Podosome and Invadopodia Dynamics	76
4.3.1	Podosome and Invadopodia Architecture	77
4.3.2	Molecular Mechanisms of PTA Assembly	78
4.3.2.1	Actin Regulatory Proteins	78
4.3.2.2	Rho GTPases	79
4.3.2.3	Signaling Molecules	79
4.3.3	Molecular Mechanisms of PTA Disassembly	80
4.4	Summary	80
	Acknowledgments	81
	References	81

5	**Integrin Trafficking**	**89**
	Susan E. LaFlamme, Feng Shi, and Jane Sottile	
5.1	Introduction	89

5.2	Historical Perspective	*89*
5.3	Clathrin versus Lipid Rafts	*90*
5.4	Internalization of Occupied versus Unoccupied Integrins	*91*
5.5	Regulation of Integrin Trafficking	*92*
5.6	The Role of Integrin Cytoplasmic Domains	*94*
5.7	Integrin-Dependent Endocytosis of Microbial Pathogens	*95*
5.8	Regulation of Cell Adhesion and Migration by Integrin Trafficking	*96*
5.9	The Regulation of Integrin Trafficking by Cell Adhesion	*96*
5.10	Integrin Trafficking and ECM Remodeling	*98*
5.11	Conclusions	*100*
	Acknowledgments	*101*
	References	*101*

6 Hemidesmosomes and their Components: Adhesion versus Signaling in Health and Disease *109*

Kristina Kligys, Kevin Hamill, and Jonathan C. R. Jones

6.1	Introduction	*109*
6.2	The Structural Components of the Hemidesmosome	*110*
6.2.1	The Plaque Proteins	*110*
6.2.1.1	Plectin	*110*
6.2.1.2	BP230	*112*
6.2.2	Membrane Elements	*112*
6.2.2.1	α6β4 Integrin	*112*
6.2.2.2	BP180	*114*
6.2.2.3	CD151	*115*
6.3	Laminin-332	*115*
6.4	Hemidesmosome Assembly and Disassembly	*116*
6.5	Hemidesmosomes and Disease	*118*
6.5.1	Inherited Skin Diseases Involving Hemidesmosome Proteins	*118*
6.5.1.1	Disease Involving α6β4 Integrin	*118*
6.5.1.2	Disease Involving Plectin	*119*
6.5.1.3	Disease Involving BP180	*119*
6.5.1.4	Disease Involving CD151	*120*
6.5.1.5	Disease Involving Laminin-332	*120*
6.5.2	Autoimmmune Diseases Associated with Hemidesmosomes	*121*
6.5.2.1	BP and Related Diseases	*121*
6.5.2.2	Cicatricial Pemphigoid	*122*
6.5.2.3	Autoimmune Diseases Involving α6β4 Integrin	*122*
6.6	α6β4 Integrin and Laminin-332 Expression in Cancer	*123*
6.7	Signaling by the α6β4 Integrin in Pathological States and Wound Healing	*123*
6.8	Laminin-332, α6β4 Integrin and Migration	*125*
6.9	Conclusions	*126*
	References	*126*

7	**Cell Matrix Adhesion in Three Dimensions** *135*
	Patricia J. Keely
7.1	Introduction *135*
7.2	Model 3D Matrices for Investigations of Cell Behavior *136*
7.2.1	Matrigel/Reconstituted Basement Membrane (rBM) *136*
7.2.2	Collagen Gels *136*
7.2.3	Fibrin Gels *138*
7.2.4	Cell-Derived Fibronectin-Based Matrices *138*
7.2.5	Engineered 3D Matrices and Microfluidic Chambers *138*
7.3	Cell Behavior and Adhesive Structures Differ between 3D and 2D Matrices *139*
7.4	"Compliancy" or Elastic Modulus Determines Cellular Response to ECMs *140*
7.4.1	Cellular Contractility and the Response to Matrix Compliance *141*
7.4.2	Rho Regulates the Cytoskeleton and Contractility *141*
7.4.3	Rho/ROCK, Contractility, and Focal Adhesion Formation *142*
7.4.4	Focal Adhesions as Transducers of Biophysical Signals *143*
7.5	A Model for Cellular Response to 3D Matrices Varying in Compliance *144*
	Acknowledgments *146*
	References *146*

Part Two   Cell–Cell Junctions

8	**Armadillo Repeat Proteins at Epithelial Adherens Junctions** *153*
	Laura J. Lewis-Tuffin and Panos Z. Anastasiadis
8.1	Introduction *153*
8.2	β-Catenin *156*
8.3	Plakoglobin *158*
8.4	p120-Catenin *159*
8.5	The Role of Armadillo Repeat Proteins and AJs in Cancer *163*
8.6	Conclusions *165*
	References *165*

9	**Signaling To and Through The Endothelial Adherens Junction** *169*
	Deana M. Ferreri and Peter A. Vincent
9.1	Introduction *169*
9.2	Cadherins *170*
9.2.1	VE-Cadherin Extracellular Domain *173*
9.2.2	Interaction of Cadherin Cytoplasmic Domain and Catenins *174*
9.2.2.1	VE-Cadherin Juxtamembrane Region and p120 *174*
9.2.2.2	VE-Cadherin Catenin Binding Domain *176*
9.3	Phosphorylation and Junction Assembly/Disassembly *179*
9.3.1	Tyrosine Phosphorylation of Endothelial AJ Proteins *180*
9.3.2	Serine/Threonine Phosphorylation of Endothelial AJ Proteins *182*
9.4	Small GTPases and Junction Assembly *183*

9.4.1	Rho GTPases *183*
9.4.2	Rap1 GTPase *186*
9.5	Conclusions *187*
	References *187*

10 Gap Junctions: Connexin Functions and Roles in Human Disease *197*
Michael Koval

10.1	Introduction *197*
10.2	Connexin Structure and Assembly *198*
10.3	Interactions Between Different Connexins *199*
10.4	Connexin Binding Proteins and Phosphorylation *201*
10.5	Channel Permeability and Signaling *202*
10.6	Connexins in Human Disease *203*
10.7	Role of Connexins in Vessel Inflammation and Atherosclerosis *206*
10.8	Conclusions *207*
	References *208*

11 Tight Junctions in Simple and Stratified Epithelium *217*
Cara J. Gottardi and Carien M. Niessen

11.1	Introduction *217*
11.2	Ultrastructure of Tight Junctions *218*
11.3	Tight Junctions and Epithelial Barrier Function *219*
11.3.1	Transmembrane Components *219*
11.3.2	Scaffolding Proteins *221*
11.3.3	Actin and Related Cytoskeleton Proteins *221*
11.4	Regulation of Tight Junction Barrier Function *222*
11.5	Tight Junctions and Epithelial Polarity *222*
11.5.1	Par3/Par6/aPKC Complex *223*
11.5.2	Crbs/Patj/Pals *223*
11.5.3	Fence Function *224*
11.6	Signaling from Tight Junctions: Coupling Junction Maturation to Transcription-Mediated Differentiation *225*
11.7	Tight Junctions in Stratifying Epithelia *225*
11.8	Tight Junctions and Disease *227*
11.9	Concluding Remarks *228*
	Acknowledgments *229*
	References *229*

12 Desmosomes in Development and Disease *235*
Ansgar Schmidt and Peter J. Koch

12.1	Introduction *235*
12.2	Molecular Composition of Desmosomes, Disease Associations, and Animal Models *236*
12.2.1	Desmosomal Cadherins *236*
12.2.2	Desmoplakin *240*
12.3	Plakoglobin *240*

12.4	Plakophilins *242*
12.5	Accessory Desmosomal Proteins *243*
12.6	Desmosomal Proteins in Embryonic Development *243*
12.7	Concluding Remarks *244*
	Acknowledgments *245*
	References *245*

13 Cadherin Trafficking and Junction Dynamics *251*
Christine M. Chiasson and Andrew P. Kowalczyk

13.1	Introduction *251*
13.2	Exocytosis and Polarized Sorting of Adherens Junction Proteins *252*
13.3	Endocytosis of Adherens Junction Proteins *255*
13.4	Catenin Regulation of Cadherin Endocytosis *258*
13.5	Rho GTPase Regulation of Cadherin Endocytosis *261*
13.6	Co-Regulation of Cadherin and Receptor Tyrosine Kinase Function by Endocytosis *262*
13.7	Conclusions *264*
	Acknowledgments *266*
	References *266*

Part Three   Cell–Matrix and Cell–Cell Crosstalk

14 Crosstalk Between Cell–Cell and Cell–Matrix Adhesion *273*
C. Michael DiPersio

14.1	Introduction *273*
14.1.1	Coordinate Regulation of Cell–Cell and Cell–Matrix Adhesion *273*
14.1.2	Crosstalk Between Integrins and Cadherins *274*
14.2	Mechanisms of Integrin–Cadherin Crosstalk *275*
14.2.1	Extracellular Proteolysis *275*
14.2.2	Cross-Regulation of Gene Expression *277*
14.2.3	Changes in Intracellular Force Generation *278*
14.2.4	Integrin–Cadherin Associations at Sites of Cell Adhesion *279*
14.2.5	Intracellular Signal Transduction Pathways *281*
14.2.5.1	FAK: An Integrin Effector that Signals to Cadherins *282*
14.2.5.2	Fer: A Tyrosine Kinase that Translocates between Cadherins and Integrins *282*
14.2.5.3	Rap1: A Central Regulator of Integrin–Cadherin Crosstalk Pathways? *283*
14.2.5.4	The Epidermis as a Model for Investigating Cadherin–Integrin Crosstalk *285*
14.3	Future Prospects *286*
	Acknowledgments *287*
	References *287*

Index *295*

List of Contributors

Panos Z. Anastasiadis
Mayo Clinic
Department of Cancer Biology
4500 San Pablo Road
Jacksonville, FL 32224
USA

Asoka Banno
University of California, San Diego
Department of Medicine
9500 Gilman Drive, MC 0726
San Diego, CA 92093-0726
USA

Keefe T. Chan
University of Wisconsin-Madison
Department of Molecular and Cellular Pharmacology
1550 Linden Drive
Madison, WI 53706
USA

Christine M. Chiasson
Emory University School of Medicine
Departments of Cell Biology and Dermatology
615 Michael Street
Atlanta
GA 30322
USA

Christa L. Cortesio
University of Wisconsin-Madison
Department of Biomolecular Chemistry
1550 Linden Drive
Madison, WI 53706
USA

C. Michael DiPersio
Albany Medical College
Center for Cell Biology and Cancer Research
47 New Scotland Avenue
Albany, NY 12208
USA

Deana M. Ferreri
Albany Medical College
Center for Cardiovascular Sciences
47 New Scotland Avenue
Albany, NY 12208
USA

Mark H. Ginsberg
University of California, San Diego
Department of Medicine
9500 Gilman Drive
San Diego, CA 92093-0726
USA

Cara J. Gottardi
Northwestern University
Feinberg School of Medicine
Department of Medicine
Division of Pulmonary and
  Critical Care
240 East Huron St.
Chicago, IL 60611
USA

Jun-Lin Guan
University of Michigan Medical
  School
Department of Internal Medicine
Division of Molecular Medicine
  and Genetics
and
Department of Cell and
  Developmental Biology
Ann Arbor, MI 48109-2200
USA

Kevin Hamill
Northwestern University
Feinberg School of Medicine
Department of Cell and Molecular
  Biology
303 E. Chicago Avenue
Chicago, IL 60611
USA

Anna Huttenlocher
University of Wisconsin-Medison
1550 Linden Drive
Madison, WI 53706
USA

Jonathan C.R. Jones
Northwestern University
Feinberg School of Medicine
Department of Cell and Molecular
  Biology
303 E. Chicago Avenue
Chicago, IL 60611
USA

Patricia J. Keely
University of Wisconsin
Department of Pharmacology
Paul C. Carbone Comprehensive
  Cancer Center
1300 University Ave
Madison, WI 53704
USA

Kristina Kligys
Northwestern University
Feinberg School of Medicine
Department of Cell and Molecular
  Biology
303 E. Chicago Avenue
Chicago, IL 60611
USA

Peter J. Koch
University of Colorado at Denver and
  Health Science Center
Department of Dermatology
12800 East 19th Avenue
Aurora, CO 80045
USA

Michael Koval
Emory University School of Medicine
Department of Medicine
Division of Pulmonary, Allergy and
  Critical Care Medicine
Atlanta, GA 30322
USA

Andrew P. Kowalczyk
Emory University School of Medicine
Departments of Cell Biology and
  Dermatology
615 Michael Street
Atlanta, GA 30322
USA

List of Contributors

Susan E. LaFlamme
Albany Medical College
Center for Cell Biology and
   Cancer Research
47 New Scotland Avenue
Albany, NY 12208
USA

Laura J. Lewis-Tuffin
Mayo Clinic
Department of Cancer Biology
4500 San Pablo Road
Jacksonville, FL 32224
USA

Carien M. Niessen
University of Cologne
Center for Molecular Medicine
Joseph Stelzmannstrasse 9
50931 Köln
Germany

Ansgar Schmidt
Philipps-University of Marburg
   Medical School
Institute of Pathology
Baldingerstrasse
35033 Marburg
Germany

Feng Shi
University of Rochester School of
   Medicine and Dentistry
Aab Cardiovascular Research
   Institute
221 Bailey Road
West Henrietta, NY 14586
USA

Jane Sottile
University of Rochester School of
   Medicine and Dentistry
Aab Cardiovascular Research Institute
221 Bailey Road
West Henrietta, NY 14586
USA

David A. Tumbarello
Cambridge Research Institute
Li Ka Shing Centre
Robinson Way
Cambridge
CB2 0RE
UK

Christopher E. Turner
SUNY Upstate Medical University
Department of Cell and Developmental
   Biology
750 East Adams St.
Syracuse, NY 13210
USA

Peter A. Vincent
Albany Medical College
Center for Cardiovascular Sciences
47 New Scotland Avenue
Albany, NY 12208
USA

Youngdong Yoo
University of Michigan Medical School
Department of Internal Medicine
Division of Molecular Medicine and
   Genetics
and
Department of Cell and Developmental
   Biology
Ann Arbor, MI 48109-2200
USA

Part One   Cell–Matrix Junctions

1
The Ins and Outs of Integrin Signaling

Asoka Banno and Mark H. Ginsberg

1.1
General Overview

1.1.1
Integrin Receptors

Integrins are adhesion receptors found in the metazoa, from the simplest sponges and cnidaria to the most complex mammals [1]. They are glycosylated heterodimers composed of non-covalently associated type I transmembrane α and β subunits [2–4]. Each integrin subunit contains a large extracellular domain (N-terminus of >700 residues), a single transmembrane domain (>20 residues), and a generally short cytoplasmic domain (C-terminus of 13–70 residues). Integrin transmembrane domains are thought to begin after an extracellular proline residue, whereas the boundary between the transmembrane and the cytoplasmic domains is less clear [5]. Membrane-proximal regions of the cytoplasmic tails are well conserved in most human integrin α and β subunits, and the boundary between the transmembrane and cytoplasmic domains is predominantly assumed to lie between the conserved W/Y and K/R residues [6, 7]. This conserved K/R residue is usually followed by a stretch of four to six apolar residues, resulting in the K/R residue being flanked by hydrophobic regions [6, 7]. The conserved sequences following K/R residues are GFFKR in α subunits and LLxxxHDRRE in β subunits [6, 7]. These apolar residues are in turn followed by a strongly polar sequence.

In humans, 18 α and eight β subunits have been identified, and these can form at least 24 different heterodimers. In general, each of the 24 integrins has a distinct, non-redundant function, binding to a specific repertoire of cell surface, extracellular matrix (ECM), and soluble protein ligands [1]. The specificity of integrin function is mainly determined by the combination of α and β subunits in each integrin, and by the cellular repertoire of the integrin expression and the functional state of those integrins, in combination with the availability of specific integrin ligands [1, 4]. In addition, the same integrin molecule can have differing ligand specificity, depending on the cell type on which they are found [8]. Furthermore, alternative splicing of extracellular and cytoplasmic domains adds to the potential complexity of integrin structure and function [2, 4, 9].

Since the recognition of the integrin family and the introduction of the term "integrin" about 20 years ago, they have been studied extensively and have become relatively well understood adhesion receptors [1, 10]. Playing central roles in cell migration and cell–ECM adhesions and controlling cell differentiation, proliferation, and apoptosis, integrins are an essential contributor in development, immune responses, leukocyte trafficking, hemostasis, and numerous human diseases such as cancer and autoimmune diseases [1, 2].

1.1.2
Functions

Cell attachment and responses to the ECM are an important requirement for the development of a multicellular organism. Integrins serve as transmembrane mechanical links between the ECM outside of a cell and the cytoskeleton inside of the cell [1, 11]. In this way, they connect ECM ligands (e.g., fibronectin, vitronectin, collagen, laminin) to actin microfilaments [1, 8, 10, 12, 13] or other cytoskeletons. Integrin–ECM interaction leads to the occupancy and clustering of integrins, which in turn promotes the recruitment of cytoskeletal and cytoplasmic proteins such as talin, paxillin, and α-actinin to form focal complexes and focal adhesions [4]. Focal complexes and focal adhesions are dynamic in that they are continually assembled, disassembled, and translocated during cell proliferation, spreading, polarization, and migration [14]. Within adhesion complexes, the cytoplasmic domains of the clustered integrins recruit cytoskeletal proteins and signaling molecules into a close proximity at high concentrations, enabling integrins to initiate intracellular signaling cascades. These signaling events enable integrins to regulate a variety of cellular behaviors [15, 16]. This process is referred to as "outside-in" integrin signaling, and results in the activation of protein tyrosine kinases such as focal adhesion kinase (FAK), Src-family kinases, and Abl, and a serine-threonine kinase such as AKT (PKB) [16, 17]. As an integral part of plasma membrane, connecting the cytoskeleton to the ECM, integrins play a role in mechanotransduction and mediate the transmission of mechanical stress across the plasma membrane [18, 19]. They are also capable of transducing physical forces to chemical signals into the cell, with a help of other signaling molecules that colocalize in focal adhesions [18, 20]. Changes in the balance of forces across integrins and resulting alternations in cell shape induce cells to switch between growth, differentiation, and apoptosis [19].

Many integrins are expressed and remain in the low-affinity binding state until cellular stimulation transforms them into a high-affinity form. Cells can use this transformation to modify their adhesive properties by varying the specificity and affinity of a given integrin [9, 15]. This regulation of cell adhesion by signaling from within the cell is referred to as "inside-out" signaling [1, 5, 9]. Rapid changes in integrin-mediated adhesion are often precisely regulated in time and space in biological settings such as platelet aggregation and leukocyte transmigration [9,

21]. Although "inside-out" signaling can be achieved via the modulation of integrin affinity for the ligand, or via affinity-independent mechanisms such as changes in integrin diffusion and integrin clustering, a major focus of integrin research over the past decade has been on the affinity-dependent regulation of integrin-mediated adhesion, a process that will be operationally defined here as integrin activation [1, 5, 9, 15]. Efforts at deciphering the mechanism of this form of "inside-out" signaling have produced compelling evidence that the integrin cytoplasmic domains are the targets of intracellular signals that modulate ligand binding affinity [21, 22].

As noted above, integrins are involved and play essential roles in many biological processes, and many excellent reviews have discussed how signals from these receptors regulate cellular behaviors [4, 11, 12, 14, 16, 20, 23]. Here, attention will be focused on how cells can change integrin affinity in response to developmental events or to changes in their environment, a process that is widely used in biological functions such as cell adhesion, cell migration, cellular aggregation, and leukocyte transmigration during inflammation.

1.2 Integrin Activation

1.2.1 Definition

Under physiological conditions, many integrins are in equilibrium between low- and high-affinity states [1]. The rapid process of shifting from the low-affinity state to the high-affinity state of integrins will be operationally defined here as "integrin activation". Activation – that is, the process of becoming active – is a term that requires definition depending on context. For example, in the case of many signaling receptors, "activation" refers to the occupancy of that receptor by an extracellular ligand with transmission of signaling events into the cell. Since integrins can also perform this function ("outside-in" signaling), the term "integrin activation" is sometimes encountered when referring to intracellular signals that result from integrin occupancy [6, 24]. Therefore, in order to avoid any confusion, the clear definition of the term "activation" with respect to the meaning of what is the "active" state of the integrin is recommended. Here, the term "activation" is used when referring to transition to the high-affinity state, an event that can be accomplished by the propagation of conformational changes from the integrin cytoplasmic domains to the extracellular domains [1, 5, 25, 26].

Activation and inactivation of integrins are tightly regulated. Although not all integrins have been shown to undergo extremes of activity, it is generally believed that most integrins shift between active and inactive states in a localized fashion when it is important for cells to regulate their adhesion in a temporal and spatial manner [1, 27]. The importance of integrin activation is well demonstrated in the functions of platelets and leukocytes [1, 9].

1.2.1.1 Platelets

αIIbβ3 integrins, also known as GPIIb-IIIa, are present at high density on resting circulating platelets but remain inactive under normal conditions [1, 9]. Upon platelet stimulation by agonists such as thrombin, ADP, or epinephrine acting through G protein-coupled receptors, or by von Willebrand factor signaling through its receptor (GPIb/V/IX), or by collagen binding to its receptor, GPVI, signals from within the cell activate αIIbβ3 integrin to bind to ligands such as fibrinogen, von Willebrand factor, and fibronectin [1, 9]. Binding of these multivalent ligands to activated αIIbβ3 leads to platelet aggregation. It is crucial that αIIbβ3 is inactive on resting circulating platelets, for if it were constitutively active or if its activation were to become deregulated, then unregulated platelet aggregation could lead to thrombosis [1, 9]. In contrast, defects in or a lack of αIIbβ3 integrins result in defects in hemostasis, as seen in a bleeding disorder known as Glanzmann thrombasthenia [1].

1.2.1.2 Leukocytes

β2 integrins, also known as CD11/18, on leukocytes are regulated in a very similar fashion to αIIbβ3 integrins on platelets [1, 9]. They are expressed on most white blood cells in their resting state. When the cells encounter agonists such as chemokines, the β2 integrins are rapidly activated to mediate firm adhesion to the integrins' ligands. Their ligands include counter-receptors such as Ig superfamily molecules (e.g., ICAM-1, 2, and 3), ECM proteins (e.g., fibronectin), blood-clotting proteins such as fibrinogen, and the complement pathway product, C3bi [1, 28, 29]. Intracellular adhesion molecules (ICAMs), which are major ligands of β2 integrins, are expressed on cells to allow the attachment of leukocytes to the cells. The β2 integrin–ICAM interaction mediates processes such as phagocytosis, cytotoxic killing, and efficient antigen presentation [1, 28]. Further, like αIIbβ3 integrins on platelets, it is critical that β2 integrins remain inactive on resting leukocytes, for deregulated activation or impaired deactivation of β2 integrins could cause failure in normal immune responses [30]. Similarly, a lack of or dysfunction in rapid activation of β2 integrins leads to the defective immune function seen in patients with leukocyte adhesion deficiency (LAD) who suffer from leukocytosis and the failure to recruit leukocytes to sites of infection [1, 9].

1.2.2
Structural Basis of Activation

As noted above, integrin activation is usually due primarily to conformational changes in the extracellular domains of the integrins. Because these conformational changes are believed to initiate at the cytoplasmic face of the integrin, they must somehow traverse the plasma membrane. Several key advances have provided important new insights into each step in the activation process: namely, the nature of the conformational change in the extracellular domain; the mechanism of transmission across the membrane; and how a specific protein – talin – interacting with the β cytoplasmic domain causes these long-range allosteric rearrangements.

1.2.2.1 Extracellular Rearrangements

Based on studies using crystallography, nuclear magnetic resonance (NMR), electron microscopy, and Förster resonance energy transfer (FRET), integrins appear to assume at least three conformations: (i) a bent "closed" conformation; (ii) an intermediate extended conformation with a closed headpiece; and (iii) an extended "open" conformation [3, 25, 31, 32]. Several studies have suggested that the bent conformation is a low-affinity state, whereas the extended conformations are associated with the high-affinity state of integrins [31–37]. However, the bent form can also bind ligand with high affinity [38] in some circumstances, and has led research workers to propose an alternative "deadbolt" model for "inside-out" activation, emphasizing that extended conformation is not necessary for integrins to bind to their physiological ligands [38, 39]. Likewise, the results of one study suggest that the extension of integrins may be regulated by different signaling pathways in a temporally specific manner rather than being a fundamental requirement for integrin activation [40]. Thus, it is very likely that integrins are in dynamic equilibrium among different conformational states [24, 32, 35, 41], and the global rearrangements in integrin conformation accompanying "inside-out" signaling remain incompletely understood [35, 42, 43].

In addition to the modulation of integrin affinity via conformational change within a single receptor molecule (affinity modulation), increased integrin-mediated adhesion can also be caused by receptor clustering on the cell surface (avidity modulation) [44–47]. Furthermore, in most circumstances it is likely that some combination of conformational change and receptor clustering is involved in the regulation of integrin-mediated adhesion. In one perceptive report, a monovalent antibody and a conditional dimerizer were used to isolate the relative contributions of affinity modulation and receptor clustering in $\alpha IIb\beta 3$-mediated functions [48]. In particular, these studies showed that affinity modulation and avidity modulation play complementary roles in the adhesive functions of this integrin.

1.2.2.2 Transmembrane Propagation

Integrin activation by "inside-out" signaling must involve the transmission of conformational rearrangements from the cytoplasmic domains to the extracellular domains; therefore it would be expected that the transmembrane domain of integrins is involved in the signal transduction process. These regions are highly conserved amongst each of the integrin α- and β-subunit families, and are also conserved between species [49]; indeed, mutations in this region can lead to a loss of integrin expression [50–52]. In fact, recent studies have begun to provide insight into how rearrangements within the transmembrane domain can lead to integrin activation.

Mutational analysis and molecular, computational modeling together suggested that a helical interface between the integrin α and β subunit transmembrane domains stabilizes the inactive state, and also suggest that disruption of this helical transmembrane interface leads to activation [50, 53–55]. To date, four models have been proposed to explain the mechanism underlying the proposed rearrangements in transmembrane domain and the sequential disruption of the α-β transmembrane interface: separation, pistoning, twisting, and hinging [1, 6, 7]. These all involve some changes in the orientation of the subunits relative to one another and to the membrane. Although recent findings support the separation and pistoning models, high-resolution structures of the transmembrane domains will be required to distinguish clearly between the different activation models [5, 6, 31].

The helical packing of integrin transmembrane regions is likely to depend on specific crossing angles and specific in-register side-chain arrays [55]. Furthermore, the insertion of the integrin membrane-proximal domain can vary, either shortening or lengthening the transmembrane domain and changing the number of residues buried within the lipid bilayer [3, 5, 49, 55]. Based on these data, the pistoning model suggests that intracellular activating signals could shorten the transmembrane helix, in response to which its membrane tilt angle and register with the neighboring helix change in order to avoid hydrophobic mismatch with the fixed width of the membrane bilayer. This helical mismatch may be the critical event in disruption of transmembrane interactions that stabilize the low-affinity conformation, leading to integrin activation [3, 5, 55]. This model is supported by a report showing that the majority of mutations identified by random mutagenesis throughout the transmembrane domain activated integrins most likely by disrupting or shortening the transmembrane helix [3, 55]. Changes in the length or orientation of the integrin transmembrane domain may also take place during physiological integrin activation [5].

1.2.2.3 Intracellular Rearrangements

Changes in interactions and/or in the structures of the cytoplasmic domains of integrins within the highly conserved regions play crucial roles in integrin activation via "inside-out" signaling [5–7, 56]. Indeed, there is direct experimental evidence for a change in the relationship of the integrin α and β cytoplasmic domains during integrin activation [25].

The interaction between the membrane-proximal regions of the α and β subunits is believed to occur in part through a salt bridge between a conserved Arg in the α tail and an Asp in the β tail and the hydrophobic residues immediately N-terminal to the Arg and Asp [21, 56, 57]. Indeed, this specific association between the α and β subunits is thought to prevent integrin activation by stabilizing the low-affinity state [2, 5, 26]. Mutations that disrupt this "clasp" lead to integrin activation [21, 22, 26, 56]. Integrins can be constitutively activated by deletion in the entire α subunit cytoplasmic tail or of the membrane-proximal GFFKR sequence [21, 22, 26, 56]. Deletion or certain point mutations in the membrane-proximal region of the β tail also result in integrin activation [26, 56, 58]. Furthermore, replacement of the cytoplasmic–transmembrane regions by heterodimeric coiled-coil peptides or an artificial linkage of the tails inactivates the receptor, and breakage of the coiled-coil or clasp activates integrins [58–60]. Taken together, these data strongly indicate that this hydrophobic and electrostatic interaction plays an essential role as a stabilizer in the association of the α and β membrane-proximal regions, maintaining the integrins in a low-affinity state [3, 5, 6]. The important role of α and β tail interaction is further supported by a study using FRET. In the resting states, although fluorophore-tagged α and β tails were sufficiently close together to undergo FRET, stimulation with agonists or the introduction of activating mutations to membrane-proximal α subunits led to a reduction in FRET [31]. Such reduction was ascribed to separation of the cytoplasmic domains, but could also relate to an alternation in the orientation of α and β cytoplasmic tails relative to each other, without actual separation [3, 31]. In consequence, an important goal for future studies will be to provide a more precise description of these structural rearrangements.

Deletions of mutations in the β tail that are more C-terminal (i.e., in the membrane-distal region) can block integrin activation. In particular, an NPxY/F motif of the β subunit has been identified as one of the critical sites [6, 21, 56, 61]. Tyr/Phe-to-Ala mutations in this conserved motif block integrin activation [6, 62]. It is now known (*vide infra*) that mutations in the NPxY motif perturb the binding of integrin β tails to numerous cytoskeletal and signaling proteins [17], thus accounting for their profound effects on integrin signaling. Hence, membrane-distal portions of the β tail may control integrin activation through interactions with regulatory cytoplasmic proteins and through effects on the conformation of membrane-proximal regions. The physiological integrin activation process also requires the membrane distal region of the β3 cytoplasmic domain [63].

In summary, a current model is that release of the structural constraint upon the cytoplasmic rearrangement initiates a conformational change that propagates across the membrane through rearrangements in the α and β transmembrane domains. This in turn leads to changes in the conformation of the extracellular domains that activate integrins [32, 59]. As noted above, intriguing questions remain at each step of this process, thereby explaining the continued intense research efforts to further clarify this unusual form of signal transduction.

1.2.2.4 Interactions at the Integrin Cytoplasmic Domains

The integrin cytoplasmic domains play a central role in integrin activation [21, 22], and overexpression of certain proteins that bind to the cytoplasmic tails can result in integrin activation [21, 22, 58, 59]. The binding of the adaptor, talin, to the β cytoplasmic domain has been identified as a crucial, final step in the activation of several classes of integrins [64–66]. This finding has been confirmed for β1 [67], β2 [31, 68, 69] and β3 [70] integrins; hence, talin appears to play a general role in activating multiple classes of integrins. Moreover, a study using mice harboring point mutations in the β3 cytoplasmic tail provided the first *in-vivo* evidence supporting the importance of talin binding for integrin activation in mammals [71].

A variety of other cellular proteins are reported to interact directly with integrin cytoplasmic tails, and may have roles in integrin activation. β3-endonexin binds specifically to β3 tails through membrane-proximal and -distal motifs, and can activate αIIbβ3 in Chinese hamster ovary (CHO) cells [6, 17]. However, since β3-endonexin-mediated activation depends on the presence of talin, its role may be to cooperate with talin in αIIbβ3 integrin activation in platelets [6, 64]. Cytohesins-1 and -3 bind β2 integrin tails, and their overexpression has been shown to increase adhesion [6, 17]. However, these proteins contain a SEC7 domain and act as guanine nucleotide exchangers for Arf GTPases. ARF6 can regulate Rac1 activity [72–75], and could therefore increase cell adhesion through affinity-independent processes, such as integrin clustering, rather than integrin activation [6]. Calcium- and integrin-binding protein (CIB) has been proposed to activate αIIbβ3 by binding to the αIIb tail [6, 17]. However, some argue that CIB is involved in post-receptor occupancy events rather than initial activation [6, 17] and a recent, elegant study showed that it blocks integrin activation by inhibiting talin binding [76]. Another α subunit-binding protein, paxillin, plays a key role in cell motility and focal adhesion turnover [77]. In T cells, the α4-paxillin interaction provides resistance to rupture by external shear forces, without changing the integrin affinity of talin [77]. Similarly, while not shown to bind directly to integrin tails, RapL/NORE1B has been reported to associate with αLβ2 and lead to its activation, clustering, and redistribution [78, 79]. Thus, the list of potential integrin activation regulators is long and will keep expanding. It will be of great interest to examine the activators to see if they are talin-independent, function as modulators of talin binding to the integrins, or act cooperatively with talin.

Talin As mentioned above, recent data strongly indicate that talin plays an important role in integrin activation [64–66]. Talin is a major cytoskeletal protein that colocalizes with and binds to integrins, as well as to actin and actin-binding proteins such as vinculin [2, 64–66]. It is an antiparallel homodimer of two ~270 kDa subunits, each of which consists of an N-terminal globular head domain of ~50 kDa and a C-terminal rod domain of ~220 kDa [80, 81]. The head domain contains a FERM domain with three subdomains, F1, F2, and F3, which often mediates interactions with the cytoplasmic tails of transmembrane proteins. The F2 and F3 subdomains of talin bind specifically to integrin β3 tails, although F3 shows a higher affinity than F2 [65]. In addition, the expression of F3 – but not F2 or other high-affinity β tail-binding proteins – activates αIIbβ3 integrins, implying that the major integrin-binding and activating fragment of talin lies within the 96-residue F3 subdomain [2, 65]. Knockdown of talin expression in CHO cells inhibits the activation of both β1 and β3 integrins without altering integrin expression, and this cannot be compensated for by the expression of activating molecules such as activated R-Ras or the CD98 heavy chain [64]. Furthermore, talin knockdown blocks agonist-stimulated fibrinogen binding to the megakaryocyte integrin αIIbβ3, suggesting that normal cellular activation of integrins also requires talin [64].

The talin F3 structure is very similar to that of phosphotyrosine-binding (PTB) domains, which recognize ligands containing β turns formed by NPxY motifs [65]. As discussed above, NPxY motifs are well conserved in most integrin β tails, and mutations that disrupt this motif perturb β turn formation, inhibiting talin binding and therefore interfering with integrin activation [61, 62]. Likewise, mutations within the talin PTB-like domain prevent integrin β tail binding and thereby block integrin activation [64, 82]. These data strongly support the view that the integrin β tail–talin interaction represents a general mechanism for integrin activation.

If talin binding is a final step in integrin activation, then it may be hypothesized that integrin activation is regulated by modulation of the talin binding to integrins. There are two obvious forms of such regulation: (i) alteration of the integrin tails; and (ii) the modification of talin. Tyrosine phosphorylation of the β tail NPxY motif inhibits talin binding, and such modification by Src family kinases inhibits cell adhesion and displaces integrins from talin-rich sites [2, 6]. These data suggest that the phosphorylation of integrins is an important, negative regulator of integrin activation. However, as tyrosine phosphorylation of β3 occurs after αIIbβ3 integrin activation in platelets, it is not clear whether or not such phosphorylation is important in regulating initial integrin activation [2].

The proteolytic cleavage of intact talin releases the head domain and results in an increase in talin binding affinity for the β tail, indicating that the integrin-binding site is masked in intact talin [2, 6, 68]. In fact, in platelets, the protease calpain can cleave talin to release the head domain, implying its role as a regulator of talin-mediated αIIbβ3 activation [2, 6]. However, as calpain also cleaves integrin β tails, resulting in activation blockade, the effects of calpain on integrin activation are complex [2, 6]. Alternatively, the binding of phosphatidylinositol 4,5-biphosphate (PtdInsP2) to talin has also been reported to induce a conformational change, enhancing talin's association with integrin β1 tails [2, 6]. Talin can bind to and also activate a splice variant of the PtdInsP2-producing enzyme phosphatidylinositol phosphate kinase type Iγ-90 (PIPKIγ-90), stimulating PtdInsP2 production, which in turn could promote talin–integrin interactions [2]. Src phosphorylation of PIPKIγ-90 has been said to lead to a dramatic increase in its affinity for talin and to an increase in recruitment to focal adhesions. Since c-Src is activated downstream of integrins, the phosphorylation of PIPKIγ-90 may represent a positive amplification mechanism that links c-Src signaling to integrin activation [2]. Although talin can be phosphorylated, the effects of these phosphorylations on integrin binding and activation remain unclear [2].

Another major question is how talin binding activates integrins. When the ability of activating talin fragments (i.e., F2–F3 domains together or F3 alone) to bind to different regions of the β3 cytoplasmic domain was compared to that of non-activating talin fragment (i.e., F2 domain) by monitoring the perturbation of specific NMR resonances of the β tail, the F2–F3 and F3 fragments – but not F2 – showed distinct perturbation of the membrane-proximal region of the β3 tail, suggesting the involvement of the β3 tail membrane-proximal region in talin-mediated integrin activation [83]. As noted earlier, as the interaction of the α and β membrane-proximal regions maintains the integrins in a low-affinity state [26], it seems likely that perturbations in the region result in integrin activation. NMR studies have suggested that direct disruption of α-β tail interaction by the talin head domain results in integrin activation [59]. In addition, F2–F3 and F3 also perturb the more distal region of the β3 tail [82]. The membrane-distal region of integrin β tail provides a substantial fraction of the binding energy, and has been suggested to contribute to integrin activation [84].

These data together suggest a two-step activation model: the talin head domain first recognizes the high-affinity binding site in the membrane-distal region, which provides a strong linkage between the talin and the integrin β tail, and subsequently binds to a second lower-affinity membrane proximal site that is involved in α-β association, triggering separation of the tails and integrin activation (Figure 1.1) [82, 85].

There are many other PTB domain-containing proteins that bind to integrin β tails in a similar fashion to talin [82, 86]. Such proteins include Numb, Dok-1, ICAP-1α, and Kindlins [6, 86]. However, talin was unusual in its ability to activate integrins, which led Wegener et al. to reason that there must be additional unique

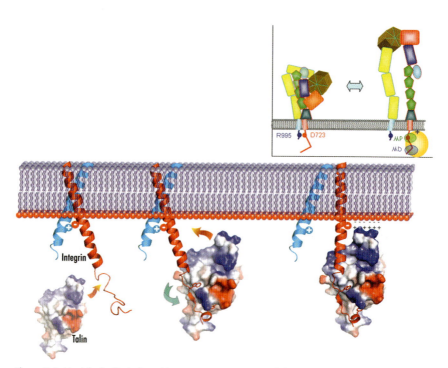

Figure 1.1 Model of talin-induced integrin activation. Left panel: The talin F3 domain (surface representation; colored by charge), freed from its autoinhibitory interactions in the full-length protein, becomes available for binding to the integrin. Center panel: F3 engages the MD part of the β3-integrin tail (red), which becomes ordered, but the α-β integrin interactions that hold the integrin in the low-affinity conformation remain intact. Right panel: In a subsequent step, F3 engages the MP portion of the β3 tail while maintaining its MD interactions. The consequences of this additional interaction are: (1) destabilization of the putative integrin salt bridge; (2) stabilization of the helical structure of the MP region; and (3) electrostatic interactions between F3 and the acidic lipid head groups. The net result is a change in the position of the transmembrane helix, which is continuous with the MP-β-tail helix. This position change causes a packing mismatch with the αIIb-transmembrane helix, separation or reorientation of the integrin tails, and activation (inset). Mutants of F3 that have compromised interactions with the MP region and other PTB domains that lack an MP-binding site stall at point B, consistent with their dominant-negative behavior. (Reprinted from Ref. [85]; © 2007, with permission from Elsevier.)

features in the integrin–talin interaction that enable talin to cause activation [85]. Indeed, a flexible loop between β strands 1 and 2 of the F3 domain of talin accepts the side chains of the membrane-proximal region of the integrin β tail. This mobile loop is absent in other PTB domain-containing proteins, which implies that this may be the unique feature that sets talin apart from other PTB domain-containing proteins in their ability to activate integrins. Furthermore, the importance of this flexible loop is supported by the data that a mutation within this loop blocks talin binding to the β membrane-proximal region and thus hinders integrin activation. In addition, when the talin F3 domain engages the β membrane-proximal region, additional favorable electrostatic contacts between the F3 and the lipid headgroups of the membrane bilayer can be made, which would further stabilize the talin–β tail association. The mutation of one of the predicted contact residues in talin F3 also blocked activation. These data led to a model in which the formation of a complex between the integrin β tail membrane proximal helix and the talin F3 domain, together with the favorable electrostatic contacts between the talin F3 and the lipid membrane, contribute significantly to the energy required to stabilize the integrin-activated state (Figure 1.1) [85]. Recent studies have shown that two other FERM domain-containing proteins, MIG-2 and radixin, interact with the cytoplasmic tails of β1 and β3 and of β2, respectively [87, 88]. Furthermore, these studies suggested that such interactions may have context-dependent roles in integrin activation [87, 88]. Thus, while talin is clearly an important player in integrin activation mechanisms, the possibility remains that other integrin tail-binding proteins can substitute for talin.

Many other PTB-containing proteins cannot activate integrins, even though they bind strongly to the membrane-distal region of the integrin β tail. These proteins may compete for talin binding, providing a potential mechanism that controls integrin avtivation. Specifically, DOK-1 and ICAP-1, compete with talin for NPxY motifs on integrins, and inhibit integrin activation [6, 85, 86]. Thus, the inhibition of talin-mediated integrin activation via competitive binding to the β tails offers a strategy to regulate cell adhesion [85, 86].

1.2.3
Regulation of Integrin Activation

As noted above, many different signaling pathways can regulate integrin activation, and talin binding to the β tail is often a final step. The challenge now is to understand how these different signaling pathways intersect with talin binding; that is, are they upstream, downstream, or do they influence activation without reference to the talin–integrin interaction? First, a few of the signaling pathways that influence activation will be discussed, after which a new, synthetic genetic strategy will be described that offers a promising approach to ordering and quantifying integrin activation pathways.

Hughes et al. found that H-Ras and its downstream effector kinase, Raf-1, suppress integrin activation in CHO cells, which is independent of *de-novo* protein synthesis and integrin phosphorylation and is mediated by the activation of the

ERK1/2 MAP kinase pathway [89, 90]. However, this suppression of integrin affinity by H-Ras is cell-type specific for, unlike in CHO cells and fibroblasts, H-Ras can promote integrin activation in other cell types such as certain hematopoietic cell lines [89]. On the other hand, R-Ras – another small GTPase that shares many common effectors with other Ras family members – generally activates integrins [91, 92]. The expression of a constitutively active variant of R-Ras converted two suspension cell lines into highly adherent cells, and the introduction of a dominant negative form of R-Ras reduced the adhesiveness of CHO cells [92].

Recent studies have identified Rap as a potent activator of integrins that is capable of inducing cell adhesion independent of PI3K [8, 78, 89]. In fact, many cytokines and growth factors promote integrin-dependent cell adhesion through the activation of Rap, and this is true with various subtypes of integrins in various cellular contexts [89]. Furthermore, Rap1 is now known to regulate all integrins that are associated with the actin cytoskeleton – that is, integrins of the β1, β2, and β3 family [93]. In patients with leukocyte adhesion deficiency (LAD)-III, who suffer from defects in leukocyte and platelet integrin activation, β1, β2, and β3 integrins are not mutated and are expressed at normal levels. Cells from these patients are impaired in their abilities to bind to integrin ligands with high affinity in response to chemoattractant signals [94, 95]. Although the expression of Rap1 and talin appears normal in these patients, an autosomal recessive mutation in CalDAG-GEFI that is a key Rap1/2 guanine exchange factor (GEF) [94] is associated with LAD-III. Similarly, $CalDAG\text{-}GEFI^{-/-}$ mice exhibit defects in activation of Rap1 as well as β1, β2, and β3 integrins and therefore an impaired inflammatory response and lack of thrombus formation [95]. These findings together have revealed CalDAG-GEFI as a critical regulator of inside-out integrin activation in human T lymphocytes, neutrophils and platelets, and has emphasized the importance of the Rap1 signaling pathway [94, 95].

During the past few years, several proteins have been identified as Rap1 effectors. RapL/NORE1B is one such protein, and has been reported to be involved in Rap1-induced integrin-mediated cell adhesion [79, 93]. Another protein clearly involved in Rap1-induced integrin-mediated cell adhesion and cell spreading is Rap1-GTP interacting adaptor molecule (RIAM) [93, 96]. The overexpression of RIAM induced the active conformation of integrins and enhanced cell adhesion, while knocking down RIAM eliminated the adhesion mediated by Rap1 [96]. In addition, Rap1 is involved in the activation of integrins by numerous cell-surface receptors [89]. For example, the crosslinking of CD98, a multi-span transmembrane protein, activates Rap to sustain αLβ2-dependent cell adhesion. The ligation of PECAM (a cell–cell adhesion molecule) in leukocytes promotes cell adhesion in a Rap-dependent manner, and a review by Kinbara et al. discusses the role of Ras GTPases in the regulation of integrin activation in greater detail [89].

The cell-type specificity of integrin affinity regulation by these signaling pathways is most likely due to variations in expression of these GTPases, or of the upstream and/or downstream elements that link them to integrin activation. Thus, as with most signaling pathways, simple linear input–output relations are not common in the regulation of integrin activation. The challenge is to discover the general rules that control activation, and to describe the steps in the pathway in sufficient quantitative detail as to enable modeling of the signaling networks that regulate integrin affinity.

Recently, Han et al. [97] ordered a pathway from agonist stimulation to integrin activation for the first time, using a synthetic approach to reconstruct an integrin activation pathway in CHO cells (Figure 1.2). The central finding of this study was that Rap1 induces the formation of an integrin activation complex containing RIAM and talin, which in turn leads to the unmasking of the integrin binding site on talin – a critical, final step in integrin activation. Moreover, these key components of the proposed pathway, Rap1, RIAM, and talin, appear to be widely used in many cellular contexts and with various integrins [97].

$\alpha IIb\beta 3$ is a prototypic integrin, and many of the principles of integrin functions established by early studies with this integrin have proved to be widely applicable across the entire integrin family [97]. However, although platelet agonists including PMA, thrombin, ADP, and collagen stimulate an increased affinity of $\alpha IIb\beta 3$ integrin on platelets for soluble ligands such as fibrinogen, von Willebrand factor, and fibronectin, these agonists have failed to activate recombinant $\alpha IIb\beta 3$ expressed in CHO cells or several other nucleated cells [21, 98]. This puzzling observation led Han and colleagues to reconstruct an integrin activation pathway in CHO cells (Figure 1.2) [97].

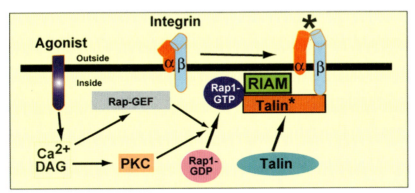

Figure 1.2 Connecting agonist stimulation to integrin activation; the core connections between agonists and integrins are depicted. Agonist receptors (e.g., G-protein-coupled receptors, tyrosine-kinase-coupled receptors) induce the formation of diacylglycerol (DAG) and increased Ca^{2+}, leading to the activation and/or translocation of active GTP-bound Rap1 to the plasma membrane via activation of protein kinase C (PKC) or a Rap guanine nucleotide exchanger (Rap-GEF). At the plasma membrane, activated Rap interacts with Rap1–GTP-interacting adaptor molecule (RIAM), leading to the recruitment of talin to form the integrin activation complex, thus unmasking the integrin binding site on talin, and leading to integrin activation. (Reprinted from Ref. [97]; © 2006, with permission from Elsevier.)

Specifically, realizing that there was something special about platelets that impacted the integrin αIIbβ3 activation process, these authors noted that platelets had a much higher concentration of talin than most other cells in the body [97]. Furthermore, protein kinase Cα (PKCα), which has been implicated as a key downstream mediator in αIIbβ3 integrin activation by many platelet agonists, was also expressed at a higher level in platelets than in αIIbβ3-expressing CHO cells [97]. Based on these two findings, the abundances of talin and PKCα were increased in CHO cells to the levels found in platelets, which resulted in PMA-stimulated activation of αIIbβ3 integrins in CHO cells [97]. This engineered system then enabled the research team to map a signaling pathway connecting agonist stimulation to integrin activation, the major components of which included Rap1, RIAM, and talin.

The results of this study also illustrated the principle that variations in cellular abundance of the components of the integrin activation complex such as talin, or in their regulators such as PKCα, can account for the cell-type specificity of integrin activation [97]. Moreover, this reconstructed model also represents a system that can now be manipulated in order to study complex signaling events involved in integrin activation. For example, it is possible to use the core pathway defined in the study of Han et al. as a template to integrate and rationalize a currently available literature, which to date has been fragmented, and to connect different agonists and signaling pathways that are known to control integrin activation [97]. Similarly, it will allow for quantitative and mutational analyses of signaling pathways that regulate integrin activation. In particular, this system may be applied to analyze contributions from each component of an integrin regulatory pathway and to establish quantitative relationships between inputs (i.e., suppressors and activators of integrins including talin, PKCα, RIAM) and outputs (integrin activation). Then, the analysis may be extended further to agonist receptors (e.g., PAR1), G-proteins (e.g., Rap1, Gαq), and tyrosine kinases (e.g., Tec kinases) that have been implicated in integrin activation. The discovery and analysis of new modulators of integrin activation should also be possible with this new system.

1.2.4
Future Research Directions

Numerous recent studies have focused their efforts on elucidating the mechanisms of integrin signaling and its regulation, as well as its role in physiological set-ups. Nevertheless, in spite of substantial progress made possible by advances in structural analyses, many questions remain to be solved by further research, as detailed in the following paragraphs.

First, the relative roles of affinity regulation and avidity regulation in the control of integrin-mediated adhesion remain uncertain in many cases. Although the measurement of integrin affinity is straightforward with monovalent ligands, it is difficult to quantify integrin clustering and the contribution that it makes to the overall strength of adhesion. The development of better methods to quantify clustering and to relate it to strength of adhesion will certainly help to elucidate this longstanding and contentious question [46, 47].

The development of approaches to analyze the conformational changes of integrins *in situ* in cell membranes also represents a major challenge. In particular, it will be enlightening to analyze tertiary structure of the integrin dimer as well as the structure of the dimer complexed with ligands and its signaling partners/effectors. Our present understanding has been driven by high-resolution structures of integrin fragments in aqueous media, the use of which has led to conflicting results such as the hybrid domain "swing-out" observed in a ligand-occupied fragment of $\alpha IIb\beta 3$ [60], which was notably absent from the ligand-occupied extracellular domain of $\alpha V\beta 3$ [99] or from the full-length $\alpha V\beta 3$ inserted into a lipid bilayer [38]. Similarly, it would be valuable to elucidate the transmission of conformational change across the plasma membrane, from inside the cell to the outside. This will certainly require high-resolution structures of integrin transmembrane domains, perhaps combined with solid-state NMR studies of full-length integrins embedded in lipid bilayers. Such structures would provide important tests of the models of integrin transmembrane domains proposed from mutational and modeling studies [50, 53–55].

Since talin binding to integrin β tails has proven to be a final step in integrin activation, a further understanding of mechanisms regulating talin binding to integrins will be an important goal. For example, it is now known that RIAM binding can unmask the integrin-binding site in talin, but what are the molecular details of this activity of RIAM? Does it require other elements? For instance, one study proposes that adaptor protein complex ADAP/SKAP-55 is a critical component of integrin activation by localizing active Rap1 and RIAM to the plasma membrane upon T-cell receptor stimulation [100]. Similarly, the integrin-binding site on talin could be unmasked through proteolytic cleavage, phospholipid binding, or phosphorylation. Are there circumstances where these mechanisms are important or is the Rap1-RIAM axis indispensable for integrin activation? Similarly, competition for the integrin β tails between talin and other PTB-containing proteins may modulate activation, but how often is this mechanism used to temper the activation process? Finally, it must be asked whether talin binding alone is sufficient for integrin activation, or are other players required even after the integrin-binding site is unmasked? It must also be emphasized that other talin-independent mechanisms of activation might exist. For example, disulfide exchange in the β subunit can activate integrins, and this can be enhanced by peptide disulfide isomerase [101]. But when might this mechanism be used to activate integrins? Likewise, interaction with membrane proteins such as urokinase plasminogen activator (uPAR) can regulate integrin signaling [102], but how does this mechanism interface with talin? The discovery that talin binding is a final common step in integrin activation has raised these and a wealth of other important questions, the elucidation of which will unravel the mechanism of an important biological process and may lead to discovery of novel therapeutic targets for the treatment of inflammation and thrombosis.

References

1 Hynes, R. O. (2002) Integrins: bidirectional, allosteric signaling machines. *Cell 110*, 673–687.
2 Campbell, I. D., and Ginsberg, M. H. (2004) The talin-tail interaction places integrin activation on FERM ground. *Trends Biochem. Sci. 29*, 429–435.
3 Ginsberg, M. H., Partridge, A., and Shattil, S. J. (2005) Integrin regulation. *Curr. Opin. Cell Biol. 17*, 509–516.
4 van der Flier, A., and Sonnenberg, A. (2001) Function and interactions of integrins. *Cell Tissue Res. 305*, 285–298.
5 Liddington, R. C., and Ginsberg, M. H. (2002) Integrin activation takes shape. *J. Cell Biol. 158*, 833–839.
6 Calderwood, D. A. (2004) Integrin activation. *J. Cell Sci. 117*, 657–666.
7 Williams, M. J., Hughes, P. E., O'Toole, T. E., and Ginsberg, M. H. (1994) The inner world of cell adhesion: integrin cytoplasmic domains. *Trends Cell Biol. 4*, 109–112.
8 Alberts, B., Johnson, A., Lewis, J., Raff, M., Roberts, K., and Walter, P. (2002) Cell junctions, cell adhesion, and the extracellular matrix, in: *Molecular Biology of the Cell*, Garland Science, New York.
9 Hynes, R. O. (1992) Integrins: versatility, modulation, and signaling in cell adhesion. *Cell 69*, 11–25.
10 Tamkun, J. W., DeSimone, D. W., Fonda, D., Patel, R. S., Buck, C., Horwitz, A. F., and Hynes, R. O. (1986) Structure of integrin, a glycoprotein involved in the transmembrane linkage between fibronectin and actin. *Cell 46*, 271–282.
11 Kumar, C. C. (1998) Signaling by integrin receptors. *Oncogene 17*, 1365–1373.
12 Wozniak, M. A., Modzelewska, K., Kwong, L., and Keely, P. J. (2004) Focal adhesion regulation of cell behavior. *Biochim. Biophys. Acta 1692*, 103–119.
13 Martin, K. H., Slack, J. K., Boerner, S. A., Martin, C. C., and Parsons, J. T. (2002) Integrin connections map: to infinity and beyond. *Science 296*, 1652–1653.
14 Zamir, E., and Geiger, B. (2001) Molecular complexity and dynamics of cell-matrix adhesions. *J. Cell Sci. 114*, 3583–3590.
15 Qin, J., Vinogradova, O., and Plow, E. F. (2004) Integrin bidirectional signaling: a molecular view. *PLoS Biol. 2*, e169.
16 Giancotti, F. G., and Ruoslahti, E. (1999) Integrin signaling. *Science 285*, 1028–1032.
17 Liu, S., Calderwood, D. A., and Ginsberg, M. H. (2000) Integrin cytoplasmic domain-binding proteins. *J. Cell Sci. 113 (Pt 20)*, 3563–3571.
18 Katsumi, A., Orr, A. W., Tzima, E., and Schwartz, M. A. (2004) Integrins in mechanotransduction. *J. Biol. Chem. 279*, 12001–12004.
19 Ingber, D. E. (1998) Cellular basis of mechanotransduction. *Biol. Bull. 194*, 323–325; discussion 325–327.
20 Alenghat, F. J., and Ingber, D. E. (2002) Mechanotransduction: all signals point to cytoskeleton, matrix, and integrins. *Sci. STKE 2002*, PE6.
21 O'Toole, T. E., Katagiri, Y., Faull, R. J., Peter, K., Tamura, R., Quaranta, V., Loftus, J. C., Shattil, S. J., and Ginsberg, M. H. (1994) Integrin cytoplasmic domains mediate inside-out signal transduction. *J. Cell Biol. 124*, 1047–1059.
22 O'Toole, T. E., Mandelman, D., Forsyth, J., Shattil, S. J., Plow, E. F., and Ginsberg, M. H. (1991) Modulation of the affinity of integrin alpha IIb beta 3 (GPIIb-IIIa) by the cytoplasmic domain of alpha IIb. *Science 254*, 845–847.
23 Orr, A. W., Helmke, B. P., Blackman, B. R., and Schwartz, M. A. (2006) Mechanisms of mechanotransduction. *Dev. Cell 10*, 11–20.
24 Humphries, M. J., McEwan, P. A., Barton, S. J., Buckley, P. A., Bella, J., and Mould, A. P. (2003) Integrin structure: heady advances in ligand binding, but activation still makes the knees wobble. *Trends Biochem. Sci. 28*, 313–320.
25 Iber, D., and Campbell, I. D. (2006) Integrin activation – the importance of a positive feedback. *Bull. Math. Biol. 68*, 945–956.

26. Hughes, P. E., Diaz-Gonzalez, F., Leong, L., Wu, C., McDonald, J. A., Shattil, S. J., and Ginsberg, M. H. (1996) Breaking the integrin hinge. A defined structural constraint regulates integrin signaling. *J. Biol. Chem.* 271, 6571–6574.
27. Lauffenburger, D. A., and Horwitz, A. F. (1996) Cell migration: a physically integrated molecular process. *Cell* 84, 359–369.
28. Koivunen, E., Ranta, T. M., Annila, A., Taube, S., Uppala, A., Jokinen, M., van Willigen, G., Ihanus, E., and Gahmberg, C. G. (2001) Inhibition of beta(2) integrin-mediated leukocyte cell adhesion by leucine-leucine-glycine motif-containing peptides. *J. Cell Biol.* 153, 905–916.
29. Li, Z. (1999) The alphaMbeta2 integrin and its role in neutrophil function. *Cell Res.* 9, 171–178.
30. Semmrich, M., Smith, A., Feterowski, C., Beer, S., Engelhardt, B., Busch, D. H., Bartsch, B., Laschinger, M., Hogg, N., Pfeffer, K., and Holzmann, B. (2005) Importance of integrin LFA-1 deactivation for the generation of immune responses. *J. Exp. Med.* 201, 1987–1998.
31. Kim, M., Carman, C. V., and Springer, T. A. (2003) Bidirectional transmembrane signaling by cytoplasmic domain separation in integrins. *Science* 301, 1720–1725.
32. Takagi, J., Petre, B. M., Walz, T., and Springer, T. A. (2002) Global conformational rearrangements in integrin extracellular domains in outside-in and inside-out signaling. *Cell* 110, 599–611.
33. Jin, M., Andricioaei, I., and Springer, T. A. (2004) Conversion between three conformational states of integrin I domains with a C-terminal pull spring studied with molecular dynamics. *Structure* 12, 2137–2147.
34. Nishida, N., Xie, C., Shimaoka, M., Cheng, Y., Walz, T., and Springer, T. A. (2006) Activation of leukocyte beta2 integrins by conversion from bent to extended conformations. *Immunity* 25, 583–594.
35. Mould, A. P., and Humphries, M. J. (2004) Regulation of integrin function through conformational complexity: not simply a knee-jerk reaction? *Curr. Opin. Cell Biol.* 16, 544–551.
36. Shimaoka, M., Lu, C., Salas, A., Xiao, T., Takagi, J., and Springer, T. A. (2002) Stabilizing the integrin alpha M inserted domain in alternative conformations with a range of engineered disulfide bonds. *Proc. Natl. Acad. Sci. USA* 99, 16737–16741.
37. Zhu, J., Boylan, B., Luo, B. H., Newman, P. J., and Springer, T. A. (2007) Tests of the extension and deadbolt models of integrin activation. *J. Biol. Chem.* 282, 11914–11920.
38. Adair, B. D., Xiong, J. P., Maddock, C., Goodman, S. L., Arnaout, M. A., and Yeager, M. (2005) Three-dimensional EM structure of the ectodomain of integrin αVβ3 in a complex with fibronectin. *J. Cell Biol.* 168, 1109–1118.
39. Xiong, J. P., Stehle, T., Goodman, S. L., and Arnaout, M. A. (2003) New insights into the structural basis of integrin activation. *Blood* 102, 1155–1159.
40. Chigaev, A., Waller, A., Zwartz, G. J., Buranda, T., and Sklar, L. A. (2007) Regulation of cell adhesion by affinity and conformational unbending of alpha4beta1 integrin. *J. Immunol.* 178, 6828–6839.
41. Luo, B. H., and Springer, T. A. (2006) Integrin structures and conformational signaling. *Curr. Opin. Cell Biol.* 18, 579–586.
42. Arnaout, M. A., Mahalingam, B., and Xiong, J. P. (2005) Integrin structure, allostery, and bidirectional signaling. *Annu. Rev. Cell. Dev. Biol.* 21, 381–410.
43. Luo, B. H., Carman, C. V., and Springer, T. A. (2007) Structural basis of integrin regulation and signaling. *Annu. Rev. Immunol.* 25, 619–647.
44. Takagi, J., and Springer, T. A. (2002) Integrin activation and structural rearrangement. *Immunol. Rev.* 186, 141–163.
45. Shimaoka, M., Takagi, J., and Springer, T. A. (2002) Conformational regulation of integrin structure and function. *Annu. Rev. Biophys. Biomol. Struct.* 31, 485–516.

46 Carman, C. V., and Springer, T. A. (2003) Integrin avidity regulation: are changes in affinity and conformation underemphasized? *Curr. Opin. Cell Biol.* 15, 547–556.

47 Bazzoni, G., and Hemler, M. E. (1998) Are changes in integrin affinity and conformation overemphasized? *Trends Biochem. Sci.* 23, 30–34.

48 Hato, T., Pampori, N., and Shattil, S. J. (1998) Complementary roles for receptor clustering and conformational change in the adhesive and signaling functions of integrin alphaIIb beta3. *J. Cell Biol.* 141, 1685–1695.

49 Stefansson, A., Armulik, A., Nilsson, I., von Heijne, G., and Johansson, S. (2004) Determination of N- and C-terminal borders of the transmembrane domain of integrin subunits. *J. Biol. Chem.* 279, 21200–21205.

50 Gottschalk, K. E., Adams, P. D., Brunger, A. T., and Kessler, H. (2002) Transmembrane signal transduction of the alpha(IIb)beta(3) integrin. *Protein Sci.* 11, 1800–1812.

51 Scott, J. P., 3rd, Scott, J. P., 2nd, Chao, Y. L., Newman, P. J., and Ward, C. M. (1998) A frameshift mutation at Gly975 in the transmembrane domain of GPIIb prevents GPIIb-IIIa expression – analysis of two novel mutations in a kindred with type I Glanzmann thrombasthenia. *Thromb. Haemost.* 80, 546–550.

52 Nurden, A. T., Breillat, C., Jacquelin, B., Combrie, R., Freedman, J., Blanchette, V. S., Schmugge, M., and Rand, M. L. (2004) Triple heterozygosity in the integrin alphaIIb subunit in a patient with Glanzmann's thrombasthenia. *J. Thromb. Haemost.* 2, 813–819.

53 Luo, B. H., Carman, C. V., Takagi, J., and Springer, T. A. (2005) Disrupting integrin transmembrane domain heterodimerization increases ligand binding affinity, not valency or clustering. *Proc. Natl. Acad. Sci. USA* 102, 3679–3684.

54 Luo, B. H., Springer, T. A., and Takagi, J. (2004) A specific interface between integrin transmembrane helices and affinity for ligand. *PLoS Biol.* 2, e153.

55 Partridge, A. W., Liu, S., Kim, S., Bowie, J. U., and Ginsberg, M. H. (2005) Transmembrane domain helix packing stabilizes integrin alphaIIbbeta3 in the low affinity state. *J. Biol. Chem.* 280, 7294–7300.

56 Hughes, P. E., O'Toole, T. E., Ylanne, J., Shattil, S. J., and Ginsberg, M. H. (1995) The conserved membrane-proximal region of an integrin cytoplasmic domain specifies ligand binding affinity. *J. Biol. Chem.* 270, 12411–12417.

57 Ma, Y. Q., Yang, J., Pesho, M. M., Vinogradova, O., Qin, J., and Plow, E. F. (2006) Regulation of integrin alphaIIbbeta3 activation by distinct regions of its cytoplasmic tails. *Biochemistry* 45, 6656–6662.

58 Lu, C., Takagi, J., and Springer, T. A. (2001) Association of the membrane proximal regions of the alpha and beta subunit cytoplasmic domains constrains an integrin in the inactive state. *J. Biol. Chem.* 276, 14642–14648.

59 Vinogradova, O., Velyvis, A., Velyviene, A., Hu, B., Haas, T., Plow, E., and Qin, J. (2002) A structural mechanism of integrin alpha(IIb)beta(3) "inside-out" activation as regulated by its cytoplasmic face. *Cell* 110, 587–597.

60 Takagi, J., Erickson, H. P., and Springer, T. A. (2001) C-terminal opening mimics 'inside-out' activation of integrin alpha5beta1. *Nat. Struct. Biol.* 8, 412–416.

61 O'Toole, T. E., Ylanne, J., and Culley, B. M. (1995) Regulation of integrin affinity states through an NPXY motif in the beta subunit cytoplasmic domain. *J. Biol. Chem.* 270, 8553–8558.

62 Ulmer, T. S., Yaspan, B., Ginsberg, M. H., and Campbell, I. D. (2001) NMR analysis of structure and dynamics of the cytosolic tails of integrin alpha IIb beta 3 in aqueous solution. *Biochemistry* 40, 7498–7508.

63 Ginsberg, M. H., Yaspan, B., Forsyth, J., Ulmer, T. S., Campbell, I. D., and Slepak, M. (2001) A membrane-distal segment of the integrin alpha IIb cytoplasmic domain regulates integrin activation. *J. Biol. Chem.* 276, 22514–22521.

64 Tadokoro, S., Shattil, S. J., Eto, K., Tai, V., Liddington, R. C., de Pereda, J. M., Ginsberg, M. H., and Calderwood, D. A.

(2003) Talin binding to integrin beta tails: a final common step in integrin activation. *Science* 302, 103–106.

65. Calderwood, D. A., Yan, B., de Pereda, J. M., Alvarez, B. G., Fujioka, Y., Liddington, R. C., and Ginsberg, M. H. (2002) The phosphotyrosine binding-like domain of talin activates integrins. *J. Biol. Chem.* 277, 21749–21758.

66. Calderwood, D. A., Zent, R., Grant, R., Rees, D. J., Hynes, R. O., and Ginsberg, M. H. (1999) The Talin head domain binds to integrin beta subunit cytoplasmic tails and regulates integrin activation. *J. Biol. Chem.* 274, 28071–28074.

67. Kuo, J. C., Wang, W. J., Yao, C. C., Wu, P. R., and Chen, R. H. (2006) The tumor suppressor DAPK inhibits cell motility by blocking the integrin-mediated polarity pathway. *J. Cell Biol.* 172, 619–631.

68. Franco, S. J., Rodgers, M. A., Perrin, B. J., Han, J., Bennin, D. A., Critchley, D. R., and Huttenlocher, A. (2004) Calpain-mediated proteolysis of talin regulates adhesion dynamics. *Nat. Cell Biol.* 6, 977–983.

69. Lim, J., Wiedemann, A., Tzircotis, G., Monkley, S. J., Critchley, D. R., and Caron, E. (2007) An essential role for talin during alpha(M)beta(2)-mediated phagocytosis. *Mol. Biol. Cell* 18, 976–985.

70. Tremuth, L., Kreis, S., Melchior, C., Hoebeke, J., Ronde, P., Plancon, S., Takeda, K., and Kieffer, N. (2004) A fluorescence cell biology approach to map the second integrin-binding site of talin to a 130-amino acid sequence within the rod domain. *J. Biol. Chem.* 279, 22258–22266.

71. Petrich, B. G., Fogelstrand, P., Partridge, A. W., Yousefi, N., Ablooglu, A. J., Shattil, S. J., and Ginsberg, M. H. (2007) The antithrombotic potential of selective blockade of talin-dependent integrin alpha(IIb)beta(3) (platelet GPIIb-IIIa) activation. *J. Clin. Invest.* 117, 2250–2259.

72. Boshans, R. L., Szanto, S., van Aelst, L., and D'Souza-Schorey, C. (2000) ADP-ribosylation factor 6 regulates actin cytoskeleton remodeling in coordination with Rac1 and RhoA. *Mol. Cell. Biol.* 20, 3685–3694.

73. Palacios, F., and D'Souza-Schorey, C. (2003) Modulation of Rac1 and ARF6 activation during epithelial cell scattering. *J. Biol. Chem.* 278, 17395–17400.

74. Santy, L. C., and Casanova, J. E. (2001) Activation of ARF6 by ARNO stimulates epithelial cell migration through downstream activation of both Rac1 and phospholipase D. *J. Cell Biol.* 154, 599–610.

75. Santy, L. C., Ravichandran, K. S., and Casanova, J. E. (2005) The DOCK180/Elmo complex couples ARNO-mediated Arf6 activation to the downstream activation of Rac1. *Curr. Biol.* 15, 1749–1754.

76. Yuan, W., Leisner, T. M., McFadden, A. W., Wang, Z., Larson, M. K., Clark, S., Boudignon-Proudhon, C., Lam, S. C., and Parise, L. V. (2006) CIB1 is an endogenous inhibitor of agonist-induced integrin alphaIIbbeta3 activation. *J. Cell Biol.* 172, 169–175.

77. Manevich, E., Grabovsky, V., Feigelson, S. W., and Alon, R. (2007) Talin1 and paxillin facilitate distinct steps in rapid VLA-4-mediated adhesion strengthening to vascular cell adhesion molecule 1. *J. Biol. Chem.* 282, 25338–25348.

78. Katagiri, K., Hattori, M., Minato, N., Irie, S., Takatsu, K., and Kinashi, T. (2000) Rap1 is a potent activation signal for leukocyte function-associated antigen 1 distinct from protein kinase C and phosphatidylinositol-3-OH kinase. *Mol. Cell. Biol.* 20, 1956–1969.

79. Katagiri, K., Maeda, A., Shimonaka, M., and Kinashi, T. (2003) RAPL, a Rap1-binding molecule that mediates Rap1-induced adhesion through spatial regulation of LFA-1. *Nat. Immunol.* 4, 741–748.

80. Critchley, D. R. (2000) Focal adhesions – the cytoskeletal connection. *Curr. Opin. Cell Biol.* 12, 133–139.

81. Rees, D. J., Ades, S. E., Singer, S. J., and Hynes, R. O. (1990) Sequence and domain structure of talin. *Nature* 347, 685–689.

82. Garcia-Alvarez, B., de Pereda, J. M., Calderwood, D. A., Ulmer, T. S., Critchley, D., Campbell, I. D., Ginsberg,

M. H., and Liddington, R. C. (2003) Structural determinants of integrin recognition by talin. *Mol. Cell* 11, 49–58.

83 Ulmer, T. S., Calderwood, D. A., Ginsberg, M. H., and Campbell, I. D. (2003) Domain-specific interactions of talin with the membrane-proximal region of the integrin beta3 subunit. *Biochemistry* 42, 8307–8312.

84 Vinogradova, O., Vaynberg, J., Kong, X., Haas, T. A., Plow, E. F., and Qin, J. (2004) Membrane-mediated structural transitions at the cytoplasmic face during integrin activation. *Proc. Natl. Acad. Sci. USA* 101, 4094–4099.

85 Wegener, K. L., Partridge, A. W., Han, J., Pickford, A. R., Liddington, R. C., Ginsberg, M. H., and Campbell, I. D. (2007) Structural basis of integrin activation by talin. *Cell* 128, 171–182.

86 Calderwood, D. A., Fujioka, Y., De Pereda, J. M., Garcia-Alvarez, B., Nakamoto, T., Margolis, B., McGlade, C. J., Liddington, R. C., and Ginsberg, M. H. (2003) Integrin beta cytoplasmic domain interactions with phosphotyrosine-binding domains: a structural prototype for diversity in integrin signaling. *Proc. Natl. Acad. Sci. USA* 100, 2272–2277.

87 Shi, X., Ma, Y. Q., Tu, Y., Chen, K., Wu, S., Fukuda, K., Qin, J., Plow, E. F., and Wu, C. (2007) The MIG-2/integrin interaction strengthens cell-matrix adhesion and modulates cell motility. *J. Biol. Chem.* 282, 20455–20466.

88 Tang, P., Cao, C., Xu, M., and Zhang, L. (2007) Cytoskeletal protein radixin activates integrin alpha(M)beta(2) by binding to its cytoplasmic tail. *FEBS Lett.* 581, 1103–1108.

89 Kinbara, K., Goldfinger, L. E., Hansen, M., Chou, F. L., and Ginsberg, M. H. (2003) Ras GTPases: integrins' friends or foes? *Nat. Rev. Mol. Cell. Biol.* 4, 767–776.

90 Hughes, P. E., Renshaw, M. W., Pfaff, M., Forsyth, J., Keivens, V. M., Schwartz, M. A., and Ginsberg, M. H. (1997) Suppression of integrin activation: a novel function of a Ras/Raf-initiated MAP kinase pathway. *Cell* 88, 521–530.

91 Scthi, T., Ginsberg, M. H., Downward, J., and Hughes, P. E. (1999) The small GTP-binding protein R-Ras can influence integrin activation by antagonizing a Ras/Raf-initiated integrin suppression pathway. *Mol. Biol. Cell* 10, 1799–1809.

92 Zhang, Z., Vuori, K., Wang, H., Reed, J. C., and Ruoslahti, E. (1996) Integrin activation by R-ras. *Cell* 85, 61–69.

93 Bos, J. L. (2005) Linking Rap to cell adhesion. *Curr. Opin. Cell Biol.* 17, 123–128.

94 Pasvolsky, R., Feigelson, S. W., Kilic, S. S., Simon, A. J., Tal-Lapidot, G., Grabovsky, V., Crittenden, J. R., Amariglio, N., Safran, M., Graybiel, A. M., Rechavi, G., Ben-Dor, S., Etzioni, A., and Alon, R. (2007) A LAD-III syndrome is associated with defective expression of the Rap-1 activator CalDAG-GEFI in lymphocytes, neutrophils, and platelets. *J. Exp. Med.* 204, 1571–1582.

95 Bergmeier, W., Goerge, T., Wang, H. W., Crittenden, J. R., Baldwin, A. C., Cifuni, S. M., Housman, D. E., Graybiel, A. M., and Wagner, D. D. (2007) Mice lacking the signaling molecule CalDAG-GEFI represent a model for leukocyte adhesion deficiency type III. *J. Clin. Invest.* 117, 1699–1707.

96 Lafuente, E. M., van Puijenbroek, A. A., Krause, M., Carman, C. V., Freeman, G. J., Berezovskaya, A., Constantine, E., Springer, T. A., Gertler, F. B., and Boussiotis, V. A. (2004) RIAM, an Ena/VASP and Profilin ligand, interacts with Rap1-GTP and mediates Rap1-induced adhesion. *Dev. Cell* 7, 585–595.

97 Han, J., Lim, C. J., Watanabe, N., Soriani, A., Ratnikov, B., Calderwood, D. A., Puzon-McLaughlin, W., Lafuente, E. M., Boussiotis, V. A., Shattil, S. J., and Ginsberg, M. H. (2006) Reconstructing and deconstructing agonist-induced activation of integrin alphaIIbbeta3. *Curr. Biol.* 16, 1796–1806.

98 O'Toole, T. E., Loftus, J. C., Du, X. P., Glass, A. A., Ruggeri, Z. M., Shattil, S. J., Plow, E. F., and Ginsberg, M. H. (1990) Affinity modulation of the alpha IIb beta 3 integrin (platelet GPIIb-IIIa) is an intrinsic property of the receptor. *Cell Regul.* 1, 883–893.

99 Xiong, J. P., Stehle, T., Zhang, R., Joachimiak, A., Frech, M., Goodman, S. L., and Arnaout, M. A. (2002) Crystal structure of the extracellular segment of integrin alphaVbeta3 in complex with an Arg-Gly-Asp ligand. *Science 296*, 151–155.

100 Menasche, G., Kliche, S., Chen, E. J., Stradal, T. E., Schraven, B., and Koretzky, G. (2007) RIAM links the ADAP/SKAP-55 signaling module to Rap1, facilitating T-cell-receptor-mediated integrin activation. *Mol. Cell. Biol. 27*, 4070–4081.

101 Lahav, J., Wijnen, E. M., Hess, O., Hamaia, S. W., Griffiths, D., Makris, M., Knight, C. G., Essex, D. W., and Farndale, R. W. (2003) Enzymatically catalyzed disulfide exchange is required for platelet adhesion to collagen via integrin alpha2beta1. *Blood 102*, 2085–2092.

102 Wei, Y., Lukashev, M., Simon, D. I., Bodary, S. C., Rosenberg, S., Doyle, M. V., and Chapman, H. A. (1996) Regulation of integrin function by the urokinase receptor. *Science 273*, 1551–1555.

2
Integrin Signaling Through Focal Adhesion Kinase
Youngdong Yoo and Jun-Lin Guan

2.1
Introduction

The extracellular matrix (ECM) is a complex mixture of polysaccharides and high-molecular-weight proteins such as laminin, fibronectin, collagen and vitronectin which are secreted by cells. The ECM provides structural support for cells and mediates cell attachment. Cellular interactions with the ECM play crucial roles for a variety of biological processes, including embryonic development, homeostasis, wound healing and malignant transformation [1, 2]. Cell adhesion to the ECM is mainly mediated by the integrin family of cell-surface receptors. Integrins are composed of non-covalently linked α and β subunits, each of which is a transmembrane glycoprotein with a single membrane-spanning segment and generally a short cytoplasmic domain [3, 4]. Eighteen α and eight β subunits, so far known to assemble into 24 integrins, have been identified in mammals, and different combinations of the α and β subunits exhibit different preferences for particular ECM molecules [4–6]. Integrins recognize environmental cues from the ECM and transmit them into signals that control a wide variety of cellular process such as cell migration, proliferation, differentiation, and survival [7–9]. Binding of the ECM to integrins induces integrin clustering and the formation of multiprotein complexes consisting of scaffolding and signaling molecules at the cytoplasmic domain of integrins [4]. Various signaling proteins and second messengers involved in integrin-induced signaling pathways have been identified, including tyrosine kinases, serine/threonine kinases, lipid mediators and small GTPases. In this chapter, attention will be focused on the role and mechanisms of the focal adhesion kinase (FAK) in integrin-mediated signal transduction.

FAK is a 125-kDa non-receptor cytoplasmic tyrosine kinase, which was initially identified more than 15 years ago based on its increased phosphorylation upon integrin-mediated cell adhesion and being a major tyrosine kinase localized to focal adhesions (hence the name, focal adhesion kinase) [10–13]. Focal adhesions (focal contacts) are formed at ECM–integrin interacting points, where signaling and cytoskeletal proteins are recruited during cell attachment, spreading and migration [14, 15]. FAK is expressed in most tissues and cell types, and in a variety of species including chicken, *Xenopus*, *Drosophila*, rodent and human, and is an evolutionarily conserved protein [13, 16–19]. In this chapter, the current understanding of the structure of FAK, the mechanisms of its activation, its downstream signaling pathways in the regulation of various cellular functions, will be discussed, and some recent studies on the role of FAK *in vivo* outlined.

2.2
Structure of FAK

FAK is composed of three major domains: (i) the N-terminal FERM (protein 4.1, ezrin, radixin and moesin homology) domain that mediates FAK interaction with various membrane proteins; (ii) a central kinase domain; and (iii) the C-terminal region containing two proline-rich (PR) domains that serve as binding sites for various SH3 domain-containing proteins. The C-terminal region also contains the focal adhesion targeting (FAT) domain which is responsible for FAK's localization to focal adhesions [14, 15] (Figure 2.1).

The N-terminal FERM domain has been shown to interact with the cytoplasmic domain of β-integrin [20]. It also mediates FAK's association with growth factor receptors, such as the epidermal growth factor (EGF) and platelet-derived growth factor (PDGF) receptors, to regulate growth factor-induced cell migration [21, 22]. Structural analysis revealed that the FERM domain contains three lobes that can serve as binding sites for lipids and proteins [23]. Recent studies have shown that the Etk protein tyrosine kinase binds to this domain and mediates the activation

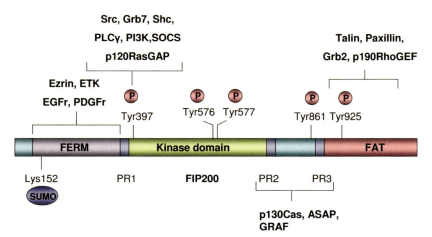

Figure 2.1 Structural domains and binding proteins of focal adhesion kinase (FAK). FAK consists of a N-terminal FERM domain, a central kinase domain, and a C-terminal FAT sequence. The proline-rich motifs (PR1, PR2, and PR3), sumoylation and tyrosine phosphorylation sites, and proteins known to interact with specific regions of FAK are also shown.

of FAK by integrin [24]. Ezrin binding to this domain can also lead to integrin-independent FAK activation [25]. The FERM domain has been proposed to modulate the kinase activity of FAK. When FAK is inactivated, the FERM domain interacts intra-molecularly with the central kinase domain, inhibiting its catalytic activity [26]. Observations that the binding of proteins such as ezrin or GEF-Trio to the FERM domain leads to enhanced kinase activity of FAK and that the binding of this domain in trans to FAK's kinase domain causes inhibition of its catalytic activity, suggest that the FERM domain may function to regulate auto-inhibitory constraints on FAK catalytic activity [25, 27, 28].

Lysine 152 in the FERM domain of FAK can be post-translationally modified by sumoylation [29]. As sumoylation is generally related to the nuclear localization of target proteins, sumo-FAK is localized in nuclear fraction of cells [29]. Observations that sumoylation increases FAK activity and that FAK can enhance gene expression and proliferation suggest that the FERM domain of FAK might serve as a signaling bridge between cell adhesion and nuclear function directly [30–32].

The central kinase domain of FAK shares sequence similarity with that of other receptor- and non-receptor tyrosine kinases [33]. There are two tyrosine residues (Y576, Y577) localized within the "catalytic loop" of the kinase domain, the phosphorylation of which is correlated with an increased catalytic activity and is important for the maximal activation of FAK [34].

Two proline-rich motifs (PR1 and PR2) in the C-terminal region function as binding sites for SH3 domain-containing proteins, including the Crk-associated substrate (Cas) adaptor protein. Cas binding to FAK is important for cell migration through activation of Rac at membrane protrusions [14, 35]. The binding of GRAF (GTPase regulator associated with FAK) and ASAP1 (Arf GTPase-activating protein containing SH3, ankyrin repeat and pleckstrin homology domain-1) to these domains regulates the actin cytoskeleton and focal contact assembly [15].

The C-terminal FAT domain contains binding sites for integrin-associated proteins such as talin and paxillin that mediate the colocalization of FAK with integrins at focal adhesions. The FAT domain sequences are both necessary and sufficient for FAK targeting to focal adhesions, although the exact mechanism has not yet been established. FAK, paxillin and talin are thought to act as links between integrins and the cytoskeleton at focal adhesions [15]. The FAT domain also binds to p190RhoGEF that activates Rho-family GTPases. The tyrosine phosphorylation of this protein by FAK has been suggested to lead to RhoA activation [36]. One intriguing observation is that the C-terminal, non-catalytic domain of FAK, termed FRNK (FAK-related non-kinase) is expressed in certain cells as an autonomous protein and functions as a negative regulator of FAK activity [37, 38]. The transcription of FRNK is controlled by an alternative promoter that resides between the kinase domain and the C-terminal domain of FAK [39]. Exogenously expressed FRNK acts as an inhibitor of FAK by displacing FAK from the focal adhesion and by interacting with signaling proteins that bind to FAK [40, 41]. However, the precise physiological function of endogenous FRNK has not yet been established.

2.3
Regulation of FAK Activity

The regulation of FAK activity can be achieved through several different mechanisms such as alternative splicing, post-translational modification (phosphorylation and dephosphorylation), and interaction with other regulatory proteins.

2.3.1
Activation of FAK

Although initial studies revealed that FAK is activated by integrin clustering, the precise mechanism of FAK activation is still not completely understood. One of the reasons for this is that multiple inputs can activate FAK [42]. The activation process of FAK requires phosphorylation on tyrosine residues, which occurs in response to integrin clustering induced by cell adhesion to the ECM. Studies showing that function-disrupting mutations in the cytoplasmic tails of the integrin β subunit impaired tyrosine phosphorylation of FAK suggest that a protein complex associating with the β tail is important for activation of FAK [12, 43–45]. Also, mutants of FAK that are unable to localize to focal adhesions fail to be phosphorylated in response to cell adhesion, suggesting that proper localization at focal adhesions is a critical requirement for FAK activation [45, 46]. The actin cytoskeleton is also essential for activation and tyrosine phosphorylation of FAK. The inhibition of actin polymerization by Cytochalasin D abolished the tyrosine phosphorylation of FAK in response to many stimuli [47, 48]. In addition, the activation of Rho family GTP-binding proteins induced tyrosine phosphorylation of FAK, whereas the inhibition of Rho blocked such activation; this suggests that the Rho-dependent regulation of actin cytoskeleton controls FAK tyrosine phosphorylation [49–52]. FAK has also been shown to undergo tyrosine phosphorylation in response to several other signaling stimuli, including phospholipids, neuropeptides, cytokines (e.g., TNF-α, hyaluronan, concanavalin A) and growth factors such as EGF, PDGF, and hepatocyte growth factor (HGF) [14, 15, 53].

Several tyrosine (Y) residues may be phosphorylated upon FAK activation, namely Y397, Y407, Y576, Y577, Y861, and Y925. Phosphorylations at these sites in response to many stimuli serve as docking sites for several classes of signaling molecule [4, 14, 26]. The best-characterized phosphorylation site is the auto-phosphorylation at Y397 residue that is located upstream of the kinase domain, which is critical for many biological functions of FAK, including cell migration, proliferation, and survival [54]. Phosphorylation at this site creates a binding site for various SH2 domain-containing proteins including Src family kinases, phospholipase C gamma (PLCγ), Shc adaptor protein, p85 subunit of phosphatidylinositol 3-kinase (PI3K), growth factor-receptor-bound protein 7 (Grb7), suppressor of cytokine signaling (SOCS), and p120 RasGAP. It is not clear whether these proteins bind differentially to phosphorylated Y397 of FAK in response to different stimuli, or whether a large pool of activated FAK with different complexes exists at the same time. Nevertheless, existing data support the idea that the binding of Src family kinases to phosphorylated Y397, which leads to the activation of both Src and FAK in the FAK/Src signaling complex, initiates downstream FAK signaling pathways.

An analysis of FAK's tyrosine phosphorylation sites has provided insight into the activation of FAK downstream pathways. Two tyrosine residues, Y576 and Y577, are located within "activation loop" of the kinase domain; phosphorylation at these sites is necessary for maximal catalytic activity of FAK. Mutations that substitute phenylalanine at these residues result in a two- to three-fold reduction of kinase activity compared with wild-type FAK [33, 34, 55]. Phosphorylation at Y861 is known to enhance the SH3 domain-mediated binding of Cas to the C-terminal proline-rich motifs of FAK [56]. This residue is also implicated in promoting the association of FAK with αvβ5 integrin signaling complexes after vascular endothelial growth factor (VEGF)-stimulation in the endothelial cells, though the exact mechanism by which it mediates this association remains unclear [57]. It has also been reported that phosphorylation at Y861 is necessary for cell migration in Ras-transformed NIH3T3 cells, and for the epithelial–mesenchymal transition of murine mammary epithelial cells [56, 58].

The phosphorylation at Y925 located in the FAT domain creates a binding site for the Grb2–SOS complex [48, 59, 60]; this association is one of the connections between FAK and Erk2/MAP kinase cascade. In addition to its function in the regulation of cell-cycle progression, Erk2 activation by FAK has been shown to phosphorylate myosin light chain kinase (MLCK) to regulate focal contact dynamics in motile cells [14, 61]. The Grb2 binding site at Y925 overlaps with that of paxillin in the FAT domain, suggesting the possibility that Grb2 binding to this site and paxillin binding to the FAT domain may be mutually exclusive [62–65]. The results of localization studies supported this idea by showing that FAK phosphorylated at Y925 is excluded from focal contacts. Notably, the FAT domain contains four helical bundles, with Y925 lying in the first helix which is known to mediate paxillin binding, supporting this idea [65, 66]. It has also been reported that this site is associated with E-cadherin deregulation in Src-induced epithelial–mesenchymal transition in colon cancer cells [67].

FAK contains four sites of potential serine phosphorylation (S722, S732, S843, and S910). The precise role of serine phosphorylation in the regulation of FAK activity is poorly understood. However, the close proximity of serine residues to protein–protein interaction sites suggests a role for serine phosphorylation in modulating the formation or stability of downstream signaling protein complexes. Serine phosphorylation has been shown to regulate FAK activity. Phosphorylation of S722 by glycogen synthase kinase 3 (GSK3) in spreading cells inhibits the kinase activity of FAK [68]. Serine 732 was shown to be phosphorylated by cyclin-dependent kinase 5 (cdk5) in post-mitotic neurons during mouse brain development. Phosphorylation at this site was shown to be important for microtubule organization, nuclear movement and neuronal migration in cultured neurons [69]. These results suggest that the phosphorylation of FAK by cdk5 regulates neuronal migration through a control of microtubule organization that is important for nuclear translocation. Phosphorylation at S843 and S910 has been reported to occur during mitosis and to be correlated with decrease of FAK activity [70, 71].

2.3.2
Inhibition of FAK Activity

FAK activity is negatively regulated by an inhibitor known as the FAK family interacting protein of 200 kDa (FIP200). FIP200, as identified by yeast two-hybridization screening, binds to the kinase domain of FAK and the other FAK family member, Pyk2, and inhibits their activity [72]. Although the inhibitory mechanism is not entirely clear, the fact that it is expressed in many tissues and cell types suggests that FIP200 plays an important role in regulating cellular processes associated with FAK signaling. An analysis of domains of FIP200 has shown that the N-terminal region of FIP200 associates with the kinase domain of FAK and inhibits catalytic activity and cellular functions of FAK such as cell migration and proliferation.

The activity of FAK has also been shown to be modulated by several other proteins. The mechanism of FAK with SOCS proteins has been shown to promote poly-ubiquitination and the degradation of FAK [73]. Protein tyrosine phosphatases (PTPs) can also regulate the activity of FAK, as tyrosine phosphorylation of FAK is an important mechanism for regulating its catalytic activity and its interaction with other signaling molecules. Several tyrosine phosphatases have been identified as regulating FAK activity, including PTP-α, Shp-2 and PTEN [74–76]. Studies using PTP-α-deficient cells revealed that this phosphatase regulates the phosphorylation of FAK at Y397 by regulating the β1-integrin-mediated stimulation of c-Src activity. Shp-2 has also been suggested to regulate the tyrosine phosphorylation of FAK. The expression of a dominant negative mutant of Shp-2 blocked the dephosphorylation of endogenous FAK in response to insulin-like growth factor 1 (IGF-1), which has been known to reduce the tyrosine phosphorylation of FAK. This observation suggests that Shp-2 inhibits FAK signaling by promoting the dephosphorylation of FAK [77, 78]. Current evidence suggests that PTEN also regulates the tyrosine phosphorylation of FAK. Indeed, the expression of PTEN in NIH-3T3 or U87-MG caused various phenotypic changes such as reduced cell spreading, motility and invasion, which are similar to those caused by FAK inactivation [79, 80]. In addition, the expression of PTEN induced a decrease of FAK phosphorylation, whereas a dominant-negative PTEN mutant led to increased tyrosine phosphorylation of FAK, suggesting that PTEN may inhibit FAK signaling by mediating its dephosphorylation [81].

2.4
Regulation of Cellular Functions by FAK Signaling Pathways

As a key mediator of integrin signaling, FAK has been shown to play important roles in the regulation of a variety of cellular functions that are dependent on integrin-mediated cell adhesion to the ECM, including cell spreading, migration, invasion, survival and proliferation. FAK regulates these cellular functions through its interactions with various other proteins and the activation of several intracellular signaling pathways.

2.4.1
Cell Adhesion and Spreading

Because of the correlation between FAK phosphorylation and cell adhesion observed since the initial discovery of FAK, it has been postulated that FAK might be involved in regulating cell adhesion to the ECM. However, there is no direct evidence to demonstrate a role for FAK in regulating integrin-mediated cell adhesion. Cultured FAK$^{-/-}$ cells show little difference in adhesion to fibronectin (FN) compared to control cells, and the overexpression of FAK in most cells does not alter cell adhesion to FN [82, 83]. Interestingly, several subsequent studies demonstrated the opposite effect of FAK on focal adhesion dynamics [85]. FAK$^{-/-}$ fibroblasts exhibit an increased number of focal adhesions, thus supporting the idea that FAK acts as a regulator of the disassembly of focal adhesions [83]. Live cell imaging analysis using fluorescently labeled focal adhesion components also supports a role for FAK in promoting focal adhesion disassembly [84, 86]. Quantitative measurement of focal adhesion turnover revealed that FAK signaling promotes focal adhesion turnover through activation of the MAP kinase signaling pathway and the proteinase calpain-2 [84, 87].

It has also been proposed that FAK controls the disassembly of focal adhesions through inhibiting Rho kinase activity. The adhesion of normal fibroblasts to FN results in the reduction of Rho activity; in contrast, the adhesion of FAK$^{-/-}$ cells does not reduce Rho activity [86, 88]. Furthermore, the activation of Rho in normal cells causes them to display similar morphology to FAK$^{-/-}$ cells, whereas inhibition of Rho activity in FAK$^{-/-}$ cells restores their morphology to that of normal cells. This suggests that the morphological differences between normal cells and FAK$^{-/-}$ cells, including the characteristic appearance of focal adhesions, are caused by different levels of Rho activity [86]. These studies support the idea that FAK regulates focal adhesion turnover by inhibiting Rho activity.

The FAK-mediated phosphorylation of paxillin might also play a role in the regulation of focal adhesion assembly by recruiting negative regulators of the focal adhesion complex. The phosphorylation sites of paxillin may serve as binding sites for CSK, an inhibitory kinase for Src and a potential negative regulator of focal adhesion complex, and PTP-PEST that dephosphorylates p130Cas [20, 89]. Grb2 association with FAK was implicated in focal adhesion turnover. As mentioned above, Y925 is located in C-terminal FAT domain, and partially overlaps with the paxillin binding site [63, 64]. Nuclear magnetic resonance (NMR) structural analysis showing that conformational rearrangement of the FAT domain is required to selectively promote either Y925 phosphorylation or paxillin association, and suggests that Grb2-binding to Y925 of FAK may negatively regulate paxillin binding by inducing a structural rearrangement, which in turn promotes the dissociation of FAK from focal contacts and subsequently leads to focal contact turnover [65, 90].

The results of several studies have suggested that FAK mediates the regulation of cell spreading on ECM proteins. FAK$^{-/-}$ cells show poor spreading when plated on FN compared to normal cells, even though the overexpression of FAK in various cultured cells does not affect morphology and cell spreading [82, 83, 91]. Furthermore, the re-expression of FAK in FAK$^{-/-}$ cells restores normal cell morphology, suggesting that FAK functions to promote cell spreading [34, 92]. Consistent with the data from FAK$^{-/-}$ cells, the overexpression of FRNK, a negative regulator of FAK, inhibited cell spreading on FN, and this is correlated with the reduced tyrosine phosphorylation of FAK [41]. Furthermore, the overexpression of FRNK resulted in reduced phosphorylation of paxillin in cells. The expression of c-Src in these cells and the consequent increase in the phosphorylation of paxillin could rescue the inhibitory effect of FRNK. These results suggest that paxillin is an important mediator of FAK regulation of cell spreading [93].

2.4.2
Cell Migration and Invasion

Perhaps the best-defined functional role for FAK signaling pathways is the regulation of cell migration, a multi-step process involving membrane protrusion through changes in the dynamics of the actin cytoskeleton and the assembly and disassembly of cell attachments. A role for FAK in regulating cell migration was first suggested by studies showing that a tyrosine kinase activity is required for endothelial cell migration, and the activity of FAK was correlated with this event [94]. More direct evidence came from a number of studies showing that the overexpression of several different dominant-negative mutants of FAK, including FRNK, FAT or the Y397F mutant, reduced cell migration [14, 53, 95, 96]. Conversely, it was demonstrated that the ability of the stable expression of FAK in various cell types to promote cell migration was dependent on the presence of Y397 and an intact Cas binding site [53, 97, 98].

Genetic studies also suggested FAK as a positive regulator of cell migration. Mouse knockout studies showed that FAK deficiency causes early embryonic lethality characterized by mesodermal defects similar to those of the FN knockout, and mesodermal cells derived from FAK null embryos showed defect in cell migration compared to those from normal embryos [83]. The re-expression of wild-type FAK, but not FAK mutants lacking kinase activity or the ability to bind to c-Src or Cas, rescued the cell migration defect of FAK$^{-/-}$ cells, suggesting that Cas phosphorylation by the FAK/Src complex is important for the regulation cell migration [34, 92]. Studies have also shown that inactivation of FAK by PTEN-induced dephosphorylation resulted in decreased cell migration in glioblastoma cell lines [79]. Interestingly, however, knockout of Shp2 or PTP-PEST–two phosphatases that can dephosphorylate FAK–led to decreased cell migration *in vitro* [75]. This suggested that dynamic processes, including repeated cycles of phosphorylation and dephosphorylation rather than increased tyrosine phosphorylation, *per se*, may be essential for the coordinated regulation of cell migration by FAK.

Recent studies also implicated FAK in controlling the chemotactic response of cells to stimuli such as growth factors. The overexpression of FAK in Madin–Darby canine kidney (MDCK) cells enhances cell migration in response to HGF, and inhibition of FAK activity by the expression of a dominant-negative mutant inhibits cell migration stimulated by urokinase-type plasminogen activator in MCF-7 cells [99, 100]. Furthermore, FAK$^{-/-}$ fibroblasts exhibit defects in cell migration in response to PDGF and EGF, which is rescued by re-expression of FAK, suggesting that FAK also is involved in regulating chemotactic cell migration [21].

Several FAK downstream pathways have been implicated in the regulation of cell migration. One well-characterized pathway involves FAK auto-phosphorylation at Y397, association of Src with phosphorylated Y397, and tyrosine phosphorylation of Cas by the FAK/Src complex, which depends on association of FAK with both Cas and Src [34, 82, 92, 101, 102]. Numerous studies have shown that Y397, the major auto-phosphorylation site of FAK, is required for FAK to enhance cell migration [34, 82]. Since Y397 is an auto-phosphorylation site, it was suggested that the kinase activity of FAK is necessary for regulating cell migration. Surprisingly, the overexpression of the kinase-dead mutant of FAK could also promote cell migration [34, 82]; however, it was found that the kinase-dead mutant can be efficiently phosphorylated at Y397 site in the cells, perhaps by endogenous FAK. This idea is supported by the observation that expression of the kinase-dead mutant in FAK$^{-/-}$ cells failed to promote cell migration, and this mutant did not show a significant level of phosphorylation in these cells [34]. Therefore, the kinase activity of FAK is important for promoting cell migration.

Formation of the FAK/Src complex by the binding of Src family kinases to Y397 is critical for FAK's ability to promote cell migration [82]. FAK-mediated recruitment of Src to focal adhesions has been suggested to control the turnover of these structures during cell migration [103]. Furthermore, fibroblasts that do not express the Src family members Src, Yes and Fyn (SYF-null cells) exhibited reduced migration on FN, although they maintained normal mitogenic response to PDGF stimulation [104]. Likewise, the expression of Csk, a Src inhibitory kinase, inhibited FAK-promoted cell migration. The critical importance of the FAK/Src/Cas pathway in cell migration has been shown in a number of cells including FAK$^{-/-}$ cells [34, 92]. The expression of a FAK mutant that is defective in binding and phosphorylating Cas failed to promote cell migration. The overexpression of a dominant-negative mutant of Cas that inhibits the recruitment of endogenous Cas to FAK also inhibits FAK-stimulated cell migration.

Tyrosine-phosphorylated Cas could associate with several SH2-containing proteins including Crk. Crk-transformed cells showed an increased association of Cas with cytoskeleton, suggesting that the Cas/Crk complex may control cell migration through the regulation of the cytoskeleton. Formation of the Cas/Crk complex has been shown to regulate membrane ruffling and cell migration through associating with DOCK180 and leading to the activation of Rac [105, 106]. A related pathway involves FAK/Src phosphorylation of another cytoskeletal adaptor molecule paxillin, which could also couple to Crk [20, 107]. Indeed, the inhibition of cell spreading by FRNK correlated with a decreased tyrosine phosphorylation of paxillin, and the overexpression of Src reversed both effects [41, 93]. It was also reported that tyrosine phosphorylation of paxillin and its association with Crk mediated the stimulation of a tumor cell line NBT-II migration on collagen [108]. However, in this system, Cas coupling to Crk played only a minor role in the regulation of cell migration.

A second pathway involves the formation of a FAK complex with PI3K. A FAK mutant that selectively disrupted the binding of PI3K, but not Src, failed to promote cell migration, although it induced the phosphorylation of Cas [109]. Likewise, inhibition of PI3K by wortmannin or LY294002 prevented FAK-stimulated cell migration. Interestingly, PI3K has also been shown to mediate α6β4-stimulated carcinoma invasion [110] as well as the regulation of cell migration by growth factor receptors [111]. One downstream effector of PI3K in the regulation of carcinoma invasion is the small GTP-binding protein, Rac, although other downstream targets may also be involved [110]. Hanks and colleagues [112] have reported recently that FAK also associates with PLCγ through Y397 to stimulate the activity of PLCγ. However, it is not yet clear whether the FAK/PLCγ complex also plays a role in the regulation of cell migration.

A third pathway in FAK regulation of cell migration involves its interaction with an adaptor molecule, Grb7, a family member of the SH2 domain-containing adaptor molecules that also include Grb10 and Grb14 [113]. Grb7 family proteins have been shown to interact with a variety of other cellular proteins including tyrosine kinase receptors and proto-oncogenes, which have been proposed to play a role in the regulation of mitogenic signaling pathways. The central domain of Grb7 family proteins is also called the GM (Grb and Mig) domain, which contains a region of approximately 300 amino acids showing high sequence homology (ca. 50% amino acid identity) to a *Caenorhabditis elegans* gene product, Mig-10 [114]. The Mig-10 protein has been shown to play a role in the long-range migration of neuronal cells during embryonic development. This in turn suggested a possible role for Grb7 in the regulation of migration of mammalian cells and, indeed, it was recently found that Grb7 interacts with FAK by binding to Y397 in an adhesion-dependent manner [115]. The overexpression of Grb7 in fibroblasts enhanced cell migration, whereas overexpression of the Grb7 SH2 domain alone inhibited cell migration. Further analysis indicated that formation of the FAK/Grb7 complex was independent of the FAK/PI3K and the FAK/Src complexes and the subsequent phosphorylation of Cas [115, 116]. It is not clear, however, whether the FAK/Grb7 pathway cooperates with these other two pathways triggered by FAK to regulate cell migration.

Actin remodeling is a crucial process in regulating cell migration. It has also been demonstrated that FAK can regulate cell migration by modulating the assembly and disassembly of actin through its effects on the Rho subfamily of small GTPases. The activity of RhoA is increased in FAK$^{-/-}$ cells and the reconstitution of FAK expression in these cells decreased the RhoA activity [86, 117, 118]. The inhibition of ROCK, a downstream effector of RhoA, could rescue the migration defect of FAK$^{-/-}$ cells [117]. Additional studies suggested that the phosphorylation and activation of p190RhoGEF could be stimulated by FAK, leading to RhoA activation [36, 119]. FAK association with GRAF, which contains GAP activity for RhoA and Cdc42, may also contribute to its regulation of actin dynamics through RhoA and/or Cdc42 [120]. Besides Rho subfamily GTPases, FAK has also been shown to interact with ASAP1, which is a GAP protein for Arf subfamily small GTPases that are involved in the regulation of actin dynamics [121]. Lastly, recent studies suggested that FAK can interact with and phosphorylate N-WASP, which acts downstream of activated Cdc42 to induce actin polymerization through interaction with the Arp2/3 complex [122]. N-WASP phosphorylation by FAK serves to maintain its cytoplasmic localization and thereby facilitate its ability to stimulate actin polymerization and cell migration. Therefore, FAK could regulate actin dynamics through multiple mechanisms which play integral roles in the promotion of cell migration in most cell types.

In addition to its well-characterized ability to promote cell migration, FAK has been shown to regulate the invasion of both normal and transformed cells by a number of mechanisms. The transformation of both $FAK^{-/-}$ and control cells with v-Src resulted in a similar stimulation of cell migration; however, v-Src transformation increased invasion of the control cells, but not the $FAK^{-/-}$ cells, suggesting a role for FAK in the regulation of cell invasion, which is distinct from its effects on cell migration [118]. Interestingly, a naturally occurring mutation in the SH3 domain of v-Src that enhances the association with and phosphorylation of FAK, promoted cell invasion [123]. The overexpression and increased phosphorylation of FAK have been demonstrated in a variety of human cancers [124], and are correlated with the invasive phenotype and metastasis [125, 126]. Additionally, the dominant-negative inhibition of FAK activity leads to a reduced v-src-stimulated cell invasion and blocked metastasis in nude mice [127]. Antisense-mediated FAK inhibition also decreases matrix metalloproteinase (MMP) expression in carcinoma cells, suggesting that FAK enhances invasion by promoting the expression of enzymes (MMPs) that have matrix-degrading activities [128]. Biochemically, it has been demonstrated that FAK promotes the assembly of a v-Src-Cas-Crk-DOCK180 complex that in turn activates Rac1 and JNK, resulting in the stimulation of MMP2 and MMP9 expression [118]. Recently, it has been reported that FAK regulates the surface expression of MT1-MMPs by controlling its endocytosis [129]. FAK has been shown to interact with endophilin A2 through its second proline-rich domain, and functions as a scaffolding protein to mediate the Src phosphorylation of endophilin A2 at Tyr315. The phosphorylation of endophilin A2 at Tyr315 inhibits its interaction with dynamin and thereby decreases the endocytosis of MT1-MMP. The resulting increased surface expression of MT1-MMP enhances the cellular degradation of the ECM.

In another study, it was shown that the activity of the FAK/src complex promotes the disruption of homotypic adhesion mediated by E-cadherin-based adherens junctions; the loss of the integrity of these junctions is associated with the malignant and invasive phenotype of carcinomas [130]. Also, the expression of an FAK mutant at the phosphorylation site inhibited the src-mediated disruption of E-cadherin-based contacts in carcinoma, suggesting the requirement of a phosphorylation-dependent FAK signaling in the disruption of these contacts [131].

A role of FAK in regulating cell invasion was also demonstrated in non-cancer cells [132]. During pregnancy, an invasion of cytotrophoblasts from the fetus into the maternal uterus is crucial in order for the fetus to gain access to blood vessels. It has been shown that the invasive fetal cells exhibit high levels of expression of FAK and its phosphorylation at Y397 residue. The inhibition of FAK by anti-sense treatment of the cytotrophoblasts reduced the invasive ability of these cells, which suggested a role for FAK in regulating the cell invasive phenotype in normal cells.

2.4.3
Cell Survival and Proliferation

It has long been known that cell adhesion to the ECM provides survival signals to many adherent cells. Upon detachment from the ECM, both epithelial cells and endothelial cells undergo apoptosis via a process termed anoikis [133]. Several studies have implicated FAK signaling pathways in the regulation of anoikis and other apoptotic responses. The expression of CD2-FAK, a constitutive active FAK mutant through membrane-targeting, prevented the detachment-induced anoikis of epithelial cells [133, 134]. Conversely, the inhibition of FAK activity by treatment with anti-sense oligonucleotide, the microinjection of an anti-FAK monoclonal antibody, or expression of the isolated C-terminal FAT domain that acts as dominant-negative mutant, leads to apoptosis, suggesting an important role for FAK in cell survival [135–137].

Further mutational studies of FAK demonstrated that the major autophosphorylation site, Y397, and the catalytic activity of FAK are necessary for protecting cells from anoikis [133, 138, 139]. In addition, FAK association with PI3K and Cas are important for FAK to inhibit apoptosis induced by ultraviolet irradiation [133, 138, 139]. FAK/PI3K interaction and PI3K activity was shown to be necessary to protect the endothelial cells from adenosine- and homocysteine-induced apoptosis [140]. It has been suggested that the protein kinase Akt, which is known to act as an inhibitor of apoptosis by regulating various cell death machinery proteins, may be the downstream target of the FAK/PI3K-mediated cell survival pathway [141]. Cas is also suggested to be involved in a FAK-mediated cell survival pathway [142]. The results of another study showed that FAK Y925, which is known to bind to Grb2 and to link FAK to the RAS-MAP kinase pathway, is also necessary to inhibit apoptosis [139]. A recent study showed that the association of FAK with receptor-interacting protein (RIP), a major component of the death-receptor complex that interacts with both FAS and tumor necrosis factor, mediates the ability of FAK to suppress apoptosis [143]. The suppression of p53 and the enhancement of NFκB activity by FAK also were implicated in FAK-mediated suppression of apoptosis [137, 139, 144].

Besides a positive role for cell survival, FAK has been shown to promote cell proliferation in a number of studies. A role for FAK in the regulation of cell proliferation was first suggested by showing that the inhibition of FAK activity by FRNK or a monoclonal antibody against FAK led to decreased DNA synthesis in endothelial cells [40, 135]. Conversely, an inducible expression of FAK leads to increased cell-cycle progression [145]. FAK increases DNA synthesis and accelerates the G_1/S transition by enhancing cyclin D1 and repressing Cdk inhibitor, p21 expression, whereas a dominant-negative mutant of FAK inhibits these steps. These effects depend on the phosphorylation of Y397 and require FAK association with c-Src and PI3-kinase at this site [146]. Furthermore, it has been demonstrated that FAK regulates the expression of KLF8, a transcription factor that was shown to enhance the transcription of cyclin D1 by directly binding to an upstream element in its promoter in NIH-3T3 cells. SiRNA-mediated inhibition of KLF8 resulted in a reduced expression of cyclin D1 and cell-cycle progression, suggesting a requirement for KLF8 in FAK's ability to promote the expression of cyclin D1 and cell proliferation [32]. Recent studies showed that the overexpression of FAK in smooth muscle cells leads to an increased cell proliferation through the increased expression of S-phase kinase-associated protein2 (skp2), which is known to be necessary for the degradation of cdk inhibitor, $p27^{kip1}$; this suggests that FAK also regulates cell-cycle progression by regulating skp2 expression [147–149].

FAK has also been suggested to promote cell-cycle progression in tumor cells. For example, in glioblastoma cells it has been shown that FAK overexpression enhances cell proliferation by reducing the expression of $p27^{kip1}$ and $p21^{waf1}$, and increasing the expression of cyclin D1 and E, whereas the FAK Y397F mutant increases the expression of $p27^{kip1}$ and $p21^{waf1}$ and reduced cyclin D1 and E expression [95].

2.5
Recent Analysis of FAK Functions *In Vivo*

Consistent with the critical roles of FAK *in vitro*, as revealed in studies using cell culture systems, recent studies using FAK knockout mice have demonstrated important biological functions of FAK *in vivo*. Null mutation of FAK in mouse embryonic stem cells by homologous recombination results in the death of derived embryos at embryonic day (E) 8.5 [150]. Mutant embryos up to E7.0 were indistinguishable from normal embryos, which suggests that FAK plays an important role in an embryonic developmental event that occurs during the period between E7.0 to E8.5. One major developmental process that occurs at this stage is the coordinated movement and differentiation of mesodermal cells. Consistently, FAK-null embryos exhibit fatal mesodermal defects, including the absence of somite formation, morphologically indistinguishable notochord, a rudimentary non-beating heart (or no heart at all), and the lack of a potent circulation. These results suggest an important role for FAK in the development of axial mesodermal tissues and the cardiovascular system [150]. However, several other studies have suggested that FAK expression may not be critical for mesodermal cell differentiation. For example, hybridization analysis using markers of somitogenetic mesoderm revealed the proper expression of those markers in FAK-null embryos, suggesting that the notochord and somite could be formed in FAK knockout embryos [150]. *In-vitro* assays analyzing the potential of differentiation of embryonic stem (ES) cells derived from FAK-null embryos revealed a normal formation of cartilaginous nodules with associated ossified areas, showing a possible normal differentiation of mesodermal cells [151]. In contrast, mesodermal cells derived from FAK-null embryos at E7.5 showed defects in migration in time-lapse video microscopy [83], suggesting that FAK plays a critical role in embryonic development through the regulation of mesodermal cell migration.

The lack of a normal heart and blood vessel development of FAK-null embryos suggested a functional role for FAK in the processes of heart development and angiogenesis [150]. However, the lethal phenotype of FAK-null embryo precludes the study of FAK's role in these processes. This has been overcome by generating conditional FAK knockout mice, wherein it has been shown that mice having a cardiac-ventricle specific deletion of FAK exhibit increased eccentric cardiac hypertrophy and fibrosis in response to Ang II stimulation, suggesting that FAK is a regulator of heart hypertrophy *in vivo* [152]. However, the results of another study showed that a myocyte-specific depletion of FAK exhibits reduced concentric hypertrophy in response to biomechanical stress to transverse aortic constriction, thereby suggesting a role for FAK in sensing and transducing biomechanical stress into a hypertropic response [153]. However, the discrepancy between the findings of the two studies could be explained by differences in the timing of FAK deletion and the extent of aortic constriction imposed in the two models.

A further physiological function of FAK in vascular development and angiogenesis has also been demonstrated by recent studies of the specific deletion of FAK in endothelial cells [154]. The inactivation of FAK results in defective angiogenesis in the embryo, yolk sac and placenta, leading to embryonic lethality as a consequence of the severe vascular developmental defects. Further studies showed that reduced embryonic cell (EC) survival upon FAK inactivation contributed to the defective vascular phenotype in the developing embryos. In addition, the deletion of FAK in ECs reduced EC migration in response to fibronectin and VEGF in wound-closure assays, and FAK-null EC migration defects could be rescued by the re-expression of FAK. The results of these studies suggested an important role for FAK in angiogenesis and vascular development through the regulation of multiple endothelial cell activities.

A strong expression of FAK in the neurons of the developing brain implies a role for FAK in neuronal development. It has been demonstrated that a tissue-specific deletion of FAK in neuronal and glial cell precursors results in severe cortical dysplasia resembling type II lissencephaly, thus implicating FAK in cortical neuron development [155]. FAK's role in the epithelial tissues has also been addressed in studies using mice having a keratinocyte-specific deletion of FAK [156]. These mutant mice exhibited defects in the hair cycle, sebaceous glands and epidermis, suggesting that the proliferation and/or migration of multi-potent epithelial stem cells in the bulge may be impaired in the absence of FAK signaling, though the underlying mechanism involved is not clear.

2.6
Conclusions

A variety of studies have demonstrated that FAK functions as an integrator of signals from several different components of the extracellular environment, such as ECM molecules and growth factors, by acting as both a scaffolding protein and a tyrosine kinase. The characterization of FAK-interacting proteins and reconstitution studies using various mutant forms of FAK have provided an important basis to delineate the mechanism of FAK signaling. In addition, the generation of FAK knockout mice model systems will continue to provide significant insights into the physiological function of FAK *in vivo*. Although many biochemical and physiological functions of FAK have currently been elucidated, much remains to be done before the mechanisms of FAK signaling are fully understood.

Acknowledgment

The studies described in this chapter were supported by NIH grants GM52890 and GM48050 to J.-L. Guan.

References

1. Hay, E.D., Extracellular matrix. *J. Cell Biol.* 1981, *91* (*3 Pt. 2*), 205s–223s.
2. Juliano, R.L. and S. Haskill, Signal transduction from the extracellular matrix. *J. Cell Biol.* 1993, *120*, 577–585.
3. Clark, E.A. and J.S. Brugge, Integrins and signal transduction pathways: the road taken. *Science* 1995, *268*, 233–239.
4. Hynes, R.O., Integrins: bidirectional, allosteric signaling machines. *Cell* 2002, *110*(6), 673–687.
5. Hynes, R.O., Integrins: a family of cell surface receptors. *Cell* 1987, *48*, 549–554.
6. Hynes, R.O., Integrins: versatility, modulation, and signaling in cell adhesion. *Cell* 1992, *69*, 11–25.
7. Assoian, R.K., Anchorage-dependent cell cycle progression. *J. Cell Biol.* 1997, *136*, 14.
8. Lauffenburger, D.A. and A.F. Horwitz, Cell migration: a physically integrated molecular process. *Cell* 1996, *84*, 359–369.
9. Ruoslahti, E. and J.C. Reed, Anchorage dependence, integrins, and apoptosis. *Cell* 1994, *77*, 477–478.
10. Guan, J.L. and D. Shalloway, Regulation of focal adhesion-associated protein tyrosine kinase by both cellular adhesion and oncogenic transformation. *Nature* 1992, *358*, 690–692.
11. Kornberg, L.J., et al., Signal transduction by integrins: increased protein tyrosine phosphorylation caused by clustering of beta 1 integrins. *Proc. Natl. Acad. Sci. USA* 1991, *88*, 8392–8396.
12. Guan, J.L., J.E. Trevithick, and R.O. Hynes, Fibronectin/integrin interaction induces tyrosine phosphorylation of a 120-kDa protein. *Cell Regul.* 1991, *2*, 951–964.
13. Schaller, M.D., et al., pp125FAK a structurally distinctive protein-tyrosine kinase associated with focal adhesions. *Proc. Natl. Acad. Sci. USA* 1992, *89*, 5192–5196.
14. Hanks, S.K., et al., Focal adhesion kinase signaling activities and their implications in the control of cell survival and motility. *Front. Biosci.* 2003, *8*, 982–996.
15. Parsons, J.T., Focal adhesion kinase: the first ten years. *J. Cell Sci.* 2003, *116* (*Pt. 8*), 1409–1416.

16 Hanks, S.K., et al., Focal adhesion protein-tyrosine kinase phosphorylated in response to cell attachment to fibronectin. *Proc. Natl. Acad. Sci. USA* 1992, *89*, 8487–8491.

17 Weiner, T.M., et al., Expression of focal adhesion kinase gene and invasive cancer. *Lancet* 1993, *342*, 1024–1025.

18 Hens, M.D. and D.W. DeSimone, Molecular analysis and developmental expression of the focal adhesion kinase pp125FAK in *Xenopus laevis*. *Dev. Biol.* 1995, *170*, 274–288.

19 Zhang, X., C.V. Wright, and S.K. Hanks, Cloning of a *Xenopus laevis* cDNA encoding focal adhesion kinase (FAK) and expression during early development. *Gene* 1995, *160*, 219–222.

20 Schaller, M.D. and J.T. Parsons, pp125FAK-dependent tyrosine phosphorylation of paxillin creates a high-affinity binding site for Crk. *Mol. Cell. Biol.* 1995, *15*, 2635–2645.

21 Sieg, D.J., et al., FAK integrates growth-factor and integrin signals to promote cell migration. *Nat. Cell Biol.* 2000, *2*, 249–256.

22 Dunty, J.M., et al., FERM domain interaction promotes FAK signaling. *Mol. Cell. Biol.* 2004, *24*, 5353–5368.

23 Sun, C.X., V.A. Robb, and D.H. Gutmann, Protein 4.1 tumor suppressors: getting a FERM grip on growth regulation. *J. Cell Sci.* 2002, *115* (Pt. 21), 3991–4000.

24 Chen, R., et al., Regulation of the PH-domain-containing tyrosine kinase Etk by focal adhesion kinase through the FERM domain. *Nat. Cell Biol.* 2001, *3*, 439–444.

25 Poullet, P., et al., Ezrin interacts with focal adhesion kinase and induces its activation independently of cell-matrix adhesion. *J. Biol. Chem.* 2001, *276*, 37686–37691.

26 Cooper, L.A., T.L. Shen, and J.L. Guan, Regulation of focal adhesion kinase by its amino-terminal domain through an autoinhibitory interaction. *Mol. Cell. Biol.* 2003, *23*, 8030–8041.

27 Medley, Q.G., et al., Signaling between focal adhesion kinase and trio. *J. Biol. Chem.* 2003, *278*, 13265–13270.

28 Toutant, M., et al., Alternative splicing controls the mechanisms of FAK autophosphorylation. *Mol. Cell. Biol.* 2002, *22*, 7731–7743.

29 Kadare, G., et al., PIAS1-mediated sumoylation of focal adhesion kinase activates its autophosphorylation. *J. Biol. Chem.* 2003, *278*, 47434–47440.

30 Jones, G. and G. Stewart, Nuclear import of N-terminal FAK by activation of the FcepsilonRI receptor in RBL-2H3 cells. *Biochem. Biophys. Res. Commun.* 2004, *314*, 39–45.

31 McKean, D.M., et al., FAK induces expression of Prx1 to promote tenascin-C-dependent fibroblast migration. *J. Cell Biol.* 2003, *161*, 393–402.

32 Zhao, J., et al., Identification of transcription factor KLF8 as a downstream target of focal adhesion kinase in its regulation of cyclin D1 and cell cycle progression. *Mol. Cell* 2003, *11*, 1503–1515.

33 Calalb, M.B., T.R. Polte, and S.K. Hanks, Tyrosine phosphorylation of focal adhesion kinase at sites in the catalytic domain regulates kinase activity: a role for Src family kinases. *Mol. Cell. Biol.* 1995, *15*, 954–963.

34 Owen, J.D., et al., Induced focal adhesion kinase (FAK) expression in FAK-null cells enhances cell spreading and migration requiring both auto- and activation loop phosphorylation sites and inhibits adhesion-dependent tyrosine phosphorylation of Pyk2. *Mol. Cell. Biol.* 1999, *19*, 4806–4818.

35 Chodniewicz, D. and R.L. Klemke, Regulation of integrin-mediated cellular responses through assembly of a CAS/Crk scaffold. *Biochim. Biophys. Acta* 2004, *1692*, 63–76.

36 Zhai, J., et al., Direct interaction of focal adhesion kinase with p190RhoGEF. *J. Biol. Chem.* 2003, *278*, 24865–24873.

37 Taylor, J.M., et al., Selective expression of an endogenous inhibitor of FAK regulates proliferation and migration of vascular smooth muscle cells. *Mol. Cell. Biol.* 2001, *21*, 1565–1572.

38 Schaller, M.D., C.A. Borgman, and J.T. Parsons, Autonomous expression of a noncatalytic domain of the focal

adhesion-associated protein tyrosine kinase pp125FAK. *Mol. Cell. Biol.* 1993, *13*, 785–791.

39. Nolan, K., J. Lacoste, and J.T. Parsons, Regulated expression of focal adhesion kinase-related nonkinase, the autonomously expressed C-terminal domain of focal adhesion kinase. *Mol. Cell. Biol.* 1999, *19*, 6120–6129.

40. Gilmore, A.P. and L.H. Romer, Inhibition of focal adhesion kinase (FAK) signaling in focal adhesions decreases cell motility and proliferation. *Mol. Biol. Cell* 1996, *7*, 1209–1224.

41. Richardson, A. and T. Parsons, A mechanism for regulation of the adhesion-associated protein tyrosine kinase pp125FAK. *Nature* 1996, *380*, 538–540.

42. Cary, L.A. and J.L. Guan, Focal adhesion kinase in integrin-mediated signaling. *Front. Biosci.* 1999, *4*, D102–D113.

43. Akiyama, S.K., et al., Transmembrane signal transduction by integrin cytoplasmic domains expressed in single-subunit chimeras. *J. Biol. Chem.* 1994, *269*, 15961–15964.

44. Leong, L., et al., Integrin signaling: roles for the cytoplasmic tails of alpha IIb beta 3 in the tyrosine phosphorylation of pp125FAK. *J. Cell Sci.* 1995, *108* (Pt.12), 3817–3825.

45. Lukashev, M.E., D. Sheppard, and R. Pytela, Disruption of integrin function and induction of tyrosine phosphorylation by the autonomously expressed beta 1 integrin cytoplasmic domain. *J. Biol. Chem.* 1994, *269*, 18311–18314.

46. Bockholt, S.M. and K. Burridge, Cell spreading on extracellular matrix proteins induces tyrosine phosphorylation of tensin. *J. Biol. Chem.* 1993, *268*, 14565–14567.

47. Lipfert, L., et al., Integrin-dependent phosphorylation and activation of the protein tyrosine kinase pp125FAK in platelets. *J. Cell Biol.* 1992, *119*, 905–912.

48. Schlaepfer, D.D., K.C. Jones, and T. Hunter, Multiple Grb2-mediated integrin-stimulated signaling pathways to ERK2/mitogen-activated protein kinase: summation of both c-Src- and focal adhesion kinase-initiated tyrosine phosphorylation events. *Mol. Cell. Biol.* 1998, *18*, 2571–2585.

49. Barry, S.T. and D.R. Critchley, The RhoA-dependent assembly of focal adhesions in Swiss 3T3 cells is associated with increased tyrosine phosphorylation and the recruitment of both pp125FAK and protein kinase C-delta to focal adhesions. *J. Cell Sci.* 1994, *107* (Pt. 7), 2033–2045.

50. Flinn, H.M. and A.J. Ridley, Rho stimulates tyrosine phosphorylation of focal adhesion kinase, p130 and paxillin. *J. Cell Sci.* 1996, *109* (Pt. 5), 1133–1141.

51. Seckl, M.J., et al., Guanosine 5'-3-O-(thio)triphosphate stimulates tyrosine phosphorylation of p125FAK and paxillin in permeabilized Swiss 3T3 cells. Role of p21rho. *J. Biol. Chem.* 1995, *270*, 6984–6990.

52. Wang, F., et al., Sphingosine 1-phosphate stimulates rho-mediated tyrosine phosphorylation of focal adhesion kinase and paxillin in Swiss 3T3 fibroblasts. *Biochem. J.* 1997, *324* (Pt. 2), 481–488.

53. Schlaepfer, D.D. and S.K. Mitra, Multiple connections link FAK to cell motility and invasion. *Curr. Opin. Genet. Dev.* 2004, *14*, 92–101.

54. Schaller, M.D. and J.T. Parsons, Focal adhesion kinase and associated proteins. *Curr. Opin. Cell Biol.* 1994, *6*, 705–710.

55. Maa, M.C. and T.H. Leu, Vanadate-dependent FAK activation is accomplished by the sustained FAK Tyr-576/577 phosphorylation. *Biochem. Biophys. Res. Commun.* 1998, *251*, 344–349.

56. Lim, Y., et al., Phosphorylation of focal adhesion kinase at tyrosine 861 is crucial for Ras transformation of fibroblasts. *J. Biol. Chem.* 2004, *279*, 29060–29065.

57. Eliceiri, B.P., et al., Src-mediated coupling of focal adhesion kinase to integrin alpha(v)beta5 in vascular endothelial growth factor signaling. *J. Cell Biol.* 2002, *157*, 149–160.

58. Nakamura, K., et al., Different modes and qualities of tyrosine phosphorylation of Fak and Pyk2 during epithelial-mesenchymal transdifferentiation and

cell migration: analysis of specific phosphorylation events using site-directed antibodies. *Oncogene* 2001, *20*, 2626–2635.

59 Chen, Q., et al., Integrin-mediated cell adhesion activates mitogen-activated protein kinases. *J. Biol. Chem.* 1994, *269*, 26602–26605.

60 Schlaepfer, D.D., et al., Integrin-mediated signal transduction linked to Ras pathway by GRB2 binding to focal adhesion kinase. *Nature* 1994, *372*, 786–791.

61 Ridley, A.J., et al., Cell migration: integrating signals from front to back. *Science* 2003, *302*, 1704–1709.

62 Hayashi, I., K. Vuori, and R.C. Liddington, The focal adhesion targeting (FAT) region of focal adhesion kinase is a four-helix bundle that binds paxillin. *Nat. Struct. Biol.* 2002, *9*, 101–106.

63 Katz, B.Z., et al., Targeting membrane-localized focal adhesion kinase to focal adhesions: roles of tyrosine phosphorylation and SRC family kinases. *J. Biol. Chem.* 2003, *278*, 29115–29120.

64 Liu, G., C.D. Guibao, and J. Zheng, Structural insight into the mechanisms of targeting and signaling of focal adhesion kinase. *Mol. Cell. Biol.* 2002, *22*, 2751–2760.

65 Prutzman, K.C., et al., The focal adhesion targeting domain of focal adhesion kinase contains a hinge region that modulates tyrosine 926 phosphorylation. *Structure* 2004, *12*, 881–891.

66 Turner, C.E., Paxillin and focal adhesion signalling. *Nat. Cell Biol.* 2000, *2*, E231–E236.

67 Brunton, V.G., et al., Identification of Src-specific phosphorylation site on focal adhesion kinase: dissection of the role of Src SH2 and catalytic functions and their consequences for tumor cell behavior. *Cancer Res.* 2005, *65*, 1335–1342.

68 Bianchi, M., et al., Regulation of FAK Ser-722 phosphorylation and kinase activity by GSK3 and PP1 during cell spreading and migration. *Biochem. J.* 2005, *391* (Pt. 2), 359–370.

69 Xie, Z., et al., Serine 732 phosphorylation of FAK by Cdk5 is important for microtubule organization, nuclear movement, and neuronal migration. *Cell* 2003, *114*, 469–482.

70 Ma, A., et al., Serine phosphorylation of focal adhesion kinase in interphase and mitosis: a possible role in modulating binding to p130(Cas). *Mol. Biol. Cell* 2001, *12*, 1–12.

71 Yamakita, Y., et al., Dissociation of FAK/p130(CAS)/c-Src complex during mitosis: role of mitosis-specific serine phosphorylation of FAK. *J. Cell Biol.* 1999, *144*, 315–324.

72 Abbi, S., et al., Regulation of focal adhesion kinase by a novel protein inhibitor FIP200. *Mol. Biol. Cell* 2002, *13*, 3178–3191.

73 Liu, E., J.F. Cote, and K. Vuori, Negative regulation of FAK signaling by SOCS proteins. *EMBO J.* 2003, *22*, 5036–5046.

74 von Wichert, G., et al., Force-dependent integrin-cytoskeleton linkage formation requires downregulation of focal complex dynamics by Shp2. *EMBO J.* 2003, *22*, 5023–5035.

75 Yu, D.H., et al., Protein-tyrosine phosphatase Shp-2 regulates cell spreading, migration, and focal adhesion. *J. Biol. Chem.* 1998, *273*, 21125–21131.

76 Zeng, L., et al., PTP alpha regulates integrin-stimulated FAK autophosphorylation and cytoskeletal rearrangement in cell spreading and migration. *J. Cell Biol.* 2003, *160*, 137–146.

77 Manes, S., et al., Concerted activity of tyrosine phosphatase SHP-2 and focal adhesion kinase in regulation of cell motility. *Mol. Cell. Biol.* 1999, *19*, 3125–3135.

78 Miao, H., et al., Activation of EphA2 kinase suppresses integrin function and causes focal-adhesion-kinase dephosphorylation. *Nat. Cell Biol.* 2000, *2*, 62–69.

79 Tamura, M., et al., Inhibition of cell migration, spreading, and focal adhesions by tumor suppressor PTEN. *Science* 1998, *280*, 1614–1617.

80 Tamura, M., et al., Tumor suppressor PTEN inhibition of cell invasion, migration, and growth: differential

involvement of focal adhesion kinase and p130Cas. *Cancer Res.* 1999, *59*, 442–449.

81 Tamura, M., et al., PTEN interactions with focal adhesion kinase and suppression of the extracellular matrix-dependent phosphatidylinositol 3-kinase/Akt cell survival pathway. *J. Biol. Chem.* 1999, *274*, 20693–20703.

82 Cary, L.A., J.F. Chang, and J.L. Guan, Stimulation of cell migration by overexpression of focal adhesion kinase and its association with Src and Fyn. *J. Cell Sci.* 1996, *109 (Pt. 7)*, 1787–1794.

83 Ilic, D., et al., Reduced cell motility and enhanced focal adhesion contact formation in cells from FAK-deficient mice. *Nature* 1995, *377*, 539–544.

84 Webb, D.J., et al., FAK-Src signalling through paxillin, ERK and MLCK regulates adhesion disassembly. *Nat. Cell Biol.* 2004, *6*, 154–161.

85 Fincham, V.J., J.A. Wyke, and M.C. Frame, v-Src-induced degradation of focal adhesion kinase during morphological transformation of chicken embryo fibroblasts. *Oncogene* 1995, *10*, 2247–2252.

86 Ren, X.D., et al., Focal adhesion kinase suppresses Rho activity to promote focal adhesion turnover. *J. Cell Sci.* 2000, *113 (Pt. 20)*, 3673–3678.

87 Franco, S., B. Perrin, and A. Huttenlocher, Isoform specific function of calpain 2 in regulating membrane protrusion. *Exp. Cell Res.* 2004, *299*, 179–187.

88 Ren, X.D., W.B. Kiosses, and M.A. Schwartz, Regulation of the small GTP-binding protein Rho by cell adhesion and the cytoskeleton. *EMBO J.* 1999, *18*, 578–585.

89 Shen, Y., et al., Direct association of protein-tyrosine phosphatase PTP-PEST with paxillin. *J. Biol. Chem.* 1998, *273*, 6474–6481.

90 Gao, G., et al., NMR solution structure of the focal adhesion targeting domain of focal adhesion kinase in complex with a paxillin LD peptide: evidence for a two-site binding model. *J. Biol. Chem.* 2004, *279*, 8441–8451.

91 Hildebrand, J.D., M.D. Schaller, and J.T. Parsons, Identification of sequences required for the efficient localization of the focal adhesion kinase, pp125FAK, to cellular focal adhesions. *J. Cell Biol.* 1993, *123*, 993–1005.

92 Sieg, D.J., C.R. Hauck, and D.D. Schlaepfer, Required role of focal adhesion kinase (FAK) for integrin-stimulated cell migration. *J. Cell Sci.* 1999, *112 (Pt. 16)*, 2677–2691.

93 Richardson, A., et al., Inhibition of cell spreading by expression of the C-terminal domain of focal adhesion kinase (FAK) is rescued by coexpression of Src or catalytically inactive FAK: a role for paxillin tyrosine phosphorylation. *Mol. Cell. Biol.* 1997, *17*, 6906–6914.

94 Romer, L.H., et al., Tyrosine kinase activity, cytoskeletal organization, and motility in human vascular endothelial cells. *Mol. Biol. Cell* 1994, *5*, 349–361.

95 Ding, Q., et al., p27Kip1 and cyclin D1 are necessary for focal adhesion kinase regulation of cell cycle progression in glioblastoma cells propagated *in vitro* and *in vivo* in the scid mouse brain. *J. Biol. Chem.* 2005, *280*, 6802–6815.

96 Hauck, C.R., et al., Inhibition of focal adhesion kinase expression or activity disrupts epidermal growth factor-stimulated signaling promoting the migration of invasive human carcinoma cells. *Cancer Res.* 2001, *61*, 7079–7090.

97 Natarajan, M., et al., HEF1 is a necessary and specific downstream effector of FAK that promotes the migration of glioblastoma cells. *Oncogene* 2006, *25*, 1721–1732.

98 Wang, D., et al., p125 focal adhesion kinase promotes malignant astrocytoma cell proliferation *in vivo*. *J. Cell Sci.* 2000, *113 (Pt. 23)*, 4221–4230.

99 Lai, J.F., et al., Involvement of focal adhesion kinase in hepatocyte growth factor-induced scatter of Madin-Darby canine kidney cells. *J. Biol. Chem.* 2000, *275*, 7474–7480.

100 Nguyen, D.H., et al., Urokinase-type plasminogen activator stimulates the Ras/Extracellular signal-regulated kinase (ERK) signaling pathway and MCF-7 cell migration by a mechanism that requires focal adhesion kinase, Src, and Shc. Rapid dissociation of GRB2/Sps-Shc complex is associated with the transient

phosphorylation of ERK in urokinase-treated cells. *J. Biol. Chem.* 2000, *275*, 19382–19388.
101 Klemke, R.L., et al., CAS/Crk coupling serves as a "molecular switch" for induction of cell migration. *J. Cell Biol.* 1998, *140*, 961–972.
102 Cary, L.A., et al., Identification of p130Cas as a mediator of focal adhesion kinase- promoted cell migration. *J. Cell Biol.* 1998, *140*, 211–221.
103 Fincham, V.J. and M.C. Frame, The catalytic activity of Src is dispensable for translocation to focal adhesions but controls the turnover of these structures during cell motility. *EMBO J.* 1998, *17*, 81–92.
104 Klinghoffer, R.A., et al., Src family kinases are required for integrin but not PDGFR signal transduction. *EMBO J.* 1999, *18*, 2459–2471.
105 Cheresh, D.A., J. Leng, and R.L. Klemke, Regulation of cell contraction and membrane ruffling by distinct signals in migratory cells. *J. Cell Biol.* 1999, *146*, 1107–1116.
106 Cho, S.Y. and R.L. Klemke, Extracellular-regulated kinase activation and CAS/Crk coupling regulate cell migration and suppress apoptosis during invasion of the extracellular matrix. *J. Cell Biol.* 2000, *149*, 223–236.
107 Birge, R.B., et al., Identification and characterization of a high-affinity interaction between v-Crk and tyrosine-phosphorylated paxillin in CT10-transformed fibroblasts. *Mol. Cell. Biol.* 1993, *13*, 4648–4656.
108 Petit, V., et al., Phosphorylation of tyrosine residues 31 and 118 on paxillin regulates cell migration through an association with CRK in NBT-II cells. *J. Cell Biol.* 2000, *148*, 957–970.
109 Reiske, H.R., et al., Requirement of phosphatidylinositol 3-kinase in focal adhesion kinase- promoted cell migration. *J. Biol. Chem.* 1999, *274*, 12361–12366.
110 Shaw, L.M., et al., Activation of phosphoinositide 3-OH kinase by the alpha6beta4 integrin promotes carcinoma invasion. *Cell* 1997, *91*, 949–960.
111 Kundra, V., et al., Regulation of chemotaxis by the platelet-derived growth factor receptor- beta. *Nature* 1994, *367*, 474–476.
112 Zhang, X., et al., Focal adhesion kinase promotes phospholipase C-gamma1 activity. *Proc. Natl. Acad. Sci. USA* 1999, *96*, 9021–9026.
113 Shen, T.L. and J.L. Guan, Grb7 in intracellular signaling and its role in cell regulation. *Front. Biosci.* 2004, *9*, 192–200.
114 Manser, J., C. Roonprapunt, and B. Margolis, C. elegans cell migration gene mig-10 shares similarities with a family of SH2 domain proteins and acts cell nonautonomously in excretory canal development. *Dev. Biol.* 1997, *184*, 150–164.
115 Han, D.C. and J.L. Guan, Association of focal adhesion kinase with Grb7 and its role in cell migration. *J. Biol. Chem.* 1999, *274*, 24425–24430.
116 Han, D.C., T.L. Shen, and J.L. Guan, Role of Grb7 targeting to focal contacts and its phosphorylation by focal adhesion kinase in regulation of cell migration. *J. Biol. Chem.* 2000, *275*, 28911–28917.
117 Chen, B.H., et al., Roles of Rho-associated kinase and myosin light chain kinase in morphological and migratory defects of focal adhesion kinase-null cells. *J. Biol. Chem.* 2002, *277*, 33857–33863.
118 Hsia, D.A., et al., Differential regulation of cell motility and invasion by FAK. *J. Cell Biol.* 2003, *160*, 753–767.
119 Schlaepfer, D.D., S.K. Mitra, and D. Ilic, Control of motile and invasive cell phenotypes by focal adhesion kinase. *Biochim. Biophys. Acta* 2004, *1692*, 77–102.
120 Hildebrand, J.D., J.M. Taylor, and J.T. Parsons, An SH3 domain-containing GTPase-activating protein for Rho and Cdc42 associates with focal adhesion kinase. *Mol. Cell. Biol.* 1996, *16*, 3169–3178.
121 Liu, Y., et al., The association of ASAP1, an ADP ribosylation factor-GTPase activating protein, with focal adhesion kinase contributes to the process of focal adhesion assembly. *Mol. Biol. Cell* 2002, *13*, 2147–2156.

122 Wu, X., et al., Focal adhesion kinase regulation of N-WASP subcellular localization and function. *J. Biol. Chem.* 2004, *279*, 9565–9576.

123 Hauck, C.R., et al., v-Src SH3-enhanced interaction with focal adhesion kinase at beta 1 integrin-containing invadopodia promotes cell invasion. *J. Biol. Chem.* 2002, *277*, 12487–12490.

124 Owens, L.V., et al., Overexpression of the focal adhesion kinase (p125FAK) in invasive human tumors. *Cancer Res.* 1995, *55*, 2752–2755.

125 Cance, W.G., et al., Immunohistochemical analyses of focal adhesion kinase expression in benign and malignant human breast and colon tissues: correlation with preinvasive and invasive phenotypes. *Clin. Cancer Res.* 2000, *6*, 2417–2423.

126 Kornberg, L.J., Focal adhesion kinase and its potential involvement in tumor invasion and metastasis. *Head Neck* 1998, *20*, 745–752.

127 Hauck, C.R., et al., FRNK blocks v-Src-stimulated invasion and experimental metastases without effects on cell motility or growth. *EMBO J.* 2002, *21*, 6289–6302.

128 Shibata, K., et al., Both focal adhesion kinase and c-Ras are required for the enhanced matrix metalloproteinase 9 secretion by fibronectin in ovarian cancer cells. *Cancer Res.* 1998, *58*, 900–903.

129 Wu, X., et al., FAK-mediated src phosphorylation of endophilin A2 inhibits endocytosis of MT1-MMP and promotes ECM degradation. *Dev. Cell* 2005, *9*, 185–196.

130 Irby, R.B. and T.J. Yeatman, Increased Src activity disrupts cadherin/catenin-mediated homotypic adhesion in human colon cancer and transformed rodent cells. *Cancer Res.* 2002, *62*, 2669–2674.

131 Avizienyte, E., et al., Src-induced de-regulation of E-cadherin in colon cancer cells requires integrin signalling. *Nat. Cell Biol.* 2002, *4*, 632–638.

132 Ilic, D., et al., Plasma membrane-associated pY397FAK is a marker of cytotrophoblast invasion *in vivo* and *in vitro*. *Am. J. Pathol.* 2001, *159*, 93–108.

133 Frisch, S.M., et al., Control of adhesion-dependent cell survival by focal adhesion kinase. *J. Cell Biol.* 1996, *134*, 793–799.

134 Chan, P.Y., et al., A transmembrane-anchored chimeric focal adhesion kinase is constitutively activated and phosphorylated at tyrosine residues identical to pp125FAK. *J. Biol. Chem.* 1994, *269*, 20567–20574.

135 Hungerford, J.E., et al., Inhibition of pp125FAK in cultured fibroblasts results in apoptosis. *J. Cell Biol.* 1996, *135*, 1383–1390.

136 Xu, L.H., et al., Attenuation of the expression of the focal adhesion kinase induces apoptosis in tumor cells. *Cell Growth Differ.* 1996, *7*, 413–418.

137 Ilic, D., et al., Extracellular matrix survival signals transduced by focal adhesion kinase suppress p53-mediated apoptosis. *J. Cell Biol.* 1998, *143*, 547–560.

138 Chan, P.C., et al., Suppression of ultraviolet irradiation-induced apoptosis by overexpression of focal adhesion kinase in Madin-Darby canine kidney cells. *J. Biol. Chem.* 1999, *274*, 26901–26906.

139 Sonoda, Y., et al., Anti-apoptotic role of focal adhesion kinase (FAK). Induction of inhibitor-of-apoptosis proteins and apoptosis suppression by the overexpression of FAK in a human leukemic cell line, HL-60. *J. Biol. Chem.* 2000, *275*, 16309–16315.

140 Bellas, R.E., et al., FAK blunts adenosine-homocysteine-induced endothelial cell apoptosis: requirement for PI 3-kinase. *Am. J. Physiol. Lung Cell. Mol. Physiol.* 2002, *282*, L1135–L1142.

141 Vivanco, I. and C.L. Sawyers, The phosphatidylinositol 3-Kinase AKT pathway in human cancer. *Nat. Rev. Cancer* 2002, *2*, 489–501.

142 Almeida, E.A., et al., Matrix survival signaling: from fibronectin via focal adhesion kinase to c-Jun NH(2)-terminal kinase. *J. Cell Biol.* 2000, *149*, 741–754.

143 Kurenova, E., et al., Focal adhesion kinase suppresses apoptosis by binding to the death domain of receptor-interacting protein. *Mol. Cell. Biol.* 2004, *24*, 4361–4371.

144 Deveraux, Q.L. and J.C. Reed, IAP family proteins – suppressors of apoptosis. *Genes Dev.* 1999, *13*, 239–252.

145 Zhao, J.H., H. Reiske, and J.L. Guan, Regulation of the cell cycle by focal adhesion kinase. *J. Cell Biol.* 1998, *143*, 1997–2008.

146 Zhao, J., R. Pestell, and J.L. Guan, Transcriptional activation of cyclin D1 promoter by FAK contributes to cell cycle progression. *Mol. Biol. Cell* 2001, *12*, 4066–4077.

147 Bond, M., G.B. Sala-Newby, and A.C. Newby, Focal adhesion kinase (FAK)-dependent regulation of S-phase kinase-associated protein-2 (Skp-2) stability. A novel mechanism regulating smooth muscle cell proliferation. *J. Biol. Chem.* 2004, *279*, 37304–37310.

148 Carrano, A.C., et al., SKP2 is required for ubiquitin-mediated degradation of the CDK inhibitor p27. *Nat. Cell Biol.* 1999, *1*, 193–199.

149 Sutterluty, H., et al., p45SKP2 promotes p27Kip1 degradation and induces S phase in quiescent cells. *Nat. Cell Biol.* 1999, *1*, 207–214.

150 Furuta, Y., et al., Mesodermal defect in late phase of gastrulation by a targeted mutation of focal adhesion kinase, FAK. *Oncogene* 1995, *11*, 1989–1995.

151 Ilic, D., et al., Focal adhesion kinase is not essential for *in vitro* and *in vivo* differentiation of ES cells. *Biochem. Biophys. Res. Commun.* 1995, *209*, 300–309.

152 Peng, X., et al., Inactivation of focal adhesion kinase in cardiomyocytes promotes eccentric cardiac hypertrophy and fibrosis in mice. *J. Clin. Invest.* 2006, *116*, 217–227.

153 DiMichele, L.A., et al., Myocyte-restricted focal adhesion kinase deletion attenuates pressure overload-induced hypertrophy. *Circ. Res.* 2006, *99*, 636–645.

154 Shen, T.L., et al., Conditional knockout of focal adhesion kinase in endothelial cells reveals its role in angiogenesis and vascular development in late embryogenesis. *J. Cell Biol.* 2005, *169*, 941–952.

155 Beggs, H.E., et al., FAK deficiency in cells contributing to the basal lamina results in cortical abnormalities resembling congenital muscular dystrophies. *Neuron* 2003, *40*, 501–514.

156 Essayem, S., et al., Hair cycle and wound healing in mice with a keratinocyte-restricted deletion of FAK. *Oncogene* 2006, *25*, 1081–1089.

3
The Paxillin Family and Tissue Remodeling
David A. Tumbarello and Christopher E. Turner

3.1
Focal Adhesions and the Paxillin Superfamily

Cell adhesion is essential for the maintenance of tissue architecture and the regulation of cell behavior associated with differentiation, morphogenesis, growth, and survival [1]. The remodeling of distinct intercellular adhesion and extracellular matrix (ECM) adhesion complexes is necessary for the proper development of the organism, as well as for normal tissue repair. In addition, at different stages of embryonic development – such as mesoderm formation during gastrulation and neural crest cell migration – dynamic changes in tissue and cell architecture are required, which are intimately associated with the transition to a motile phenotype [2, 3]. Alternatively, inappropriate signaling events may occur that disrupt the normal tissue architecture, thereby leading to a variety of pathologies, which include atherosclerosis, tumorigenesis, cancer metastasis, and tissue fibrosis [4–6]. Biologically, these processes occur through modulations in cell-to-cell and cell-to-ECM interactions triggered by various external cues. These include adhesion to different ECM components as well as the binding of various soluble growth factors to their appropriate receptors that allow for cell migration and morphogenesis to proceed. Fundamental to this process is the linkage of the internal actin cytoskeleton to the extracellular environment; this is referred to as a focal adhesion.

3.1.1
Focal Adhesions

Focal adhesions are comprised of a complex of intracellular structural and signaling proteins that are linked to the extracellular environment by members of the heterodimeric transmembrane integrin receptor family that bind to a number of different extracellular ligands, including laminin, vitronectin, fibronectin, and collagen [7, 8] (Figure 3.1). Integrins are generally considered to be inactive in their non-ligand bound state, but become active and cluster upon ligand binding or inside-out signaling events, thereby resulting in an increase in their avidity for the ECM and strengthening of adhesion [8, 9]. This results in the activation of

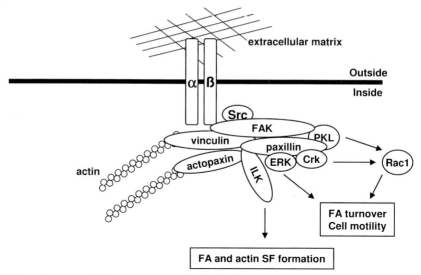

Figure 3.1 A simplified representation of a focal adhesion (FA). SF, stress fibers. For details of other abbreviations, see the text.

outside-in signaling events that regulate actin cytoskeleton organization, focal adhesion formation, gene expression, and cell survival [10]. Integrins provide links from the ECM to the internal actin cytoskeleton, with the exception of the α6β4 integrin heterodimer, which is present in the hemidesmosomes of epidermal cells and is linked to the internal intermediate filament system [8, 9].

Downstream of integrin-mediated cell adhesion, multiple signaling networks and regulatory mechanisms exist that function to maintain the dynamic nature of focal adhesions and the actin cytoskeleton, essential for proper cell adhesion and migration. One of the critical components of focal adhesions is the multi-domain adapter/scaffold protein *paxillin*. It serves to coordinate integrin-mediated signal transduction through the recruitment and activation of a variety of these downstream signaling networks, including those mediated by the SH2-SH3 adapter Crk, the integrin-linked kinase (ILK), the ARF-GAP PKL/GIT2, and the p21 Rho GTPase family [11]. Paxillin has been identified as one of the earliest proteins recruited to focal adhesions following integrin-mediated adhesion [12], and is an essential component in embryonic development as paxillin knockout mice die before embryonic day (E) 9.5. Furthermore, fibroblasts harvested from these null embryos exhibit major defects in integrin signaling and cell migration [13], the likely cause of the embryonic lethality. Thus, the ability of paxillin to process signals from integrin activation in order to modulate downstream targets involved in actin-cytoskeletal remodeling and focal adhesion turnover position it as an important component involved in the regulation of tissue morphogenesis [11].

3.1.2
The Paxillin Superfamily

The predominant form of paxillin, paxillin α, encodes a ubiquitously expressed 68 kDa protein that was first identified as a substrate for the Rous sarcoma virus v-Src oncoprotein [14–16]. Moreover, it was found subsequently also to be a member of a larger superfamily that consists of paxillin α, β, γ, and δ, and the homologues Hic-5 and Leupaxin [16–20] (Figure 3.2). The paxillin β and γ isoforms are the products of alternative splicing, resulting from the insertion of 34 and 48 amino acids, respectively, between amino acids 277 and 278 [19]. Paxillin β exhibits a restricted expression in monocytic cells of the adult organism, while paxillin γ is predominantly expressed in cancer cells [19, 21]. Paxillin δ protein expression results from alternative translation at an internal methionine within the full-length paxillin mRNA transcript [18]. Paxillin δ lacks the first 132 amino acids of paxillin, and it encodes a protein of approximately 46 kDa. It is enriched in cells of epithelial origin and its expression is downregulated in response to transforming growth factor (TGF) β1-induced epithelial–mesenchymal transformation (EMT) in the NMuMG cell line [18].

Hic-5 (hydrogen peroxide-inducible clone-5) was originally identified in a screen for TGFβ1 and hydrogen peroxide-inducible genes [20]. Hic-5 exhibits a more restricted tissue expression pattern compared to its family member paxillin, which suggests that it may play a distinct role in regulating cellular phenotype [22]. Hic-5

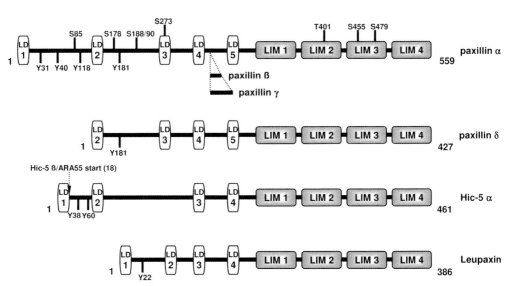

Figure 3.2 The paxillin superfamily. The leucine-rich (LD) motif and LIM (lin-11, isl-1, mec-3) domain regions are represented along with tyrosine (Y), serine (S), and threonine (T) phosphorylation sites.

is primarily expressed in smooth muscle and mesenchymal tissue layers, such as those surrounding the intestines, bronchial airways, and the metanephric kidney [22–24]. In contrast, Hic-5 expression is absent in the epithelium of the stomach, colon, skin, and mammary gland, while paxillin is present in these tissue layers [22]. Hic-5 expression is also restricted to the stromal cell layers of the prostate gland that have undergone smooth muscle differentiation, and to myoepithelial cells of the mammary, sweat, and salivary glands [22, 25–27]. Interestingly, with regards to mammary gland development, Hic-5 expression is elevated in the virgin mammary gland, while during pregnancy its expression decreases, followed by an elevation of expression during later stages of weaning [28]. Furthermore, Hic-5 protein expression is upregulated and is necessary for TGFβ1-induced EMT in a kidney proximal tubule epithelial cell line [29]. Interestingly, paxillin-null fibroblasts derived from knockout mice exhibit elevated levels of Hic-5 expression, while paxillin-null ES cells do not express Hic-5 [13, 30, 31]. In addition, in paxillin-null fibroblasts, Hic-5 is unable to fully complement paxillin function, but re-expression of Hic-5 in paxillin-null embryonic stem (ES) cells rescues their cell spreading and focal adhesion kinase (FAK) phosphorylation defect [13, 31, 32]. Conversely, a forced expression of Hic-5 in fibroblasts inhibited cell spreading and integrin-dependent FAK activation [33], suggesting a possible antagonistic relationship between paxillin and Hic-5. Consequently, the interrelationship between paxillin and Hic-5 function with regards to integrin-mediated signaling may not only be complex, but is also quite likely to be important in regulating distinct downstream signaling pathways. Of interest at this point, there is increasing evidence suggesting that the differential expression of paxillin family members–and in particular the full-length paxillin α, paxillin δ and Hic-5 isoforms within a given cell population–plays an important role in defining cell morphology, cytoarchitecture and integrin signaling, for example during the process of EMT (see below) [18, 29]. Lastly, Leupaxin has a restricted expression in cells of hematopoietic origin [17], and functions as an adapter protein in the osteoclast podosome complex [34].

3.1.3
Structure of Paxillin

Paxillin α is evolutionarily conserved across species including human, mouse, frog, zebrafish, fly, slime mold, and yeast [35–41]. The human paxillin gene is comprised of 11 exons that are localized to chromosome 12q24 [35]. The paxillin α isoform can be divided into two halves: the amino terminus (aa 1–313) contains five leucine-rich (LD) motifs, multiple potential SH2-binding tyrosine residues (Y31, Y40, Y88, Y118, Y181), several serine and threonine phosphorylation sites, and a proline-rich region potentially functioning as a SH3-binding site for Src **(see Figure 3.2)** [11, 42–47]. The LD motifs provide docking sites for a variety of structural and signaling proteins, which include vinculin, actopaxin, the ARF-GAP paxillin kinase linker (PKL/GIT2) and the related GIT1, ILK, FAK and its family member PYK2/RAFTK/CAKß [15, 48–52]. Upon tyrosine phosphorylation, several of the potential SH2 binding sites have been shown directly to bind Crk, Csk, the Csk homologous kinase (Chk), the p85 subunit of phosphatidylinositol kinase-3 (PI3K), the MAPK family member ERK, and p120RasGAP [53–57], and indirectly to modulate FAK binding to the LD motifs [58], while serine phosphorylation of the amino terminus (aa 273) regulates interactions with GIT1 [59]. The carboxy terminus of paxillin α (aa 314–559), consisting of four double-zinc finger motifs known as lin-11, isl-1, mec-3 (LIM) domains, is necessary for paxillin focal adhesion targeting, and also contains binding sites for tubulin, the syndecan-binding protein syndesmos, and the protein tyrosine phosphatase PTP-PEST [52, 60–63]. Serine/threonine phosphorylation of the carboxy terminus plays an important role in regulating paxillin recruitment to focal adhesions to facilitate cell migration [64]. It is through the spatial and temporal modulation of these multiple interactions that paxillin most likely mediates the cellular response to changes in the extracellular environment. Importantly, paxillin δ lacks several amino-terminal functional domains when compared to paxillin α, which include the LD1 motif and the primary SH2 binding sites, tyrosine 31 and tyrosine 118. Therefore, when coexpressed with full-length paxillin, as is observed in epithelial cells, paxillin δ might serve as a competitive inhibitor to suppress integrin signaling [18].

3.1.4
Structure of Hic-5

Hic-5 was originally cloned as a protein of 444 amino acids from a mouse osteoblastic cell line [20]. This protein was independently isolated as an androgen receptor coactivator and named ARA55 [65]. Subsequently, a longer Hic-5 cDNA encoding an extra 17 amino acids at the amino terminus was isolated from a mouse E12.5 cDNA library [36]. This full-length Hic-5 is now designated Hic-5 α, and the shorter isoform Hic-5 β/ARA55 [11, 66]. The human Hic-5 gene consists of 11 exons localized to chromosome 16p11-12, and encodes a protein with a molecular weight of 55 kDa [67]. The general domain structure of Hic-5 α, while similar to paxillin α, comprises an amino-terminal region containing only four LD motifs (three LD motifs for the Hic-5 β/ARA55 isoform) as well as two potential SH2-binding consensus sites (Y38 and Y60), and a carboxy-terminal region containing four LIM domains [36]. The amino-terminal LD motifs have been shown to interact directly with FAK, PYK2, actopaxin, PKL, GIT1, and vinculin [48, 50, 68–71]. In addition, the two tyrosine residues at amino acid positions 38 and 60, when phosphorylated, can function as SH2-binding sites for Crk and Csk [36, 72, 73]. Like paxillin α, the carboxy-terminal LIM domains have been shown to be necessary for focal adhesion targeting and binding to PTP-PEST, but contain additional binding sites for the TGFβ1 receptor effector Smad3, the heat shock protein 27 (Hsp27), tumor necrosis factor receptor-associated factor (TRAF) 4, the peroxisome proliferator-activated receptor gamma (PPARγ), the dopamine transporter (DAT), and syndesmos [61, 74–79].

3.2
The Paxillin Superfamily and Tissue Remodeling

3.2.1
Epithelial–Mesenchymal Transformation

Epithelial cells have the ability to convert to a mesenchymal or fibroblast-like cell phenotype through the process of EMT, which requires an intricate spatiotemporal coordination of signal transduction events [80]. Physiologically, EMT occurs during various stages of embryogenesis (including gastrulation) where the epiblast gives rise to the primary mesenchyme that migrates throughout the embryo and eventually forms the mesodermal layer. EMT also results in the generation of neural crest cells from the neuroepithelia; these migrate throughout the embryo, eventually differentiating into many different cell types that are essential for organ, craniofacial, and cardiac development [81, 82]. Additionally, the reverse process – referred to as mesenchymal–epithelial transformation [83] – occurs during developmental processes such as somitogenesis and kidney development [84]. EMT is also a common event contributing to a variety of pathologies including cancer metastasis and tissue fibrosis [4, 5]. Specifically, EMT is associated with the advanced stages of breast and colorectal cancer, and has been implicated in the progression of kidney tubulointerstitial fibrosis [5, 85–87].

Epithelial cells exhibit apical–basal polarity, characterized by an asymmetric distribution of cellular components, in order to provide the correct spatial distribution of proteins necessary for membrane trafficking and to maintain tissue structure and function [88]. The actin cytoskeleton is a crucial component of mature, polarized epithelial cells, contributing to tissue integrity through the generation of a circumferential actin ring associated with cell–cell adherens junctions [89, 90]. In contrast to epithelial cells, mesenchymal or fibroblast cells lack apical–basal polarity, and have an inability to form efficient intercellular junctions [91]. In two-dimensional (2D)-cell culture, mesenchymal cells adopt a flattened, spindle-shaped morphology that contributes to their ability to migrate as single cells [84, 92].

In order for epithelial cells to progress successfully through an EMT, they must undergo morphologic changes that result in the loss of epithelial apical–basal polarity and a gain of mesenchymal front-to-back polarity (Figure 3.3). Various growth factors (including TGFβ1) and ECM components (e.g. collagen) have the ability to initiate this process in distinct cell types [82]. EMT is characterized by the dissolution of intercellular junctions through the internalization and downregulation of various proteins present in tight junctions, such as zonula occludens (ZO)-1, and adherens junctions, such as E-cadherin. Additionally, remodeling of the actin cytoskeleton occurs, which is characterized by the loss of a circumferentially arranged actin cytoskeleton and the establishment of a stress fiber-oriented cytoskeleton. The latter is characterized by the formation of parallel actin bundles that traverse the cell body and terminate in robust and prominent focal adhesions that mediate cell–ECM adhesion to promote a motile phenotype.

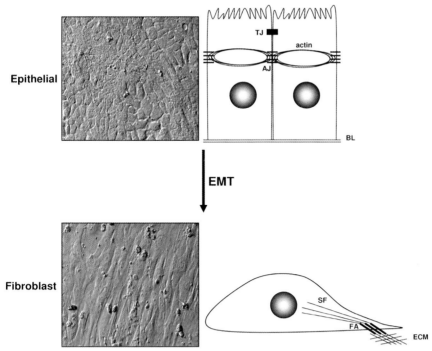

Figure 3.3 Epithelial–mesenchymal transformation (EMT). Left: Hoffmann modulation phase-contrast images of NMuMG cells either left untreated or treated for 48 h with TGFβ1 to induce EMT [18]. Right: A schematic representation of epithelial and fibroblast phenotypes. AJ, adherens junctions; BL, basal lamina; ECM, extracellular matrix; FA, focal adhesions; SF, actin stress fibers; TJ, tight junctions.

With regards to the paxillin family expression profile and EMT (Figure 3.4), paxillin α is expressed in both epithelial and mesenchymal cells, although tyrosine phosphorylation of paxillin α increases during EMT [18, 46]. Indeed, the high levels of paxillin α and FAK phosphorylation observed in the developing embryo are indicative of the increased cell migration and major tissue remodeling events that are occurring at that time [93, 94]. Paxillin δ and Hic-5 expression levels are reciprocally regulated during EMT, with paxillin δ exhibiting elevated levels in epithelial cells to potentially suppress integrin signaling and cell migration [18], and Hic-5 becoming elevated following EMT to promote cell motility through regulation of focal adhesion and actin stress fiber formation [18, 29].

3.2.2
Integrin-Mediated Signaling and EMT

Integrin-mediated signaling via cell–ECM adhesions is essential in facilitating an EMT, as well as maintaining the mesenchymal phenotype [82, 95, 96]. Alterations in integrin expression have been documented to occur during the transition to a

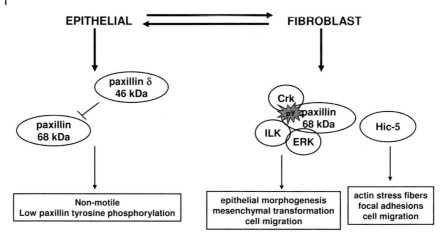

Figure 3.4 Paxillin expression dictates the cellular phenotype. In epithelial cells, full-length paxillin α signaling through its tyrosine phosphorylation sites is suppressed by the elevated expression of the 46 kDa paxillin δ isoform. During epithelial–mesenchymal transition (EMT) and the development of a mesenchymal/fibroblast phenotype, paxillin δ is downregulated and full-length paxillin α becomes tyrosine-phosphorylated, allowing signaling to occur through ILK, ERK, and Crk to stimulate cell migration, epithelial morphogenesis, and mesenchymal transformation. Additionally, Hic-5 protein expression is induced and functions to stimulate cell migration as well as focal adhesion and actin stress fiber formation.

mesenchymal phenotype, including upregulation of the α5β1 integrin receptor accompanied by the upregulation of its ligand, fibronectin, and a downregulation of epithelial-associated ECM receptors that reside at the basal surface, such as α6β1 and α6β4 [97–99]. Specifically, with regards to integrin function, the overexpression of a membrane-targeted β1 integrin into epithelial β1 null cells (GE11) stimulated the dissolution of intercellular junctions and the formation of cell–ECM adhesions, subsequently inducing a cell-scattering phenotype [100]. Alternatively, knockdown of β1 integrin expression inhibited Collagen I-induced EMT in the normal murine mammary gland epithelial (NMuMG) cell line [101]. Furthermore, with regards to TGFβ1-induced EMT in NMuMG cells, there exists (potentially) a TGFβ1 receptor-independent pathway necessary for EMT progression, which requires the activation of the beta1 integrin and subsequent activation of the MAPK family member p38 [102]. However, recent evidence suggests that this process may be more complicated, as it was later shown that both kinase-inactive TGFβ1 type I and II receptors block p38 phosphorylation and EMT, suggesting a potential TGFβ1 receptor-dependent component to this mechanism [103]. Regardless, modulation in integrin-mediated adhesion appears to be an essential component to EMT progression.

There are multiple proteins downstream of integrin activation that play a role in providing distinct signaling pathways necessary for EMT. For example, there are various kinases associated with focal adhesions (most via interactions with paxillin) that become activated, including FAK, PI3K, the Src family kinases, and ILK [104–106]. The Src family kinases have been shown to play an active role in regulating integrin- and cadherin-mediated adhesion. For example, elevated levels of Src activity and its peripheral localization along with phosphorylated myosin in colon carcinoma cells led to the formation of integrin-mediated adhesive complexes and the disruption of E-cadherin-based cell adherens junctions, characterized by a transition to a motile, mesenchymal phenotype [96, 107]. PI3K-dependent activation of the serine/threonine kinase Akt (also known as protein kinase B) is involved in E-cadherin downregulation and the promotion of EMT [108]. Furthermore, PI3K/Akt activation, along with activation of the p21 Rho GTPase family member RhoA, are suggested to play primary roles in mediating the morphological transformation associated with TGFβ1-induced EMT in the NMuMG cell line [105, 109]. Interestingly, ILK has been shown to be activated by PI3K, and functions to phosphorylate Akt on serine residue 473; this suggests that ILK may play a crucial role in the PI3K/Akt-dependent cytoskeletal remodeling associated with EMT [110]. In addition, ILK expression is capable of facilitating the dedifferentiation of epithelial cells to a mesenchymal cell phenotype [111], potentially through a LEF1/β-catenin-dependent mechanism to stimulate mesenchymal gene transcription [112] and through the stimulation of Snail expression, subsequently inducing the transcriptional repression of the E-cadherin gene [113, 114]. Finally, the remodeling of cell–cell and cell–ECM adhesions and the associated cytoskeleton that occurs during EMT, regulated by many of these upstream signaling cascades, is mediated through their influence on the activity of the Rho family of small GTPase comprising Cdc42, Rac and RhoA [115, 116].

Cdc42 is essential for the development and maintenance of epithelial apical–basal polarity and membrane trafficking, with Cdc42-dependent regulation of the Par6, Par3, and atypical protein kinase C (aPKC) complex being important for adherens and tight junction formation [117]. Rac1 performs multiple and somewhat conflicting roles. It has been implicated in the formation and maintenance of epithelial cell–cell adhesion as well as their disassembly [116], and also in the stimulation of lamellipodia formation to promote hepatocyte growth factor (HGF)-induced cell scattering [118]. The contribution of RhoA activity to tissue morphology is equally complex. For instance, the RhoA effectors ROCK and mDia have opposing effects on intercellular junctions. Specifically, ROCK activation of actomyosin contractility promotes the disassembly of cell–cell adherens junctions and the development of a contractile mesenchymal phenotype. In contrast, the formin family member, mDia, promotes the maintenance of adherens junctions and facilitates the proper localization of adherens junction proteins to sites of cell–cell contact [119]. Conversely, there is evidence to suggest that upon E-cadherin adhesion, RhoA activity is suppressed through the activation of p190RhoGAP [120], most likely via a Rac1 and p120 catenin-dependent mechanism [121]. In addition to its direct effects on the actin cytoskeleton, TGFβ1-stimulated, RhoA-dependent regulation of gene expression is also essential for a successful EMT [109, 122]. It is also clear that there is a delicate spatial and temporal relationship between RhoA and Rac1 activity that functions to regulate the balance between epithelial and fibroblast phenotypes [123]. The paxillin family members, through their ability to associate with many of these aforementioned signaling molecules, provide a platform to choreograph these events.

3.2.3
Paxillins α and δ in EMT

Paxillin is tyrosine-phosphorylated in response to integrin-mediated adhesion to multiple ECM ligands, such as fibronectin and collagen I, as well as in response to stimulation by multiple soluble agonists, such as TGFβ1, epidermal growth factor (EGF), bombesin, angiotensin II, and activin A [124–129]. The major tyrosine phosphorylation sites of paxillin reside at amino acid positions 31, 40, 88, 118, and 181 [11, 45, 46]. Following phosphorylation by the FAK/Src complex, the tyrosine 31 and 118 amino acid residues mediate interactions with the SH2 domain-containing proteins, Crk, Csk, ERK, and p120RasGAP [55, 56, 128, 130, 131]. Of the remaining tyrosine residues at amino acid positions 40, 88, and 181, only tyrosine 40 has been shown to have the ability to act as a potential SH2-binding site through an *in-vitro* interaction with the p85 subunit of PI3K [55].

In the NMuMG mammary epithelial cell line, paxillin is highly phosphorylated at tyrosine 31 and 118 during TGFβ1-induced EMT [46]. In addition, FAK and the related kinase Pyk2 both exhibit an increase in kinase activity and in tyrosine phosphorylation of various docking sites during this process [104], although the increase in Pyk2 phosphorylation was greater than that of FAK [104]. This raises the possibility that FAK and Pyk2 mediate different downstream signaling pathways, potentially via differential adapter function and phosphorylation of substrates, including paxillin [104]. Additionally, the ability of paxillin and Hic-5 to bind Pyk2 and FAK with differential affinities [71, 132] may further contribute to the divergent effects of these kinases downstream of integrin activation, such as their impact on cell migration and proliferation [133–136].

Importantly, paxillin phosphorylation has been implicated in processes associated with cancer progression, as well as epithelial morphogenesis. Specifically, an increase in its tyrosine phosphorylation correlated with a highly metastatic phenotype (analogous in many ways to EMT) in osteosarcoma cells, and the observed tyrosine phosphorylation was essential for the stimulation of cell migration in these cells [137]. In addition, paxillin Y31 and Y118 phosphorylation is also required for hepatoma cancer and Nara bladder tumor cell migration [128, 138]. Interestingly, paxillin Y118 phosphorylation may be important in epithelial morphogenesis through a direct interaction with ERK [56]. In response to HGF, the association of tyrosine 118-phosphorylated paxillin with ERK is necessary for ERK-induced paxillin serine 83 phosphorylation [56, 139]. In turn, the serine 83 phosphorylation of paxillin is required for HGF-induced cell migration and epithelial branching morphogenesis, as well as for NGF-induced neurite outgrowth in PC12 cells [139, 140]. In addition, perturbation of the paxillin/ERK association in Madin–Darby cancer kidney (MDCK) cells inhibits HGF-induced cell spreading and branching morphogenesis in three-dimensional (3D) collagen gels [56]. Therefore, paxillin phosphorylation and the subsequent activation of downstream signals may be an important component to facilitating cell migration and epithelial remodeling.

The adhesion-dependent phosphorylation of paxillin on Y31 and Y118 promotes the recruitment to focal adhesions of the SH2-SH3 adapter Crk [141], and is necessary for Crk-dependent cell spreading and collagen-induced cell migration [18, 128, 142]. Crk most likely mediates this process through its interaction with the DOCK180/ELMO complex that, in turn, functions as an atypical GEF for Rac1 [143–146]. The simultaneous recruitment of the PKL/PIX/PAK complex through the paxillin LD4 motif also appears to be important in facilitating this Crk-induced Rac1 activation [142].

In addition, tyrosine-phosphorylated paxillin can interact with other SH2 domain-containing proteins, suggesting a much more complex regulation of downstream signaling. For example, in NMuMG cells stimulated with TGFβ1, tyrosine-phosphorylated paxillin competes with the RhoA GAP, p190RhoGAP, for binding to the SH2 domain of the Ras GTPase-activating protein, p120RasGAP. The association of paxillin with p120RasGAP is thought to result in the displacement and subsequent activation of p190RhoGAP to inactivate RhoA [55]. Interestingly, in contrast to wild-type paxillin, a paxillin Y31F and Y118F phosphorylation mutant localized to abnormally large focal adhesions at the leading edge of a wounded monolayer of NMuMG cells and stimulated more robust actin stress fibers, indicative of a RhoA-like phenotype [46]. Thus, it was suggested that the induction of paxillin tyrosine phosphorylation at the leading edge of a wound may contribute to cell migration through a local inhibition of RhoA activity [55]. Alternatively, p120RasGAP – potentially through its ability to inactivate Ras – has been shown to be required for directed cell migration [147], which suggests that tyrosine-phosphorylated paxillin may be regulating migration through the direct recruitment of p120RasGAP to the leading edge. The precise function of a tyrosine-phosphorylated paxillin/p120RasGAP complex during cell migration remains to be determined. In any case, paxillin tyrosine phosphorylation appears to be a necessary component for the regulation of cell migration through potential interactions with Crk and p120RasGAP. Once again, an elevated expression of paxillin δ, which lacks the Y31 and Y118 phosphorylation sites, may be important in maintaining a sedentary epithelial phenotype through suppression of the paxillin/Crk/p120RasGAP signaling pathways [18].

The direct association of ILK with paxillin, via interaction with the LD1 motif, is required for ILK's proper localization to focal adhesions [48, 49]. In the human kidney proximal tubule epithelial (HKC) cell line, ILK expression is necessary for TGFβ1-induced EMT, and an increase in ILK expression is associated with chronic renal fibrosis [106]. ILK may also contribute to the development of stress fibers during EMT through stimulating RhoA, via a FAK-dependent process [148]. Due to ILK's established role as a mediator of EMT, the ability of paxillin to recruit ILK to focal adhesions may be an important regulatory mechanism for the modulation of cellular phenotype and the activation of ILK signaling. Conversely, paxillin δ, which lacks the ILK binding LD1 motif present in paxillin α, is elevated in epithelial cells where it may function to antagonize ILK function and, in so doing, suppress integrin signaling and help sustain the epithelial morphology.

3.2.4
Hic-5 and EMT

Although several *in-vitro* studies have indicated that Hic-5 may antagonize paxillin function [33, 36, 68], the fact that Hic-5 upregulation during EMT coincides with an increase in integrin signaling, including elevated paxillin phosphorylation, suggests a more cooperative role for these two proteins [18]. Further evidence for this is derived from the observation that the upregulated Hic-5 colocalizes with phosphorylated paxillin in the focal adhesions of the resulting fibroblasts and promotes their cell migration [18, 29; D. A. Tumbarello and C. E. Turner, unpublished observations].

As with paxillin, Hic-5 is integral to the Rho-mediated events associated with EMT. In kidney proximal tubule epithelial cells, RhoA activation is necessary for some of the TGFβ1-induced changes in gene expression during EMT, including the upregulation of the smooth muscle cell protein alpha-smooth muscle actin (α-sm actin) [122]. This RhoA- and ROCK-dependent process is mediated by a serum response factor (SRF)-induced mechanism [122]. RhoA and ROCK activation are also necessary for the induction of Hic-5 expression in response to TGFβ1 [29]. Although not formally tested, this also likely occurs via SRF, as the Hic-5 promoter contains a conserved CArG serum response element [149]. In turn, the upregulated Hic-5 is required for the RhoA- and ROCK-dependent formation of robust focal adhesions and stress fibers in these cells [29], indicating the potential for a feed-forward interrelationship between Hic-5 and RhoA (Figure 3.5). Importantly, the RhoA-dependent induction of smooth muscle cell proteins, such as α-sm actin and possibly Hic-5 in non-muscle cells (subsequently identified as myofibroblasts) may contribute to the pathologies associated with kidney tubulointerstitial fibrosis [5], by promoting increased cell contraction and tension to influence the remodeling of the ECM [150].

The manner in which Hic-5 stimulates RhoA activity is unclear, although Hic-5 phosphorylation is likely to be involved. Several studies have indicated that tyrosine-phosphorylated Hic-5 preferentially associates with the SH2 domain of the carboxy-terminal Src kinase (Csk) [36, 72, 73]. As Csk is a negative regulator of Src kinase activity [151], Hic-5 may locally regulate Rho GTPase activity through a recruitment of Csk to focal adhesions and suppression of Src/FAK dependent activation of p190RhoGAP [152].

In contrast to paxillin, Hic-5 likely plays an important gene regulatory role in cell morphology. It has been suggested that, with regards to TGFβ1-induced EMT, Smad-dependent pathways are predominantly involved in growth inhibitory signaling through modulations of cell cycle-associated proteins, while the alternative pathways, involving the mitogen-activated protein kinase (MAPK) family, Rho GTPases and protein kinase B/Akt signaling, are involved in the morphological transformation [105, 109, 153]. Interestingly, Hic-5 associates with the TGFβ1

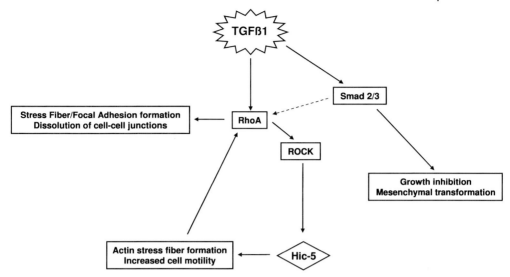

Figure 3.5 The role of Hic-5 in epithelial–mesenchymal transition (EMT) is RhoA/ROCK-dependent. TGFβ1 stimulation independently activates Smad-dependent pathways involved in growth inhibition and EMT along with a RhoA/ROCK signaling pathway leading to the upregulation of Hic-5 protein expression. Hic-5 stimulates actin stress fiber and focal adhesion formation, potentially in a RhoA/ROCK-dependent manner, indicating a positive feedback relationship. The dotted lines indicate a potential pathway.

receptor type I effector protein Smad3 and inhibits its activity [76]; this potentially indicates a role for Hic-5 in negatively regulating certain aspects of TGFβ1 signaling. Furthermore, it has been suggested that during cancer progression there is a switch from the tumor suppressor function of TGFβ1 to a pro-oncogenic function [87, 154, 155]. Therefore, pathways may exist downstream of TGFβ1 activation that promote tumorigenesis and EMT, whilst at the same time inhibiting Smad-dependent growth inhibition. In this regard, the induction of Hic-5 expression and subsequent association with Smad3 may function to suppress TGFβ1-induced growth arrest, while promoting the morphologic changes associated with TGFβ1-induced EMT. However, while these data indicate a proactive role for Hic-5 in EMT, it should be noted that Hic-5 also interacts with and regulates PPARγ and the LEF/TCF complex, which have been suggested to play a role in stimulating an epithelial program [79, 156].

Interestingly, Hic-5 also shuttles between focal adhesions and the nucleus to act as a transcriptional coactivator of glucocorticoid receptor (GR) and androgen receptor (AR) signaling, as well as a transcriptional cofactor for c-fos and p21 gene expression [65, 157–159]. The AR plays an important role in prostate development, as well as contributing to the progression of prostate cancer [160–162]. During prostate development, Hic-5 β/ARA55 is only present in stromal cells surrounding mitotically active, non-canalized acini, which have not completed arborization and ductal differentiation, and are specifically exhibiting smooth muscle differentiation [27]. This may indicate a role for Hic-5 in contributing to ductal morphogenesis and smooth muscle differentiation. With respect to prostate cancer, Hic-5 β/ARA55 is present at variable levels in prostate cancer samples, and a higher expression is associated with a shorter overall survival in hormone refractory prostate cancer [65, 163, 164]. Therefore, Hic-5 β/ARA55 may be associated with the progression from androgen-dependent to androgen-independent prostate cancer, which is resistant to anti-androgen therapy, by promoting downstream transactivation in the absence of hormone [160].

In summary, modulation in cell–ECM signaling features prominently in the morphogenic events associated with EMT, both in the physiologic context of tissue remodeling during embryonic development and in the pathology of tissue fibrosis and tumor cell metastasis. Evidence has been presented that the paxillin family, through differential expression and modulation of downstream effector pathways, plays a significant role in directing cells towards a particular phenotype. It will be interesting to determine in the future if selective manipulation of these proteins – for example the suppression of Hic-5 in fibrotic tissue or forced expression of paxillin δ in tumor cells – might ameliorate these conditions.

References

1 Gumbiner, B. M. (1996) Cell adhesion: the molecular basis of tissue architecture and morphogenesis. *Cell* 84, 345–357.

2 Gumbiner, B. M. (1992) Epithelial morphogenesis. *Cell* 69, 385–387.

3 Birchmeier, C., and Birchmeier, W. (1993) Molecular aspects of mesenchymal-epithelial interactions. *Annu. Rev. Cell Biol.* 9, 511–540.

4 Thiery, J. P. (2002) Epithelial-mesenchymal transitions in tumour progression. *Nature Rev. Cancer 2*, 442–454.

5 Kalluri, R., and Neilson, E. G. (2003) Epithelial-mesenchymal transition and its implications for fibrosis. *J. Clin. Invest.* 112, 1776–1784.

6 Thyberg, J., Hedin, U., Sjolund, M., Palmberg, L., and Bottger, B. A. (1990) Regulation of differentiated properties and proliferation of arterial smooth muscle cells. *Arteriosclerosis 10*, 966–990.

7 Wayner, E. A., and Carter, W. G. (1987) Identification of multiple cell adhesion receptors for collagen and fibronectin in human fibrosarcoma cells possessing unique alpha and common beta subunits. *J. Cell Biol.* 105, 1873–1884.

8 Hynes, R. O. (2002) Integrins: bidirectional, allosteric signaling machines. *Cell* 110, 673–87.

9 van der Flier, A., and Sonnenberg, A. (2001) Function and interactions of integrins. *Cell Tissue Res.* 305, 285–298.

10 Schwartz, M. A., Schaller, M. D., and Ginsberg, M. H. (1995) Integrins: emerging paradigms of signal transduction. *Annu. Rev. Cell Dev. Biol.* 11, 549–599.

11 Brown, M. C., and Turner, C. E. (2004) Paxillin: adapting to change. *Physiol. Rev.* 84, 1315–1339.

12 Zaidel-Bar, R., Cohen, M., Addadi, L., and Geiger, B. (2004) Hierarchical assembly of cell-matrix adhesion complexes. *Biochem. Soc. Trans. 32*, 416–420.

13 Hagel, M., George, E. L., Kim, A., Tamimi, R., Opitz, S. L., Turner, C. E., Imamoto, A., and Thomas, S. M. (2002) The adaptor protein paxillin is essential for normal development in the mouse

and is a critical transducer of fibronectin signaling. *Mol. Cell. Biol.* 22, 901–915.

14. Glenney, J. R.Jr, ., and Zokas, L. (1989) Novel tyrosine kinase substrates from Rous sarcoma virus-transformed cells are present in the membrane skeleton. *J. Cell Biol.* 108, 2401–2408.

15. Turner, C. E., and Miller, J. T. (1994) Primary sequence of paxillin contains putative SH2 and SH3 domain binding motifs and multiple LIM domains: identification of a vinculin and pp125Fak-binding region. *J. Cell Sci.* 107, 1583–1591.

16. Turner, C. E., Glenney, J. R., Jr., and Burridge, K. (1990) Paxillin: a new vinculin-binding protein present in focal adhesions. *J. Cell Biol.* 111, 1059–1068.

17. Lipsky, B. P., Beals, C. R., and Staunton, D. E. (1998) Leupaxin is a novel LIM domain protein that forms a complex with PYK2. *J. Biol. Chem.* 273, 11709–11713.

18. Tumbarello, D. A., Brown, M. C., Hetey, S. E., and Turner, C. E. (2005) Regulation of paxillin family members during epithelial-mesenchymal transformation: a putative role for paxillin delta. *J. Cell Sci.* 118, 4849–4863.

19. Mazaki, Y., Hashimoto, S., and Sabe, H. (1997) Monocyte cells and cancer cells express novel paxillin isoforms with different binding properties to focal adhesion proteins. *J. Biol. Chem.* 272, 7437–7444.

20. Shibanuma, M., Mashimo, J., Kuroki, T., and Nose, K. (1994) Characterization of the TGF beta 1-inducible hic-5 gene that encodes a putative novel zinc finger protein and its possible involvement in cellular senescence. *J. Biol. Chem.* 269, 26767–26774.

21. Mazaki, Y., Uchida, H., Hino, O., Hashimoto, S., and Sabe, H. (1998) Paxillin isoforms in mouse. Lack of the gamma isoform and developmentally specific beta isoform expression. *J. Biol. Chem.* 273, 22435–22441.

22. Yuminamochi, T., Yatomi, Y., Osada, M., Ohmori, T., Ishii, Y., Nakazawa, K., Hosogaya, S., and Ozaki, Y. (2003) Expression of the LIM proteins paxillin and Hic-5 in human tissues. *J. Histochem. Cytochem.* 51, 513–521.

23. Brunskill, E. W., Witte, D. P., Yutzey, K. E., and Potter, S. S. (2001) Novel cell lines promote the discovery of genes involved in early heart development. *Dev. Biol.* 235, 507–520.

24. Kim-Kaneyama, J. R., Suzuki, W., Ichikawa, K., Ohki, T., Kohno, Y., Sata, M., Nose, K., and Shibanuma, M. (2005) Uni-axial stretching regulates intracellular localization of Hic-5 expressed in smooth-muscle cells in vivo. *J. Cell Sci.* 118, 937–949.

25. Nessler-Menardi, C., Jotova, I., Culig, Z., Eder, I. E., Putz, T., Bartsch, G., and Klocker, H. (2000) Expression of androgen receptor coregulatory proteins in prostate cancer and stromal-cell culture models. *Prostate* 45, 124–131.

26. Li, P., Yu, X., Ge, K., Melamed, J., Roeder, R. G., and Wang, Z. (2002) Heterogeneous expression and functions of androgen receptor co-factors in primary prostate cancer. *Am. J. Pathol.* 161, 1467–1474.

27. Cai, G., Huang, H., Shapiro, E., Zhou, H., Yeh, S., Melamed, J., Greco, M. A., and Lee, P. (2005) Expression of androgen receptor associated protein 55 (ARA55) in the developing human fetal prostate. *J. Urol.* 173, 2190–2193.

28. Gao, Z., and Schwartz, L. M. (2005) Identification and analysis of Hic-5/ARA55 isoforms: Implications for integrin signaling and steroid hormone action. *FEBS Lett.* 579, 5651–5657.

29. Tumbarello, D. A., and Turner, C. E. (2007) Hic-5 contributes to epithelial-mesenchymal transformation through a RhoA/ROCK-dependent pathway. *J. Cell Physiol.* 211, 736–747.

30. Jamieson, J. S., Tumbarello, D. A., Halle, M., Brown, M. C., Tremblay, M. L., and Turner, C. E. (2005) Paxillin is essential for PTP-PEST-dependent regulation of cell spreading and motility: a role for paxillin kinase linker. *J. Cell Sci.* 118, 5835–5847.

31. Wade, R., Bohl, J., and Vande Pol, S. (2002) Paxillin null embryonic stem cells are impaired in cell spreading and

tyrosine phosphorylation of focal adhesion kinase. *Oncogene* 21, 96–107.

32 Wade, R., and Vande Pol, S. (2006) Minimal features of paxillin that are required for the tyrosine phosphorylation of focal adhesion kinase. *Biochem. J.* 393, 565–573.

33 Nishiya, N., Tachibana, K., Shibanuma, M., Mashimo, J. I., and Nose, K. (2001) Hic-5-reduced cell spreading on fibronectin: competitive effects between paxillin and Hic-5 through interaction with focal adhesion kinase. *Mol. Cell. Biol.* 21, 5332–5345.

34 Gupta, A., Lee, B. S., Khadeer, M. A., Tang, Z., Chellaiah, M., Abu-Amer, Y., Goldknopf, J., and Hruska, K. A. (2003) Leupaxin is a critical adaptor protein in the adhesion zone of the osteoclast. *J. Bone Miner. Res.* 18, 669–685.

35 Salgia, R., Li, J. L., Lo, S. H., Brunkhorst, B., Kansas, G. S., Sobhany, E. S., Sun, Y., Pisick, E., Hallek, M., Ernst, T., and et al. (1995) Molecular cloning of human paxillin, a focal adhesion protein phosphorylated by P210BCR/ABL. *J. Biol. Chem.* 270, 5039–5047.

36 Thomas, S. M., Hagel, M., and Turner, C. E. (1999) Characterization of a focal adhesion protein, Hic-5, that shares extensive homology with paxillin. *J. Cell Sci.* 112, 181–190.

37 Ogawa, M., Hiraoka, Y., Taniguchi, K., Sakai, Y., and Aiso, S. (2001) mRNA sequence of the Xenopus laevis paxillin gene and its expression. *Biochim. Biophys. Acta* 1519, 235–240.

38 Crawford, B. D., Henry, C. A., Clason, T. A., Becker, A. L., and Hille, M. B. (2003) Activity and distribution of paxillin, focal adhesion kinase, and cadherin indicate cooperative roles during zebrafish morphogenesis. *Mol. Biol. Cell* 14, 3065–3081.

39 Wheeler, G. N., and Hynes, R. O. (2001) The cloning, genomic organization and expression of the focal contact protein paxillin in Drosophila. *Gene* 262, 291–299.

40 Glockner, G., Eichinger, L., Szafranski, K., Pachebat, J. A., Bankier, A. T., Dear, P. H., Lehmann, R., Baumgart, C., Parra, G., Abril, J. F., Guigo, R., Kumpf, K., Tunggal, B., Cox, E., Quail, M. A., Platzer, M., Rosenthal, A., and Noegel, A. A. (2002) Sequence and analysis of chromosome 2 of Dictyostelium discoideum. *Nature* 418, 79–85.

41 Mackin, N. A., Sousou, T. J., and Erdman, S. E. (2004) The PXL1 gene of *Saccharomyces cerevisiae* encodes a paxillin-like protein functioning in polarized cell growth. *Mol. Biol. Cell* 15, 1904–1917.

42 Turner, C. E. (2000) Paxillin and focal adhesion signalling. *Nat. Cell Biol.* 2, E231–E236.

43 Turner, C. E. (2000) Paxillin interactions. *J. Cell Sci.* 113 (Pt. 23), 4139–4140.

44 Weng, Z., Taylor, J. A., Turner, C. E., Brugge, J. S., and Seidel-Dugan, C. (1993) Detection of Src homology 3-binding proteins, including paxillin, in normal and v-Src-transformed Balb/c 3T3 cells. *J. Biol. Chem.* 268, 14956–14963.

45 Schaller, M. D., and Schaefer, E. M. (2001) Multiple stimuli induce tyrosine phosphorylation of the Crk-binding sites of paxillin. *Biochem. J.* 360, 57–66.

46 Nakamura, K., Yano, H., Uchida, H., Hashimoto, S., Schaefer, E., and Sabe, H. (2000) Tyrosine phosphorylation of paxillin alpha is involved in temporospatial regulation of paxillin-containing focal adhesion formation and F-actin organization in motile cells. *J. Biol. Chem.* 275, 27155–27164.

47 Tumbarello, D. A., Brown, M. C., and Turner, C. E. (2002) The paxillin LD motifs. *FEBS Lett.* 513, 114–118.

48 Nikolopoulos, S. N., and Turner, C. E. (2000) Actopaxin, a new focal adhesion protein that binds paxillin LD motifs and actin and regulates cell adhesion. *J. Cell Biol.* 151, 1435–1448.

49 Nikolopoulos, S. N., and Turner, C. E. (2001) Integrin-linked kinase (ILK) binding to paxillin LD1 motif regulates ILK localization to focal adhesions. *J. Biol. Chem.* 276, 23499–23505.

50 Turner, C. E., Brown, M. C., Perrotta, J. A., Riedy, M. C., Nikolopoulos, S. N., McDonald, A. R., Bagrodia, S., Thomas, S., and Leventhal, P. S. (1999) Paxillin LD4 motif binds PAK and PIX through a novel 95-kD ankyrin repeat, ARF-GAP

protein: A role in cytoskeletal remodeling. *J. Cell Biol. 145*, 851–863.

51. Tachibana, K., Sato, T., D'Avirro, N., and Morimoto, C. (1995) Direct association of pp125FAK with paxillin, the focal adhesion-targeting mechanism of pp125FAK. *J. Exp. Med. 182*, 1089–1099.

52. Brown, M. C., Perrotta, J. A., and Turner, C. E. (1996) Identification of LIM3 as the principal determinant of paxillin focal adhesion localization and characterization of a novel motif on paxillin directing vinculin and focal adhesion kinase binding. *J. Cell Biol. 135*, 1109–1123.

53. Birge, R. B., Fajardo, J. E., Reichman, C., Shoelson, S. E., Songyang, Z., Cantley, L. C., and Hanafusa, H. (1993) Identification and characterization of a high-affinity interaction between v-Crk and tyrosine-phosphorylated paxillin in CT10-transformed fibroblasts. *Mol. Cell. Biol. 13*, 4648–4656.

54. Sabe, H., Kurosaki, T., Takata, M., and Hanafusa, H. (1994) Analysis of the binding of the Src homology 2 domain of Csk to tyrosine-phosphorylated proteins in the suppression and mitotic activation of c-Src. *Mol. Cell. Biol. 14*, 7306–7313.

55. Tsubouchi, A., Sakakura, J., Yagi, R., Mazaki, Y., Schaefer, E., Yano, H., and Sabe, H. (2002) Localized suppression of RhoA activity by Tyr31/118-phosphorylated paxillin in cell adhesion and migration. *J. Cell Biol. 159*, 673–683.

56. Ishibe, S., Joly, D., Zhu, X., and Cantley, L. G. (2003) Phosphorylation-dependent paxillin-ERK association mediates hepatocyte growth factor-stimulated epithelial morphogenesis. *Mol. Cell 12*, 1275–1285.

57. Grgurevich, S., Mikhael, A., and McVicar, D. W. (1999) The Csk homologous kinase, Chk, binds tyrosine phosphorylated paxillin in human blastic T cells. *Biochem. Biophys. Res. Commun. 256*, 668–675.

58. Zaidel-Bar, R., Milo, R., Kam, Z., and Geiger, B. (2007) A paxillin tyrosine phosphorylation switch regulates the assembly and form of cell-matrix adhesions. *J. Cell Sci. 120*, 137–148.

59. Nayal, A., Webb, D. J., Brown, C. M., Schaefer, E. M., Vicente-Manzanares, M., and Horwitz, A. R. (2006) Paxillin phosphorylation at Ser273 localizes a GIT1-PIX-PAK complex and regulates adhesion and protrusion dynamics. *J. Cell Biol. 173*, 587–589.

60. Herreros, L., Rodriguez-Fernandez, J. L., Brown, M. C., Alonso-Lebrero, J. L., Cabanas, C., Sanchez-Madrid, F., Longo, N., Turner, C. E., and Sanchez-Mateos, P. (2000) Paxillin localizes to the lymphocyte microtubule organizing center and associates with the microtubule cytoskeleton. *J. Biol. Chem. 275*, 26436–26440.

61. Denhez, F., Wilcox-Adelman, S. A., Baciu, P. C., Saoncella, S., Lee, S., French, B., Neveu, W., and Goetinck, P. F. (2002) Syndesmos, a syndecan-4 cytoplasmic domain interactor, binds to the focal adhesion adaptor proteins paxillin and Hic-5. *J. Biol. Chem. 277*, 12270–12274.

62. Shen, Y., Schneider, G., Cloutier, J. F., Veillette, A., and Schaller, M. D. (1998) Direct association of protein-tyrosine phosphatase PTP-PEST with paxillin. *J. Biol. Chem. 273*, 6474–6481.

63. Sanchez-Garcia, I., and Rabbitts, T. H. (1994) The LIM domain: a new structural motif found in zinc-finger-like proteins. *Trends Genet. 10*, 315–320.

64. Brown, M. C., Perrotta, J. A., and Turner, C. E. (1998) Serine and threonine phosphorylation of the paxillin LIM domains regulates paxillin focal adhesion localization and cell adhesion to fibronectin. *Mol. Biol. Cell 9*, 1803–1816.

65. Fujimoto, N., Yeh, S., and Chang, C. (1999) Cloning and characterization of androgen receptor coactivator, ARA55, in human prostate. *J. Biol. Chem. 274*, 8316–8321.

66. Mashimo, J., Shibanuma, M., Satoh, H., Chida, K., and Nose, K. (2000) Genomic structure and chromosomal mapping of the mouse hic-5 gene that encodes a focal adhesion protein. *Gene 249*, 99–103.

67. Zhang, J., Zhang, L. X., Meltzer, P. S., Barrett, J. C., and Trent, J. M. (2000) Molecular cloning of human Hic-5, a potential regulator involved in signal

transduction and cellular senescence. *Mol. Carcinogen.* 27, 177–183.

68 Nishiya, N., Shirai, T., Suzuki, W., and Nose, K. (2002) Hic-5 interacts with GIT1 with a different binding mode from paxillin. *J. Biochem. (Tokyo)* 132, 279–289.

69 Hagmann, J., Grob, M., Welman, A., van Willigen, G., and Burger, M. M. (1998) Recruitment of the LIM protein hic-5 to focal contacts of human platelets. *J. Cell Sci.* 111, 2181–2188.

70 Fujita, H., Kamiguchi, K., Cho, D., Shibanuma, M., Morimoto, C., and Tachibana, K. (1998) Interaction of Hic-5, A senescence-related protein, with focal adhesion kinase. *J. Biol. Chem.* 273, 26516–26521.

71 Matsuya, M., Sasaki, H., Aoto, H., Mitaka, T., Nagura, K., Ohba, T., Ishino, M., Takahashi, S., Suzuki, R., and Sasaki, T. (1998) Cell adhesion kinase beta forms a complex with a new member, Hic-5, of proteins localized at focal adhesions. *J. Biol. Chem.* 273, 1003–1014.

72 Ishino, M., Aoto, H., Sasaski, H., Suzuki, R., and Sasaki, T. (2000) Phosphorylation of Hic-5 at tyrosine 60 by CAKbeta and Fyn. *FEBS Lett.* 474, 179–183.

73 Hetey, S. E., Lalonde, D. P., and Turner, C. E. (2005) Tyrosine-phosphorylated Hic-5 inhibits epidermal growth factor-induced lamellipodia formation. *Exp. Cell Res.* 311, 147–156.

74 Nishiya, N., Iwabuchi, Y., Shibanuma, M., Cote, J. F., Tremblay, M. L., and Nose, K. (1999) Hic-5, a paxillin homologue, binds to the protein-tyrosine phosphatase PEST (PTP-PEST) through its LIM 3 domain. *J. Biol. Chem.* 274, 9847–9853.

75 Jia, Y., Ransom, R. F., Shibanuma, M., Liu, C., Welsh, M. J., and Smoyer, W. E. (2001) Identification and characterization of hic-5/ARA55 as an hsp27 binding protein. *J. Biol. Chem.* 276, 39911–39918.

76 Wang, H., Song, K., Sponseller, T. L., and Danielpour, D. (2005) Novel function of androgen receptor-associated protein 55/Hic-5 as a negative regulator of Smad3 signaling. *J. Biol. Chem.* 280, 5154–5162.

77 Carneiro, A. M., Ingram, S. L., Beaulieu, J. M., Sweeney, A., Amara, S. G., Thomas, S. M., Caron, M. G., and Torres, G. E. (2002) The multiple LIM domain-containing adaptor protein Hic-5 synaptically colocalizes and interacts with the dopamine transporter. *J. Neurosci.* 22, 7045–7054.

78 Wu, R. F., Xu, Y. C., Ma, Z., Nwariaku, F. E., Sarosi, G. A.Jr, ., and Terada, L. S. (2005) Subcellular targeting of oxidants during endothelial cell migration. *J. Cell Biol.* 171, 893–904.

79 Drori, S., Girnun, G. D., Tou, L., Szwaya, J. D., Mueller, E., Xia, K., Shivdasani, R. A., and Spiegelman, B. M. (2005) Hic-5 regulates an epithelial program mediated by PPARgamma. *Genes Dev.* 19, 362–375.

80 Greenburg, G., and Hay, E. D. (1982) Epithelia suspended in collagen gels can lose polarity and express characteristics of migrating mesenchymal cells. *J. Cell Biol.* 95, 333–339.

81 Fleischmajer, R., and Billingham, R. E. (Eds.), (1968) *Epithelial-mesenchymal interactions: 18th Hahnemann symposium.* Williams & Wilkins Co., Baltimore.

82 Savagner, P. (2001) Leaving the neighborhood: molecular mechanisms involved during epithelial-mesenchymal transition. *BioEssays* 23, 912–923.

83 Denis, C., Methia, N., Frenette, P. S., Rayburn, H., Ullman-Cullere, M., Hynes, R. O., and Wagner, D. D. (1998) A mouse model of severe von Willebrand disease: defects in hemostasis and thrombosis. *Proc. Natl. Acad. Sci. USA* 95, 9524–9529.

84 Thiery, J. P., and Sleeman, J. P. (2006) Complex networks orchestrate epithelial-mesenchymal transitions. *Nat. Rev. Mol. Cell. Biol.* 7, 131–142.

85 Gamallo, C., Palacios, J., Suarez, A., Pizarro, A., Navarro, P., Quintanilla, M., and Cano, A. (1993) Correlation of E-cadherin expression with differentiation grade and histological type in breast carcinoma. *Am. J. Pathol.* 142, 987–993.

86 Moll, R., Mitze, M., Frixen, U. H., and Birchmeier, W. (1993) Differential loss of E-cadherin expression in infiltrating ductal and lobular breast carcinomas. *Am. J. Pathol.* 143, 1731–1742.

87. Bierie, B., and Moses, H. L. (2006) TGF-beta and cancer. *Cytokine Growth Factor Rev. 17*, 29–40.
88. Rodriguez-Boulan, E., and Nelson, W. J. (1989) Morphogenesis of the polarized epithelial cell phenotype. *Science 245*, 718–725.
89. Owaribe, K., Kodama, R., and Eguchi, G. (1981) Demonstration of contractility of circumferential actin bundles and its morphogenetic significance in pigmented epithelium in vitro and in vivo. *J. Cell Biol. 90*, 507–514.
90. Zhang, J., Betson, M., Erasmus, J., Zeikos, K., Bailly, M., Cramer, L. P., and Braga, V. M. (2005) Actin at cell-cell junctions is composed of two dynamic and functional populations. *J. Cell Sci. 118*, 5549–5562.
91. Hay, E. D. (1995) An overview of epithelio-mesenchymal transformation. *Acta Anatom. 154*, 8–20.
92. Guarino, M., and Giordano, F. (1995) Experimental induction of epithelial-mesenchymal interconversions. *Exp. Toxicol. Pathol. 47*, 325–334.
93. Turner, C. E. (1991) Paxillin is a major phosphotyrosine-containing protein during embryonic development. *J. Cell Biol. 115*, 201–207.
94. Turner, C. E., Schaller, M. D., and Parsons, J. T. (1993) Tyrosine phosphorylation of the focal adhesion kinase pp125FAK during development: relation to paxillin. *J. Cell Sci. 105*, 637–645.
95. Thannickal, V. J., Lee, D. Y., White, E. S., Cui, Z., Larios, J. M., Chacon, R., Horowitz, J. C., Day, R. M., and Thomas, P. E. (2003) Myofibroblast differentiation by transforming growth factor-beta1 is dependent on cell adhesion and integrin signaling via focal adhesion kinase. *J. Biol. Chem. 278*, 12384–12389.
96. Avizienyte, E., Wyke, A. W., Jones, R. J., McLean, G. W., Westhoff, M. A., Brunton, V. G., and Frame, M. C. (2002) Src-induced de-regulation of E-cadherin in colon cancer cells requires integrin signalling. *Nat. Cell Biol. 4*, 632–638.
97. Maschler, S., Wirl, G., Spring, H., Bredow, D. V., Sordat, I., Beug, H., and Reichmann, E. (2005) Tumor cell invasiveness correlates with changes in integrin expression and localization. *Oncogene 24*, 2032–2041.
98. Zuk, A., and Hay, E. D. (1994) Expression of beta 1 integrins changes during transformation of avian lens epithelium to mesenchyme in collagen gels. *Dev. Dyn. 201*, 378–393.
99. Hay, E. D., and Zuk, A. (1995) Transformations between epithelium and mesenchyme: normal, pathological, and experimentally induced. *Am. J. Kidney Dis. 26*, 678–690.
100. Gimond, C., van Der Flier, A., Van Delft, S., Brakebusch, C., Kuikman, I., Collard, J. G., Fassler, R., and Sonnenberg, A. (1999) Induction of cell scattering by expression of beta1 integrins in beta1-deficient epithelial cells requires activation of members of the rho family of GTPases and downregulation of cadherin and catenin function. *J. Cell Biol. 147*, 1325–1340.
101. Shintani, Y., Wheelock, M. J., and Johnson, K. R. (2006) Phosphoinositide-3 kinase-Rac1-c-Jun NH2-terminal kinase signaling mediates collagen I-induced cell scattering and up-regulation of N-cadherin expression in mouse mammary epithelial cells. *Mol. Biol. Cell 17*, 2963–2975.
102. Bhowmick, N. A., Zent, R., Ghiassi, M., McDonnell, M., and Moses, H. L. (2001) Integrin beta 1 signaling is necessary for transforming growth factor-beta activation of p38MAPK and epithelial plasticity. *J. Biol. Chem. 276*, 46707–46713.
103. Bakin, A. V., Rinehart, C., Tomlinson, A. K., and Arteaga, C. L. (2002) p38 mitogen-activated protein kinase is required for TGFbeta-mediated fibroblastic transdifferentiation and cell migration. *J. Cell Sci. 115*, 3193–3206.
104. Nakamura, K., Yano, H., Schaefer, E., and Sabe, H. (2001) Different modes and qualities of tyrosine phosphorylation of Fak and Pyk2 during epithelial-mesenchymal transdifferentiation and cell migration: analysis of specific phosphorylation events using site-directed antibodies. *Oncogene 20*, 2626–2635.

105 Bakin, A. V., Tomlinson, A. K., Bhowmick, N. A., Moses, H. L., and Arteaga, C. L. (2000) Phosphatidylinositol 3-kinase function is required for transforming growth factor beta-mediated epithelial to mesenchymal transition and cell migration. *J. Biol. Chem. 275*, 36803–36810.

106 Li, Y., Yang, J., Dai, C., Wu, C., and Liu, Y. (2003) Role for integrin-linked kinase in mediating tubular epithelial to mesenchymal transition and renal interstitial fibrogenesis. *J. Clin. Invest. 112*, 503–516.

107 Avizienyte, E., Fincham, V. J., Brunton, V. G., and Frame, M. C. (2004) Src SH3/2 domain-mediated peripheral accumulation of Src and phospho-myosin is linked to deregulation of E-cadherin and the epithelial-mesenchymal transition. *Mol. Biol. Cell 15*, 2794–2803.

108 Larue, L., and Bellacosa, A. (2005) Epithelial-mesenchymal transition in development and cancer: role of phosphatidylinositol 3' kinase/AKT pathways. *Oncogene 24*, 7443–7454.

109 Bhowmick, N. A., Ghiassi, M., Bakin, A., Aakre, M., Lundquist, C. A., Engel, M. E., Arteaga, C. L., and Moses, H. L. (2001) Transforming growth factor-beta1 mediates epithelial to mesenchymal transdifferentiation through a RhoA-dependent mechanism. *Mol. Biol. Cell 12*, 27–36.

110 Delcommenne, M., Tan, C., Gray, V., Rue, L., Woodgett, J., and Dedhar, S. (1998) Phosphoinositide-3-OH kinase-dependent regulation of glycogen synthase kinase 3 and protein kinase B/AKT by the integrin-linked kinase. *Proc. Natl. Acad. Sci. USA 95*, 11211–11216.

111 Wu, C., Keightley, S. Y., Leung-Hagesteijn, C., Radeva, G., Coppolino, M., Goicoechea, S., McDonald, J. A., and Dedhar, S. (1998) Integrin-linked protein kinase regulates fibronectin matrix assembly, E-cadherin expression, and tumorigenicity. *J. Biol. Chem. 273*, 528–536.

112 Novak, A., Hsu, S. C., Leung-Hagesteijn, C., Radeva, G., Papkoff, J., Montesano, R., Roskelley, C., Grosschedl, R., and Dedhar, S. (1998) Cell adhesion and the integrin-linked kinase regulate the LEF-1 and beta-catenin signaling pathways. *Proc. Natl. Acad. Sci. USA 95*, 4374–4379.

113 Peinado, H., Quintanilla, M., and Cano, A. (2003) Transforming growth factor beta-1 induces snail transcription factor in epithelial cell lines: mechanisms for epithelial mesenchymal transitions. *J. Biol. Chem. 278*, 21113–21123.

114 Tan, C., Costello, P., Sanghera, J., Dominguez, D., Baulida, J., de Herreros, A. G., and Dedhar, S. (2001) Inhibition of integrin linked kinase (ILK) suppresses beta-catenin-Lef/Tcf-dependent transcription and expression of the E-cadherin repressor, snail, in APC-/- human colon carcinoma cells. *Oncogene 20*, 133–140.

115 Jaffe, A. B., and Hall, A. (2005) Rho GTPases: biochemistry and biology. *Annu. Rev. Cell Dev. Biol. 21*, 247–269.

116 Lozano, E., Betson, M., and Braga, V. M. (2003) Tumor progression: Small GTPases and loss of cell-cell adhesion. *BioEssays 25*, 452–463.

117 Etienne-Manneville, S. (2004) Cdc42–the centre of polarity. *J. Cell Sci. 117*, 1291–1300.

118 Lamorte, L., Royal, I., Naujokas, M., and Park, M. (2002) Crk adapter proteins promote an epithelial-mesenchymal-like transition and are required for HGF-mediated cell spreading and breakdown of epithelial adherens junctions. *Mol. Biol. Cell 13*, 1449–1461.

119 Sahai, E., and Marshall, C. J. (2002) ROCK and Dia have opposing effects on adherens junctions downstream of Rho. *Nat. Cell Biol. 4*, 408–415.

120 Noren, N. K., Arthur, W. T., and Burridge, K. (2003) Cadherin engagement inhibits RhoA via p190RhoGAP. *J. Biol. Chem. 278*, 13615–13618.

121 Wildenberg, G. A., Dohn, M. R., Carnahan, R. H., Davis, M. A., Lobdell, N. A., Settleman, J., and Reynolds, A. B. (2006) p120-catenin and p190RhoGAP regulate cell-cell adhesion by coordinating antagonism between Rac and Rho. *Cell 127*, 1027–1039.

122 Masszi, A., Di Ciano, C., Sirokmany, G., Arthur, W. T., Rotstein, O. D., Wang, J.,

McCulloch, C. A., Rosivall, L., Mucsi, I., and Kapus, A. (2003) Central role for Rho in TGF-beta1-induced alpha-smooth muscle actin expression during epithelial-mesenchymal transition. *Am. J. Physiol. Renal Physiol.* 284, F911–F924.

123 Zondag, G. C., Evers, E. E., ten Klooster, J. P., Janssen, L., van der Kammen, R. A., and Collard, J. G. (2000) Oncogenic Ras downregulates Rac activity, which leads to increased Rho activity and epithelial-mesenchymal transition. *J. Cell Biol.* 149, 775–782.

124 Riedy, M. C., Brown, M. C., Molloy, C. J., and Turner, C. E. (1999) Activin A and TGF-beta stimulate phosphorylation of focal adhesion proteins and cytoskeletal reorganization in rat aortic smooth muscle cells. *Exp. Cell Res.* 251, 194–202.

125 Turner, C. E., Pietras, K. M., Taylor, D. S., and Molloy, C. J. (1995) Angiotensin II stimulation of rapid paxillin tyrosine phosphorylation correlates with the formation of focal adhesions in rat aortic smooth muscle cells. *J. Cell Sci.* 108, 333–342.

126 Burridge, K., Turner, C. E., and Romer, L. H. (1992) Tyrosine phosphorylation of paxillin and pp125FAK accompanies cell adhesion to extracellular matrix: a role in cytoskeletal assembly. *J. Cell Biol.* 119, 893–903.

127 Zachary, I., Sinnett-Smith, J., Turner, C. E., and Rozengurt, E. (1993) Bombesin, vasopressin, and endothelin rapidly stimulate tyrosine phosphorylation of the focal adhesion-associated protein paxillin in Swiss 3T3 cells. *J. Biol. Chem.* 268, 22060–22065.

128 Petit, V., Boyer, B., Lentz, D., Turner, C. E., Thiery, J. P., and Valles, A. M. (2000) Phosphorylation of tyrosine residues 31 and 118 on paxillin regulates cell migration through an association with CRK in NBT-II cells. *J. Cell Biol.* 148, 957–970.

129 Tapia, J. A., Camello, C., Jensen, R. T., and Garcia, L. J. (1999) EGF stimulates tyrosine phosphorylation of focal adhesion kinase (p125FAK) and paxillin in rat pancreatic acini by a phospholipase C-independent process that depends on phosphatidylinositol 3-kinase, the small GTP-binding protein, p21rho, and the integrity of the actin cytoskeleton. *Biochim. Biophys. Acta* 1448, 486–499.

130 Schaller, M. D., and Parsons, J. T. (1995) pp125FAK-dependent tyrosine phosphorylation of paxillin creates a high-affinity binding site for Crk. *Mol. Cell. Biol.* 15, 2635–45.

131 Schaller, M. D., Hildebrand, J. D., and Parsons, J. T. (1999) Complex formation with focal adhesion kinase: A mechanism to regulate activity and subcellular localization of Src kinases. *Mol. Biol. Cell* 10, 3489–3505.

132 Osada, M., Ohmori, T., Yatomi, Y., Satoh, K., Hosogaya, S., and Ozaki, Y. (2001) Involvement of Hic-5 in platelet activation: integrin alphaIIbbeta3-dependent tyrosine phosphorylation and association with proline-rich tyrosine kinase 2. *Biochem. J.* 355, 691–697.

133 Zhao, J., Zheng, C., and Guan, J. (2000) Pyk2 and FAK differentially regulate progression of the cell cycle. *J. Cell Sci.* 113 (Pt 17), 3063–3072.

134 Lipinski, C. A., Tran, N. L., Bay, C., Kloss, J., McDonough, W. S., Beaudry, C., Berens, M. E., and Loftus, J. C. (2003) Differential role of proline-rich tyrosine kinase 2 and focal adhesion kinase in determining glioblastoma migration and proliferation. *Mol. Cancer Res.* 1, 323–332.

135 Jiang, X., Jacamo, R., Zhukova, E., Sinnett-Smith, J., and Rozengurt, E. (2006) RNA interference reveals a differential role of FAK and Pyk2 in cell migration, leading edge formation and increase in focal adhesions induced by LPA in intestinal epithelial cells. *J. Cell Physiol.* 207, 816–828.

136 Lipinski, C. A., Tran, N. L., Menashi, E., Rohl, C., Kloss, J., Bay, R. C., Berens, M. E., and Loftus, J. C. (2005) The tyrosine kinase pyk2 promotes migration and invasion of glioma cells. *Neoplasia* 7, 435–445.

137 Azuma, K., Tanaka, M., Uekita, T., Inoue, S., Yokota, J., Ouchi, Y., and Sakai, R. (2005) Tyrosine phosphorylation of paxillin affects the metastatic potential of human osteosarcoma. *Oncogene* 24, 4754–4764.

138 Iwasaki, T., Nakata, A., Mukai, M., Shinkai, K., Yano, H., Sabe, H., Schaefer, E., Tatsuta, M., Tsujimura, T., Terada, N., Kakishita, E., and Akedo, H. (2002) Involvement of phosphorylation of Tyr-31 and Tyr-118 of paxillin in MM1 cancer cell migration. *Int. J. Cancer* 97, 330–335.

139 Ishibe, S., Joly, D., Liu, Z. X., and Cantley, L. G. (2004) Paxillin serves as an ERK-regulated scaffold for coordinating FAK and Rac activation in epithelial morphogenesis. *Mol. Cell* 16, 257–267.

140 Huang, C., Borchers, C. H., Schaller, M. D., and Jacobson, K. (2004) Phosphorylation of paxillin by p38MAPK is involved in the neurite extension of PC-12 cells. *J. Cell Biol.* 164, 593–602.

141 Valles, A. M., Beuvin, M., and Boyer, B. (2004) Activation of Rac1 by paxillin-Crk-DOCK180 signaling complex is antagonized by Rap1 in migrating NBT-II cells. *J. Biol. Chem.* 279, 44490–44496.

142 Lamorte, L., Rodrigues, S., Sangwan, V., Turner, C. E., and Park, M. (2003) Crk associates with a multimolecular Paxillin/GIT2/beta-PIX complex and promotes Rac-dependent relocalization of Paxillin to focal contacts. *Mol. Biol. Cell* 14, 2818–2831.

143 Hasegawa, H., Kiyokawa, E., Tanaka, S., Nagashima, K., Gotoh, N., Shibuya, M., Kurata, T., and Matsuda, M. (1996) DOCK180, a major CRK-binding protein, alters cell morphology upon translocation to the cell membrane. *Mol. Cell Biol.* 16, 1770–1776.

144 Brugnera, E., Haney, L., Grimsley, C., Lu, M., Walk, S. F., Tosello-Trampont, A. C., Macara, I. G., Madhani, H., Fink, G. R., and Ravichandran, K. S. (2002) Unconventional Rac-GEF activity is mediated through the Dock180-ELMO complex. *Nat. Cell Biol.* 4, 574–582.

145 Gumienny, T. L., Brugnera, E., Tosello-Trampont, A. C., Kinchen, J. M., Haney, L. B., Nishiwaki, K., Walk, S. F., Nemergut, M. E., Macara, I. G., Francis, R., Schedl, T., Qin, Y., Van Aelst, L., Hengartner, M. O., and Ravichandran, K. S. (2001) CED-12/ELMO, a novel member of the CrkII/Dock180/Rac pathway, is required for phagocytosis and cell migration. *Cell* 107, 27–41.

146 Grimsley, C. M., Kinchen, J. M., Tosello-Trampont, A. C., Brugnera, E., Haney, L. B., Lu, M., Chen, Q., Klingele, D., Hengartner, M. O., and Ravichandran, K. S. (2004) Dock180 and ELMO1 proteins cooperate to promote evolutionarily conserved Rac-dependent cell migration. *J. Biol. Chem.* 279, 6087–6097.

147 Kulkarni, S. V., Gish, G., van der Geer, P., Henkemeyer, M., and Pawson, T. (2000) Role of p120 Ras-GAP in directed cell movement. *J. Cell Biol.* 149, 457–470.

148 Khyrul, W. A., LaLonde, D. P., Brown, M. C., Levinson, H., and Turner, C. E. (2004) The integrin-linked kinase regulates cell morphology and motility in a rho-associated kinase-dependent manner. *J. Biol. Chem.* 279, 54131–54139.

149 Sun, Q., Chen, G., Streb, J. W., Long, X., Yang, Y., Stoeckert, C. J.Jr., and Miano, J. M. (2006) Defining the mammalian CArGome. *Genome Res.* 16, 197–207.

150 Wang, J., Zohar, R., and McCulloch, C. A. (2006) Multiple roles of alpha-smooth muscle actin in mechanotransduction. *Exp. Cell Res.* 312, 205–214.

151 Cooper, J. A., Gould, K. L., Cartwright, C. A., and Hunter, T. (1986) Tyr527 is phosphorylated in pp60c-src: implications for regulation. *Science* 231, 1431–1434.

152 Roof, R. W., Haskell, M. D., Dukes, B. D., Sherman, N., Kinter, M., and Parsons, S. J. (1998) Phosphotyrosine (p-Tyr. -dependent and -independent mechanisms of p190 RhoGAP-p120 RasGAP interaction: Tyr 1105 of p190, a substrate for c-Src, is the sole p-Tyr mediator of complex formation. *Mol. Cell. Biol.* 18, 7052–7063.

153 Yue, J., and Mulder, K. M. (2001) Transforming growth factor-beta signal transduction in epithelial cells. *Pharmacol. Ther.* 91, 1–34.

154 Derynck, R., Akhurst, R. J., and Balmain, A. (2001) TGF-beta signaling in tumor suppression and cancer progression. *Nat. Genet.* 29, 117–129.

155 Wakefield, L. M., and Roberts, A. B. (2002) TGF-beta signaling: positive and negative effects on tumorigenesis. *Curr. Opin. Genet. Dev.* 12, 22–29.

156 Ghogomu, S. M., van Venrooy, S., Ritthaler, M., Wedlich, D., and Gradl, D. (2006) HIC-5 is a novel repressor of lymphoid enhancer factor/T-cell factor-driven transcription. *J. Biol. Chem.* 281, 1755–1764.

157 Yang, L., Guerrero, J., Hong, H., DeFranco, D. B., and Stallcup, M. R. (2000) Interaction of the tau2 transcriptional activation domain of glucocorticoid receptor with a novel steroid receptor coactivator, Hic-5, which localizes to both focal adhesions and the nuclear matrix. *Mol. Biol. Cell* 11, 2007–2018.

158 Kim-Kaneyama, J., Shibanuma, M., and Nose, K. (2002) Transcriptional activation of the c-fos gene by a LIM protein, Hic-5. *Biochem. Biophys. Res. Commun.* 299, 360–365.

159 Shibanuma, M., Kim-Kaneyama, J. R., Sato, S., and Nose, K. (2004) A LIM protein, Hic-5, functions as a potential coactivator for Sp1. *J. Cell Biochem.* 91, 633–645.

160 Suzuki, H., Ueda, T., Ichikawa, T., and Ito, H. (2003) Androgen receptor involvement in the progression of prostate cancer. *Endocr. Relat. Cancer* 10, 209–216.

161 Mimeault, M., and Batra, S. K. (2006) Recent advances on multiple tumorigenic cascades involved in prostatic cancer progression and targeting therapies. *Carcinogenesis* 27, 1–22.

162 Shannon, J. M., and Cunha, G. R. (1983) Autoradiographic localization of androgen binding in the developing mouse prostate. *Prostate* 4, 367–373.

163 Miyoshi, Y., Ishiguro, H., Uemura, H., Fujinami, K., Miyamoto, H., Kitamura, H., and Kubota, Y. (2003) Expression of AR associated protein 55 (ARA55. and androgen receptor in prostate cancer. *Prostate* 56, 280–286.

164 Mestayer, C., Blanchere, M., Jaubert, F., Dufour, B., and Mowszowicz, I. (2003) Expression of androgen receptor coactivators in normal and cancer prostate tissues and cultured cell lines. *Prostate* 56, 192–200.

4
Adhesion Dynamics in Motile Cells
Christa L. Cortesio, Keefe T. Chan, and Anna Huttenlocher

4.1
Introduction

Cell adhesion and its dynamic regulation during cell motility are central to fundamental biological processes, including the regulated movements of cells during embryonic development, immune responses, wound healing, and tissue regeneration. Dysregulated cell adhesion and motility contribute to the pathogenesis of disease states such as cancer, inflammatory disease, and atherosclerotic disease. It is not surprising, therefore, that there is substantial interest in understanding the molecular mechanisms that govern cell adhesion regulation and dynamics during cell motility, and in exploiting these mechanisms for therapeutic benefits in a wide range of disease processes. Recently, considerable progress has been made in our understanding of the molecular mechanisms that regulate adhesion assembly and disassembly in migrating cells; however, an understanding of the temporal and spatial modulation of adhesion dynamics during motility remains a challenge. In this chapter, recent progress in understanding the mechanisms of adhesion formation and turnover in motile cells in the context of two very different types of adhesion structures, podosomes and focal adhesions, is discussed.

   Cell migration requires a dynamic interaction between the extracellular matrix (ECM) and the actin cytoskeleton that can be separated into distinct steps. In order to migrate, cells initially extend a protrusion at the leading edge through the localized polymerization of actin and subsequent formation of integrin-mediated adhesions. After stabilization of adhesions at the leading edge, cells generate the traction forces required for cell translocation. The final stage of the migratory cycle involves rear retraction and release of adhesions at the rear of the cell. These classic steps are representative of the migratory cycle of fibroblasts, and these principles generally apply to many other cell types. However, leukocytes tend to display a more gliding movement and generally do not form the organized types of adhesion structures observed in fibroblasts. General principles govern the migration of many different cell types in that adhesions at the front of the cell must be strong in order to develop traction forces required for cell movement, whilst adhesions at the rear must be weak enough to allow for rear retraction. Therefore, an

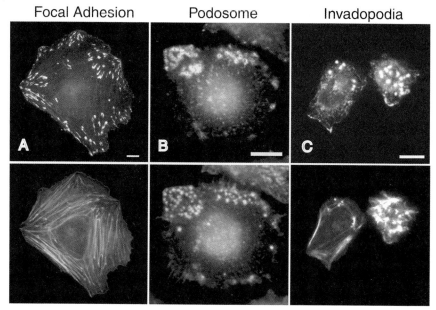

Figure 4.1 Representative images of focal adhesions, podosomes and invadopodia. (A) Immunofluorescent image of a HeLa cell co-stained with an anti-paxillin antibody (top panel) and rhodamine phalloidin (bottom panel). (B) Immunofluorescent image of a macrophage cell co-stained with an anti-vinculin antibody (top panel) and rhodamine phalloidin (bottom panel). (C) Immunofluorescent image of an MTLn3 breast cancer cell co-stained with an anti-cortactin antibody (top panel) and rhodamine phalloidin (bottom panel). All scale bars = 10 μm.

asymmetry in the strength or dynamics of adhesions between the front and rear of the cell is important for efficient cell migration.

Adherent cells such as fibroblasts commonly form integrin-mediated adhesions known as focal adhesions (Figure 4.1A). Focal adhesions are generally found in less-motile cells, and function to tether actin-containing stress fibers. In contrast to focal adhesions, highly motile and invasive cells form specialized integrin-mediated adhesions known as *podosomes* and *invadopodia* (Figure 4.1B and C). Podosomes have been observed in cells of monocytic lineage such as dendritic cells, macrophages and osteoclasts, while related structures known as invadopodia are found in transformed fibroblasts and a number of carcinoma cell lines [1]. In this chapter, the structural and regulatory components that contribute to the dynamics of focal adhesions and podosomes in motile cells, are reviewed.

4.2
Focal Adhesion Dynamics

Most adherent cell types form organized integrin-mediated adhesions with the ECM that serve to link the actin cytoskeleton and the ECM. The most extensively characterized cell–matrix adhesions–focal adhesions–contain clusters of αβ heterodimeric integrin receptors associated with large complexes of structural and signaling proteins linked to the actin cytoskeleton [2, 3]. Focal adhesions were first identified by electron microscopy as electron-dense submembrane plaques associated with actin filament bundles [4, 5]. Focal adhesions are localized to both central and peripheral regions of the cell, and generally are found at the ends of actin-containing stress fibers. These adhesion structures undergo dynamic cycles of assembly and disassembly, with variable durations lasting generally between 10 and 60 min [6–8]. In contrast, focal complexes are early adhesions that form at the cell periphery and are often more dynamic than focal adhesions. Focal adhesions are typically formed by cultured cells on rigid surfaces; however, analogous structures have been observed *in vivo* [9–11]. Here, some of the mechanisms involved in the dynamic regulation of focal adhesions will be highlighted, and readers are referred to relevant review articles on the regulation of focal adhesions [12–15] and their role in mechanical sensing [16–18].

4.2.1
Focal Adhesion Composition

Over the past 25 years, substantial progress has been made in dissecting the molecular composition of focal adhesions, including the identification of more than 50 unique components [19–22]. Focal adhesion components can be separated into three general categories: (i) membrane-associated proteins; (ii) cytoskeletal proteins; and (iii) signaling proteins (see Table 4.1). Integrins are required for focal adhesion formation and serve to relay signals bidirectionally between the extracellular and intracellular environments [23–25]. Additional membrane-associated focal adhesion components, such as syndecans [26], modulate focal adhesion formation.

Several integrin-associated cytoskeletal proteins have been shown to function as direct bridges between integrins and the actin cytoskeleton, such as talin [27, 28], α-actinin [29], filamin [30], and tensin [31]. Other structural or adaptor proteins interact with integrin cytoplasmic tails but do not directly bind actin, including focal adhesion kinase (FAK) [32, 33], the scaffolding protein paxillin [34], and the integrin-linked kinase (ILK) [35, 36]. Furthermore, focal adhesions contain structural proteins that do not associate directly with integrins, including vinculin, which interacts with talin, α-actinin, and actin, and functions to stabilize focal adhesions and suppress cell migration [37]. Finally, focal adhesions contain signaling proteins that likely transiently associate with focal adhesions and participate in their dynamic regulation. Some of these signaling molecules include the Src tyrosine kinase [38], protein kinase C (PKC) [39], the Rho family of GTPases [40], p21-activated kinase (PAK) [41], the calcium-dependent protease, calpain 2 [42], and the tyrosine phosphatase SHP-2 [43].

Table 4.1 Adhesion components.

Common to focal adhesions and PTAs

Component	Function	Relevant reviews
Integrins ($\beta1$, $\beta3$)	Membrane-associated	A, B, C, D, E
Syndecans	Membrane-associated	
Talin	Cytoskeletal	
Paxillin	Cytoskeletal	
Tensin	Cytoskeletal	
Vinculin	Cytoskeletal	
Dynamin	Cytoskeletal	
VASP	Cytoskeletal	
c-Src	Signaling	
Rho GTPases	Signaling	
FAK/Pyk2	Signaling	

Unique to PTAs

Component	Function	Relevant reviews
Integrins ($\alpha M \beta 2$, $\alpha X \beta 2$)	Membrane-associated	F, G, H, I
Cortactin	Actin-associated	
Gelsolin	Actin-associated	
Caldesmon	Actin-associated	
Arp 2/3	Actin-regulator	
WASP/N-WASP	Actin-regulator	

[A] See Ref. [12].
[B] See Ref. [22].
[C] Bass, M.D. and Humphries, M.J. (2002). Cytoplasmic interactions of syndecan-4 orchestrate adhesion receptor and growth factor receptor signaling. *Biochem. J.* 368, 1–15.
[D] Kruchten, A.E. and McNiven, M.A. (2006). Dynamin as a mover and pincher during cell migration and invasion. *J. Cell Sci. 119*, 1683–1690.
[E] Holt, M.R., Critchley, D.R., and Brindle, N.P. (1998). The focal adhesion phosphoprotein, VASP. *Int. J. Biochem. Cell Biol. 3*, 307–311.
[F] See Ref. [99].
[G] See Ref. [104].
[H] See Ref. [1].
[I] Hai, C.M., and Gu, Z. (2006). Caldesmon phosphorylation in actin cytoskeletal remodeling. *Eur. J. Cell Biol. 85*, 305–309.

4.2.2
Mechanisms of Focal Adhesion Assembly

Focal adhesions are dynamic structures that undergo cycles of assembly and disassembly, both at the leading and trailing edges of the cell. Recently, significant progress has been made in understanding the molecular mechanisms by which focal adhesions are formed. New adhesions that contain activated integrins form preferentially at the leading edge of the cell. The initial step in focal adhesion assembly is dependent on the physical clustering of integrins and the binding of ECM ligand [44, 45]. Binding to ligand induces conformational changes that result in the association of cytoplasmic proteins with the integrin cytoplasmic tails [46]. Assembly of these protein complexes initiates signaling cascades, including the activation of Rho family GTPases, tyrosine phosphorylation, and local phospholipid production [47]. In turn, activation of these signaling pathways is critical for the modulation of focal adhesion assembly. For example, if tyrosine phosphorylation is inhibited, these multi-protein adhesion complexes do not assemble [48]. The results of recent studies have suggested that focal adhesion formation involves the sequential and hierarchical incorporation of components with distinct kinetics where $\alpha v \beta 3$ integrin, talin, paxillin and phosphotyrosine assemble into early adhesion complexes, followed later by α-actinin, vinculin, FAK, $\alpha 5 \beta 1$ integrin and tensin [6, 49, 50].

4.2.2.1 Rho GTPases

The Rho family GTPases are critical regulators of focal adhesion and stress fiber assembly [51, 52]. Rac and Cdc42 mediate the formation of early integrin clusters and focal complexes at the leading edge of the cell [53, 54]. Although integrin clustering is necessary for cell migration, large clusters are generally formed in less-motile cells. The assembly of more stable focal adhesions is regulated by Rho and the downstream activation of myosin-dependent contractility [55]. Although the requirement for Rho GTPases in focal adhesion assembly is well established, the downstream effector mechanisms that contribute to the temporal and spatial modulation of cell motility remain a subject of intense investigation [56–58]. For example, Rho effectors including Rho kinase (ROCK) play a central role in the generation of contractility both at the leading and trailing edges of cells [59, 60]. Accordingly, the strongest forces between the cell and ECM have been reported to occur at adhesions at both the cell front and in areas of rear retraction [61]. In addition to their effects on adhesion formation, Rho family GTPases are also key regulators of adhesion turnover. For example, Rac modulates adhesion disassembly through its effector PAK [62] and Rho activation is critical for rear detachment through ROCK and the generation of myosin-based contractility [63].

4.2.2.2 Talin and Phosphoinositides

One key focal adhesion component with an established role in focal adhesion assembly is the cytoskeletal protein talin [27, 64]. Talin-null cells display impaired focal adhesion assembly [65] and fail to respond to the force-induced recruitment of additional cytoskeletal proteins such as vinculin [66]. Talin has also been demonstrated to be a critical regulator of integrin activation [67–70]. An additional mechanism by which talin may regulate focal adhesion assembly is through its direct interaction with the type I gamma661 phosphatidylinositol phosphate kinase (PIPKIγ661) [71]. Recent studies have implicated phosphatidylinositol 4,5-bisphosphate (PIP_2) in focal adhesion assembly [72–74].

4.2.3
Mechanisms of Focal Adhesion Disassembly

Recently, substantial progress has been made in defining the mechanisms that regulate the disassembly of focal adhesions. Matrix proteins such as thrombospondin [75], as well as stimulation by growth factors such as platelet-derived growth factor (PDGF) [76] and epidermal growth factor (EGF) [77], have been demonstrated to induce the disassembly of focal adhesions. Furthermore, it has been shown that focal adhesion disassembly depends on microtubule dynamics, as disruption of microtubules stabilizes focal adhesions [78–80].

4.2.3.1 Calpain
The intracellular calcium-dependent proteases known as calpains [81, 82] have been shown to be important for the rear detachment phase of cell motility [83], and have been implicated in the destabilization of focal adhesions downstream of microtubules [84–86]. Calpains cleave several proteins found in focal adhesions such as FAK, paxillin, and talin [87–89], and these proteolytic events appear to be important for focal adhesion disassembly. Specifically, calpain-mediated proteolysis of talin is a rate-limiting step in focal adhesion turnover [8].

4.2.3.2 Tyrosine Phosphorylation and Contractile Machinery
In addition to well-established roles in adhesion assembly, tyrosine phosphorylation is also critical for focal adhesion disassembly. For example, infection of cells with Rous sarcoma virus induces disassembly of focal adhesions. In addition, FAK-null fibroblasts exhibit an increased number and size of peripheral focal adhesions [90]. Furthermore, expression of kinase-dead mutants of Src also promotes enlarged, peripheral adhesions, indicating that FAK and Src likely are both involved in focal adhesion disassembly. In fact, recent evidence has shown that FAK-Src signaling is required for adhesion disassembly through the regulation of myosin-light chain kinase (MLCK) and ERK [7]. FAK may also regulate adhesion disassembly through its association with calpain and ERK [91]. More recent evidence has also implicated a role for tyrosine phosphatases in adhesion disassembly. Fibroblasts deficient in the phosphatase SHP-2 have increased focal adhesions and a phenotype that resembles FAK-deficient fibroblasts [43]. Finally, substantial evidence implicates a critical role for the cellular contractile machinery in focal adhesion disassembly. For example, recent studies report that fibroblasts deficient in myosin IIA exhibit impaired adhesion disassembly and rear detachment [92, 93].

4.3
Podosome and Invadopodia Dynamics

Highly migratory and invasive cells form specialized types of integrin-mediated adhesions, referred to as a podosomes or invadopodia.

Podosomes were first identified in fibroblasts transformed by the v-Src oncogene [94], and were later referred to as invadopodia because of their matrix-degrading function [95]. Podosomes have also been observed in cells of monocytic lineage such as osteoclasts, macrophages and dendritic cells, and more recently in smooth muscle cells [96] and endothelial cells [97] treated with PKC-activating phorbol esters.

The related adhesion structures, invadopodia, are generally found in invasive cancer cells. Both podosomes and invadopodia are actin-rich adhesions that serve to link the ECM to the actin cytoskeleton and have the capacity for matrix degradation. The relationship between podosomes and invadopodia is not clearly defined in the literature, and the terms are frequently used interchangeably. Some studies suggest that podosomes are dynamic structures that represent precursors to the more stable invadopodia, which mediate focused areas of matrix degradation [98]. To be concise, and for the purpose of this chapter, when referring to both structures the term podosome-type adhesions (PTAs) will be used, as proposed in a recent review [99].

PTAs are generally found in invasive cells that are specialized to cross tissue boundaries through the localized secretion of matrix-degrading metalloproteases (MMPs) [1]. For example, podosome formation is critical for the *in-vitro* bone resorption functions of osteoclasts [100], while invadopodia appear to be important for cancer cell invasion into blood vessels, an early step in the metastatic process [101]. Furthermore, podosomes appear to be critical for the normal functioning of the immune system, since mutations in the podosome component Wiskott–Aldrich syndrome protein (WASP), are associated with human immunodeficiency [102]. Podosomes are highly dynamic adhesive structures with an average half-life of 2 to 12 min [103], while invadopodia are more stable structures that can last for periods of up to 1 hour [104]. In contrast to focal adhesions, few studies have addressed the dynamic regulation of PTAs in live cells. Hence, at this point it is relevant to discuss the present knowledge of the molecular mechanisms that regulate the dynamic formation and disassembly of PTAs. The reader is also referred to recent reviews describing other aspects of their function and regulation [99, 105, 106].

4.3.1
Podosome and Invadopodia Architecture

PTA architecture is defined by an actin-rich core, where the actin-polymerizing machinery and actin regulatory proteins function to drive membrane protrusion. This core is surrounded by a ring structure composed of signaling and adaptor proteins, including cortactin and focal adhesion proteins such as talin and paxillin. In contrast to focal adhesions, PTAs are not associated with stress fibers but are primary sites of rapid actin polymerization and contain actin regulatory proteins, including cortactin, gelsolin, WASP, and the actin-nucleating Arp 2/3 complex [106]. Furthermore, a hallmark of PTAs is their capacity for matrix degradation through the localized secretion of ECM-degrading MMPs [1].

PTAs are formed on the ventral surface of migratory cells and appear as dot-like structures [99], or they may form larger ring-like structures, termed rosettes [107]. The shapes and sizes of PTAs will differ depending on the cell type or ECM environment. Integrins, including $\beta 1$, $\beta 2$ and $\beta 3$ integrins, localize to the PTAs and are found within both the actin-rich core and the surrounding ring structure [1].

Integrin-associated cytoskeletal proteins such as paxillin [108], vinculin and talin [109] are recruited to PTAs; however, their functions in this context remain largely unknown. Protein kinases also localize to PTAs, including members of the Src kinase family [107], protein kinase C [110], Pyk2 [111] and FAK [112]. Other signaling molecules including Rho GTPases play an important role in PTA regulation [1, 113].

4.3.2
Molecular Mechanisms of PTA Assembly

Significant recent progress has been made in defining the molecular mechanisms that regulate PTA formation, including a role for intracellular signaling cascades through Src tyrosine kinases [114], PKC [97, 115], cortactin [116, 117], WASP [103, 118], and Rho GTPases [119, 120]. Extracellular factors play a crucial role in modulating PTA formation through the regulation of these signaling pathways. For example, the ECM environment and engagement of specific integrins is necessary for PTA formation, and can be further modified by soluble factors, including growth factors and cytokines, such as EGF [121], vascular endothelial growth factor (VEGF) [122] and colony-stimulating factor-1 (CSF-1) [101].

4.3.2.1 Actin Regulatory Proteins

Many actin regulatory proteins are necessary for the formation of podosomes and invadopodia (see Table 4.1). For example, the actin-related protein 2/3 (Arp 2/3) is critical for actin nucleation and is necessary for the formation of podosomes [123]. Furthermore, regulators of the Arp 2/3 complex, WASP and the non-hematopoietic homologue, neural Wiskott–Aldrich syndrome protein (N-WASP), are necessary for both the formation of podosomes [124] and invadopodia [121]. Accordingly, RNA interference (RNAi) has demonstrated a requirement for N-WASP, the WASP-interacting protein WIP and Arp2/3 in invadopodia formation of invasive breast cancer cells [121]. Similarly, a partial knockdown of WASP in dendritic cells was sufficient to inhibit podosome formation [125]. The importance of localized WASP activity for efficient podosome assembly was further emphasized by studies showing that siRNA targeted to WASP interacting protein (WIP) resulted in impaired podosome formation in macrophages, and that the WIP–WASP interaction was important for this effect [126].

Other actin regulatory proteins have been reported to play a role in PTA formation. Gelsolin is an actin-binding protein which functions to sever actin filaments, thereby creating new sites for actin nucleation. Osteoclasts from mice deficient in gelsolin are devoid of podosomes and display abnormal osteoclast activity [120]. Cortactin, an actin-binding protein and Src kinase substrate that activates Arp2/3, has emerged as a commonly used marker of PTAs and has also been reported to be necessary for PTA formation. Depletion of cortactin impairs the formation of PTAs in osteoclasts and transformed fibroblasts, and inhibits degradation of the ECM [127]. While many of the actin regulatory proteins are necessary for podosome formation, others are involved in regulating the dynamic formation of these structures. For example, the depletion of cofilin, which functions to generate new barbed ends of actin filaments, inhibits the formation of long-lived invadopodia and thereby impairs their matrix-degradation activity [121].

4.3.2.2 Rho GTPases

Members of the Rho family of GTPases are critical regulators of the actin cytoskeleton, and have been implicated in the regulation of PTA formation [128]. A signaling cascade involving Cdc42, WASP and Arp2/3 is essential for the formation of both podosomes [129] and invadopodia [121]. Furthermore, an essential role for Rac2 has been demonstrated in murine macrophages by knockout experiments of Rac isoforms. In addition, activated Rho has been localized to podosomes, and its inhibition is associated with impaired podosome formation and matrix degradation [119]. However, despite some studies having shown an essential role for Rho GTPase family members in PTA formation, there have been conflicting reports regarding the roles of specific Rho GTPase family members in the regulation of PTA assembly. For example, the expression of a constitutively active Rac disrupted podosome assembly in osteoclasts [130] but had no effect on dendritic cell podosomes [131]. Contradictory data on Cdc42 also exist, with multiple studies implicating Cdc42 as a positive regulator of podosome and invadopodia formation in endothelial cells and breast cancer cells [121], while the expression of a constitutively active Cdc42 in macrophages results in podosome disassembly [103]. These findings suggest that there is most likely a balance between the activities of Rho GTPases family members required for the formation of PTAs and for their dynamic regulation in live cells [99].

4.3.2.3 Signaling Molecules

Substantial evidence supports a requirement for tyrosine phosphorylation in the formation of PTAs. Podosomes were first identified in fibroblasts transformed by the v-Src oncogene [94]. Interestingly, excessive Src activation by the expression of transforming v-Src induces podosome formation but impairs the formation of focal adhesions. Furthermore, in contrast to focal adhesions, the activity of Src tyrosine kinases is both sufficient [95, 107] and necessary [132] for the formation of PTAs. This requirement appears to occur at an early step in podosome assembly, and most likely occurs through the phosphorylation of specific effector substrates by Src, including cortactin and scaffolding proteins, such as paxillin. In addition, recent evidence suggests that Src provides an important scaffolding role in podosome formation in osteoclasts [114].

PKC also has an important role early in the formation of podosomes, and requires c-Src as a downstream effector [97, 115]. FAK and focal adhesion kinase-related protein tyrosine kinase (Pyk2) localize to podosome-like structures, but their functions in podosome assembly have not been fully elucidated. In osteoclasts, Pyk2 functions to recruit Src kinase to sites of podosome assembly after integrin engagement [111]. Furthermore, in v-Src-transformed fibroblasts, a FAK–v-Src complex at invadopodia appears critical for tissue invasion [133].

4.3.3
Molecular Mechanisms of PTA Disassembly

Few studies have addressed the molecular mechanisms that regulate the disassembly of PTAs. When stimulated with exogenous factors such as tumor necrosis factor-alpha (TNF-α) or prostaglandin E_2, the dissolution of podosome-like structures occurs in dendritic cells [134], suggesting that exogenous factors can induce PTA disassembly. However, the molecular mechanisms that contribute to this disassembly in dendritic cells have not been identified. A specific pathway that has been implicated in dendritic cell podosome disassembly is through regulated proteolysis by the calcium-dependent intracellular proteases calpains [135]. Calpains have also been implicated in the disassembly of osteoclast podosomes [136]. Calpains cleave several proteins found in PTAs including talin, paxillin, FAK, Pyk2 and several actin-binding proteins such as cortactin and WASP [8, 99, 135], but the involvement of specific substrate cleavage in podosome disassembly has not been identified.

In addition to calpains, Src tyrosine kinases have also been implicated in the disassembly and turnover of podosomes in osteoclasts [137]. The results of recent studies have suggested that the kinase activity of Src–and not simply its adaptor functions–are essential for its effects on podosome turnover [114]. In a manner analogous to focal adhesions, Src tyrosine kinases likely affect PTA disassembly by targeting specific effector substrates. Interestingly, the turnover of podosomes in osteoclasts was inversely related to localized tyrosine phosphorylation of cortactin [114], suggesting that other Src substrates most likely function to regulate PTA turnover. The involvement of calpains downstream of Src in the regulation of focal adhesion turnover suggests that this may be an attractive candidate to mediate the disassembly of PTAs downstream of Src.

4.4
Summary

Significant progress has been made in understanding the molecular mechanisms by which adhesive structures assemble and disassemble in motile cells. It appears that many of the signaling pathways that govern focal adhesion formation and disassembly also play roles in the regulation of PTA dynamics. For example, Src kinases are clearly important for the dynamic turnover of both focal adhesions and PTAs. Src kinase activity is also necessary for PTA formation; however, its role in focal adhesion formation is less clear.

Interestingly, expression of the constitutively active v-Src kinase in cells that normally form focal adhesions promotes a switch to the formation of PTAs. A key question for future investigation will address the molecular determinants that promote this switch between the formation of firm adhesive structures through focal adhesion and the generation of invasive, matrix-degrading adhesion structures. It will also be important to provide progress in understanding how focal adhesions and PTAs are temporally and spatially regulated to mediate cell motility and invasion. Finally, an additional challenge for future investigations will be to understand how invasive adhesions form, and how these functions may be modulated to treat human disease.

Acknowledgments

The authors thank Kate Cooper for carrying out the macrophage podosome staining.

References

1. Linder, S., and Aepfelbacher, M. (2003). Podosomes: adhesion hot-spots of invasive cells. *Trends Cell Biol.* 13, 376–385.
2. Burridge, K., and Chrzanowska-Wodnicka, M. (1996). Focal adhesions, contractility, and signaling. *Annu. Rev. Cell Dev. Biol.* 12, 463–518.
3. Yamada, K.M., and Geiger, B. (1997). Molecular interactions in cell adhesion complexes. *Curr. Opin. Cell Biol.* 9, 76–85.
4. Abercrombie, M., Heaysman, J.E., and Pegrum, S.M. (1971). The locomotion of fibroblasts in culture. IV. Electron microscopy of the leading lamella. *Exp. Cell Res.* 67, 359–367.
5. Heath, J.P., and Dunn, G.A. (1978). Cell to substratum contacts of chick fibroblasts and their relation to the microfilament system. A correlated interference-reflexion and high-voltage electron-microscope study. *J. Cell Sci.* 29, 197–212.
6. Laukaitis, C.M., Webb, D.J., Donais, K., and Horwitz, A.F. (2001). Differential dynamics of alpha 5 integrin, paxillin, and alpha-actinin during formation and disassembly of adhesions in migrating cells. *J. Cell Biol.* 153, 1427–1440.
7. Webb, D.J., Donais, K., Whitmore, L.A., Thomas, S.M., Turner, C.E., Parsons, J.T., and Horwitz, A.F. (2004). FAK-Src signalling through paxillin, ERK and MLCK regulates adhesion disassembly. *Nat. Cell Biol.* 6, 154–161.
8. Franco, S.J., Rodgers, M.A., Perrin, B.J., Han, J., Bennin, D.A., Critchley, D.R., and Huttenlocher, A. (2004). Calpain-mediated proteolysis of talin regulates adhesion dynamics. *Nat. Cell Biol.* 6, 977–983.
9. Kano, Y., Katoh, K., Masuda, M., and Fujiwara, K. (1996). Macromolecular composition of stress fiber-plasma membrane attachment sites in endothelial cells in situ. *Circ. Res.* 79, 1000–1006.
10. Turner, C.E., Kramarcy, N., Sealock, R., and Burridge, K. (1991). Localization of paxillin, a focal adhesion protein, to smooth muscle dense plaques, and the myotendinous and neuromuscular junctions of skeletal muscle. *Exp. Cell Res.* 192, 651–655.
11. Cypher, C., and Letourneau, P.C. (1991). Identification of cytoskeletal, focal adhesion, and cell adhesion proteins in growth cone particles isolated from developing chick brain. *J. Neurosci. Res.* 30, 259–265.
12. Cukierman, E., Pankov, R., Stevens, D.R., and Yamada, K.M. (2001). Taking cell-matrix adhesions to the third dimension. *Science* 294, 1708–1712.
13. Geiger, B., Bershadsky, A., Pankov, R., and Yamada, K.M. (2001). Transmembrane crosstalk between the extracellular matrix–cytoskeleton crosstalk. *Nat. Rev. Mol. Cell. Biol.* 2, 793–805.
14. Yamada, K.M., Pankov, R., and Cukierman, E. (2003). Dimensions and

dynamics in integrin function. *Braz. J. Med. Biol. Res.* 36, 959–966.
15 Webb, D.J., Brown, C.M., and Horwitz, A.F. (2003). Illuminating adhesion complexes in migrating cells: moving toward a bright future. *Curr. Opin. Cell Biol.* 15, 614–620.
16 Bershadsky, A., Kozlov, M., and Geiger, B. (2006). Adhesion-mediated mechanosensitivity: a time to experiment, and a time to theorize. *Curr. Opin. Cell Biol.* 18, 472–481.
17 Chen, C.S., Tan, J., and Tien, J. (2004). Mechanotransduction at cell-matrix and cell-cell contacts. *Annu. Rev. Biomed. Eng.* 6, 275–302.
18 Wehrle-Haller, B., and Imhof, B. (2002). The inner lives of focal adhesions. *Trends Cell Biol.* 12, 382–389.
19 Zamir, E., and Geiger, B. (2001). Components of cell-matrix adhesions. *J. Cell Sci.* 114, 3577–3579.
20 Romer, L.H., Birukov, K.G., and Garcia, J.G. (2006). Focal adhesions: paradigm for a signaling nexus. *Circ. Res.* 98, 606–616.
21 Cohen, M., Joester, D., Geiger, B., and Addadi, L. (2004). Spatial and temporal sequence of events in cell adhesion: from molecular recognition to focal adhesion assembly. *Chembiochem* 5, 1393–1399.
22 Wozniak, M.A., Modzelewska, K., Kwong, L., and Keely, P.J. (2004). Focal adhesion regulation of cell behavior. *Biochim Biophys Acta* 1692, 103–119.
23 Hynes, R.O. (2002). Integrins: bidirectional, allosteric signaling machines. *Cell* 110, 673–687.
24 Hood, J.D., and Cheresh, D.A. (2002). Role of integrins in cell invasion and migration. *Nat. Rev. Cancer* 2, 91–100.
25 Luo, B.H., and Springer, T.A. (2006). Integrin structures and conformational signaling. *Curr. Opin. Cell Biol.* 18, 579–586.
26 Couchman, J.R., and Woods, A. (1999). Syndecan-4 and integrins: combinatorial signaling in cell adhesion. *J. Cell Sci.* 112 (Pt 20), 3415–3420.
27 Nayal, A., Webb, D.J., and Horwitz, A.F. (2004). Talin: an emerging focal point of adhesion dynamics. *Curr. Opin. Cell Biol.* 16, 94–98.
28 Critchley, D.R. (2005). Genetic, biochemical and structural approaches to talin function. *Biochem. Soc. Trans.* 33, 1308–1312.
29 Otey, C.A., and Carpen, O. (2004). Alpha-actinin revisited: a fresh look at an old player. *Cell Motil. Cytoskeleton* 58, 104–111.
30 Popowicz, G.M., Schleicher, M., Noegel, A.A., and Holak, T.A. (2006). Filamins: promiscuous organizers of the cytoskeleton. *Trends Biochem. Sci.* 31, 411–419.
31 Lo, S.H. (2004). *Tensin. Int. J. Biochem. Cell Biol.* 36, 31–34.
32 Mitra, S.K., Hanson, D.A., and Schlaepfer, D.D. (2005). Focal adhesion kinase: in command and control of cell motility. *Nat. Rev. Mol. Cell. Biol.* 6, 56–68.
33 Parsons, J.T. (2003). Focal adhesion kinase: the first ten years. *J. Cell Sci.* 116, 1409–1416.
34 Brown, M.C., and Turner, C.E. (2004). Paxillin: adapting to change. *Physiol. Rev.* 84, 1315–1339.
35 Legate, K.R., Montanez, E., Kudlacek, O., and Fassler, R. (2006). ILK, PINCH and parvin: the tIPP of integrin signalling. *Nat. Rev. Mol. Cell. Biol.* 7, 20–31.
36 Wu, C. (2001). ILK interactions. *J. Cell Sci.* 114, 2549–2550.
37 Ziegler, W.H., Liddington, R.C., and Critchley, D.R. (2006). The structure and regulation of vinculin. *Trends Cell Biol.* 16, 453–460.
38 Frame, M.C. (2004). Newest findings on the oldest oncogene; how activated src does it. *J. Cell Sci.* 117, 989–998.
39 Woods, A., and Couchman, J.R. (1992). Protein kinase C involvement in focal adhesion formation. *J. Cell Sci.* 101 (Pt 2), 277–290.
40 Ridley, A.J. (1994). Signal transduction through the GTP-binding proteins Rac and Rho. *J. Cell Sci. Suppl.* 18, 127–131.
41 Zhao, Z.S., Manser, E., Loo, T.H., and Lim, L. (2000). Coupling of PAK-interacting exchange factor PIX to GIT1 promotes focal complex disassembly. *Mol. Cell. Biol.* 20, 6354–6363.

42 Beckerle, M.C., Burridge, K., DeMartino, G.N., and Croall, D.E. (1987). Colocalization of calcium-dependent protease II and one of its substrates at sites of cell adhesion. *Cell* 51, 569–577.

43 Yu, D.H., Qu, C.K., Henegariu, O., Lu, X., and Feng, G.S. (1998). Protein-tyrosine phosphatase Shp-2 regulates cell spreading, migration, and focal adhesion. *J. Biol. Chem.* 273, 21125–21131.

44 Miyamoto, S., Akiyama, S.K., and Yamada, K.M. (1995). Synergistic roles for receptor occupancy and aggregation in integrin transmembrane function. *Science* 267, 883–885.

45 Galbraith, C.G., Yamada, K.M., and Sheetz, M.P. (2002). The relationship between force and focal complex development. *J. Cell Biol.* 159, 695–705.

46 Kim, M., Carman, C.V., and Springer, T.A. (2003). Bidirectional transmembrane signaling by cytoplasmic domain separation in integrins. *Science* 301, 1720–1725.

47 Ridley, A.J., Schwartz, M.A., Burridge, K., Firtel, R.A., Ginsberg, M.H., Borisy, G., Parsons, J.T., and Horwitz, A.R. (2003). Cell migration: integrating signals from front to back. *Science* 302, 1704–1709.

48 Yap, A.S., Stevenson, B.R., Cooper, V., and Manley, S.W. (1997). Protein tyrosine phosphorylation influences adhesive junction assembly and follicular organization of cultured thyroid epithelial cells. *Endocrinology* 138, 2315–2324.

49 Zaidel-Bar, R., Ballestrem, C., Kam, Z., and Geiger, B. (2003). Early molecular events in the assembly of matrix adhesions at the leading edge of migrating cells. *J. Cell Sci.* 116, 4605–4613.

50 Zaidel-Bar, R., Cohen, M., Addadi, L., and Geiger, B. (2004). Hierarchical assembly of cell-matrix adhesion complexes. *Biochem. Soc. Trans.* 32, 416–420.

51 Clark, E.A., King, W.G., Brugge, J.S., Symons, M., and Hynes, R.O. (1998). Integrin-mediated signals regulated by members of the rho family of GTPases. *J. Cell Biol.* 142, 573–586.

52 Nobes, C.D., and Hall, A. (1999). Rho GTPases control polarity, protrusion, and adhesion during cell movement. *J. Cell Biol.* 144, 1235–1244.

53 Nobes, C.D., and Hall, A. (1995). Rho, rac, and cdc42 GTPases regulate the assembly of multimolecular focal complexes associated with actin stress fibers, lamellipodia, and filopodia. *Cell* 81, 53–62.

54 Rottner, K., Hall, A., and Small, J.V. (1999). Interplay between Rac and Rho in the control of substrate contact dynamics. *Curr. Biol.* 9, 640–648.

55 Chrzanowska-Wodnicka, M., and Burridge, K. (1996). Rho-stimulated contractility drives the formation of stress fibers and focal adhesions. *J. Cell Biol.* 133, 1403–1415.

56 Garcia-Mata, R., and Burridge, K. (2007). Catching a GEF by its tail. *Trends Cell Biol.* 17, 36–43.

57 Ridley, A.J. (2001). Rho GTPases and cell migration. *J. Cell Sci.* 114, 2713–2722.

58 Jaffe, A.B., and Hall, A. (2005). Rho GTPases: biochemistry and biology. *Annu. Rev. Cell Dev. Biol.* 21, 247–269.

59 Sturge, J., Wienke, D., and Isacke, C.M. (2006). Endosomes generate localized Rho-ROCK-MLC2-based contractile signals via Endo180 to promote adhesion disassembly. *J. Cell Biol.* 175, 337–347.

60 Amano, M., Fukata, Y., and Kaibuchi, K. (2000). Regulation and functions of Rho-associated kinase. *Exp. Cell Res.* 261, 44–51.

61 Beningo, K.A., Dembo, M., Kaverina, I., Small, J.V., and Wang, Y.L. (2001). Nascent focal adhesions are responsible for the generation of strong propulsive forces in migrating fibroblasts. *J. Cell Biol.* 153, 881–888.

62 Symons, M. (2000). Adhesion signaling: PAK meets Rac on solid ground. *Curr. Biol.* 10, R535–R537.

63 Riento, K., and Ridley, A.J. (2003). Rocks: multifunctional kinases in cell behaviour. *Nat. Rev. Mol. Cell. Biol.* 4, 446–456.

64 Critchley, D.R. (2000). Focal adhesions – the cytoskeletal connection. *Curr. Opin. Cell Biol.* 12, 133–139.

65 Priddle, H., Hemmings, L., Monkley, S., Woods, A., Patel, B., Sutton, D., Dunn, G.A., Zicha, D., and Critchley, D.R. (1998). Disruption of the talin gene compromises focal adhesion assembly in undifferentiated but not differentiated embryonic stem cells. *J. Cell Biol.* 142, 1121–1133.

66 Giannone, G., Jiang, G., Sutton, D.H., Critchley, D.R., and Sheetz, M.P. (2003). Talin1 is critical for force-dependent reinforcement of initial integrin-cytoskeleton bonds but not tyrosine kinase activation. *J. Cell Biol.* 163, 409–419.

67 Cram, E.J., and Schwarzbauer, J.E. (2004). The talin wags the dog: new insights into integrin activation. *Trends Cell Biol.* 14, 55–57.

68 Tadokoro, S., Shattil, S.J., Eto, K., Tai, V., Liddington, R.C., de Pereda, J.M., Ginsberg, M.H., and Calderwood, D.A. (2003). Talin binding to integrin beta tails: a final common step in integrin activation. *Science* 302, 103–106.

69 Calderwood, D.A., Zent, R., Grant, R., Rees, D.J., Hynes, R.O., and Ginsberg, M.H. (1999). The Talin head domain binds to integrin beta subunit cytoplasmic tails and regulates integrin activation. *J. Biol. Chem.* 274, 28071–28074.

70 Vinogradova, O., Velyvis, A., Velyviene, A., Hu, B., Haas, T., Plow, E., and Qin, J. (2002). A structural mechanism of integrin alpha(IIb)beta(3) "inside-out" activation as regulated by its cytoplasmic face. *Cell* 110, 587–597.

71 Ling, K., Doughman, R.L., Firestone, A.J., Bunce, M.W., and Anderson, R.A. (2002). Type I gamma phosphatidylinositol phosphate kinase targets and regulates focal adhesions. *Nature* 420, 89–93.

72 Cluzel, C., Saltel, F., Lussi, J., Paulhe, F., Imhof, B.A., and Wehrle-Haller, B. (2005). The mechanisms and dynamics of (alpha)v(beta)3 integrin clustering in living cells. *J. Cell Biol.* 171, 383–392.

73 Martel, V., Racaud-Sultan, C., Dupe, S., Marie, C., Paulhe, F., Galmiche, A., Block, M.R., and Albiges-Rizo, C. (2001). Conformation, localization, and integrin binding of talin depend on its interaction with phosphoinositides. *J. Biol. Chem.* 276, 21217–21227.

74 Kong, X., Wang, X., Misra, S., and Qin, J. (2006). Structural basis for the phosphorylation-regulated focal adhesion targeting of type Igamma phosphatidylinositol phosphate kinase (PIPKIgamma) by talin. *J. Mol. Biol.* 359, 47–54.

75 Orr, A.W., Pallero, M.A., Xiong, W.C., and Murphy-Ullrich, J.E. (2004). Thrombospondin induces RhoA inactivation through FAK-dependent signaling to stimulate focal adhesion disassembly. *J. Biol. Chem.* 279, 48983–48992.

76 Bornfeldt, K.E., Raines, E.W., Graves, L.M., Skinner, M.P., Krebs, E.G., and Ross, R. (1995). Platelet-derived growth factor. Distinct signal transduction pathways associated with migration versus proliferation. *Ann. N. Y. Acad. Sci.* 766, 416–430.

77 Xie, H., Pallero, M.A., Gupta, K., Chang, P., Ware, M.F., Witke, W., Kwiatkowski, D.J., Lauffenburger, D.A., Murphy-Ullrich, J.E., and Wells, A. (1998). EGF receptor regulation of cell motility: EGF induces disassembly of focal adhesions independently of the motility-associated PLCgamma signaling pathway. *J. Cell Sci.* 111 (Pt 5), 615–624.

78 Krylyshkina, O., Kaverina, I., Kranewitter, W., Steffen, W., Alonso, M.C., Cross, R.A., and Small, J.V. (2002). Modulation of substrate adhesion dynamics via microtubule targeting requires kinesin-1. *J. Cell Biol.* 156, 349–359.

79 Small, J.V., and Kaverina, I. (2003). Microtubules meet substrate adhesions to arrange cell polarity. *Curr. Opin. Cell Biol.* 15, 40–47.

80 Ezratty, E.J., Partridge, M.A., and Gundersen, G.G. (2005). Microtubule-induced focal adhesion disassembly is mediated by dynamin and focal adhesion kinase. *Nat. Cell Biol.* 7, 581–590.

81 Franco, S.J., and Huttenlocher, A. (2005). Regulating cell migration: calpains make the cut. *J. Cell Sci.* 118, 3829–3838.

82 Glading, A., Lauffenburger, D.A., and Wells, A. (2002). Cutting to the chase: calpain proteases in cell motility. *Trends Cell Biol.* 12, 46–54.

83 Huttenlocher, A., Palecek, S.P., Lu, Q., Zhang, W., Mellgren, R.L., Lauffenburger, D.A., Ginsberg, M.H., and Horwitz, A.F. (1997). Regulation of cell migration by the calcium-dependent protease calpain. *J. Biol. Chem. 272*, 32719–32722.

84 Dourdin, N., Bhatt, A.K., Dutt, P., Greer, P.A., Arthur, J.S., Elce, J.S., and Huttenlocher, A. (2001). Reduced cell migration and disruption of the actin cytoskeleton in calpain-deficient embryonic fibroblasts. *J. Biol. Chem. 276*, 48382–48388.

85 Bhatt, A., Kaverina, I., Otey, C., and Huttenlocher, A. (2002). Regulation of focal complex composition and disassembly by the calcium-dependent protease calpain. *J. Cell Sci. 115*, 3415–3425.

86 Carragher, N.O., and Frame, M.C. (2002). Calpain: a role in cell transformation and migration. *Int. J. Biochem. Cell Biol. 34*, 1539–1543.

87 Franco, S., Perrin, B., and Huttenlocher, A. (2004). Isoform specific function of calpain 2 in regulating membrane protrusion. *Exp. Cell Res. 299*, 179–187.

88 Carragher, N.O., Levkau, B., Ross, R., and Raines, E.W. (1999). Degraded collagen fragments promote rapid disassembly of smooth muscle focal adhesions that correlates with cleavage of pp125(FAK), paxillin, and talin. *J. Cell Biol. 147*, 619–630.

89 Schoenwaelder, S.M., Yuan, Y., Cooray, P., Salem, H.H., and Jackson, S.P. (1997). Calpain cleavage of focal adhesion proteins regulates the cytoskeletal attachment of integrin alphaIIbbeta3 (platelet glycoprotein IIb/IIIa) and the cellular retraction of fibrin clots. *J. Biol. Chem. 272*, 1694–1702.

90 Ilic, D., Furuta, Y., Kanazawa, S., Takeda, N., Sobue, K., Nakatsuji, N., Nomura, S., Fujimoto, J., Okada, M., and Yamamoto, T. (1995). Reduced cell motility and enhanced focal adhesion contact formation in cells from FAK-deficient mice. *Nature 377*, 539–544.

91 Carragher, N.O., Westhoff, M.A., Fincham, V.J., Schaller, M.D., and Frame, M.C. (2003). A novel role for FAK as a protease-targeting adaptor protein: regulation by p42 ERK and Src. *Curr. Biol. 13*, 1442–1450.

92 Vicente-Manzanares, M., Zareno, J., Whitmore, L., Choi, C.K., and Horwitz, A.F. (2007). Regulation of protrusion, adhesion dynamics, and polarity by myosins IIA and IIB in migrating cells. *J. Cell Biol. 176*, 573–580.

93 Even-Ram, S., Doyle, A.D., Conti, M.A., Matsumoto, K., Adelstein, R.S., and Yamada, K.M. (2007). Myosin IIA regulates cell motility and actomyosin-microtubule crosstalk. *Nat. Cell Biol. 9*, 299–309.

94 Tarone, G., Cirillo, D., Giancotti, F.G., Comoglio, P.M., and Marchisio, P.C. (1985). Rous sarcoma virus-transformed fibroblasts adhere primarily at discrete protrusions of the ventral membrane called podosomes. *Exp. Cell Res. 159*, 141–157.

95 Chen, W.T. (1989). Proteolytic activity of specialized surface protrusions formed at rosette contact sites of transformed cells. *J. Exp. Zool. 251*, 167–185.

96 Hai, C.M., Hahne, P., Harrington, E.O., and Gimona, M. (2002). Conventional protein kinase C mediates phorbol-dibutyrate-induced cytoskeletal remodeling in a7r5 smooth muscle cells. *Exp. Cell Res. 280*, 64–74.

97 Tatin, F., Varon, C., Genot, E., and Moreau, V. (2006). A signalling cascade involving PKC, Src and Cdc42 regulates podosome assembly in cultured endothelial cells in response to phorbol ester. *J. Cell Sci. 119*, 769–781.

98 Baldassarre, M., Pompeo, A., Beznoussenko, G., Castaldi, C., Cortellino, S., McNiven, M.A., Luini, A., and Buccione, R. (2003). Dynamin participates in focal extracellular matrix degradation by invasive cells. *Mol. Biol. Cell 14*, 1074–1084.

99 Linder, S. (2007). The matrix corroded: podosomes and invadopodia in extracellular matrix degradation. *Trends Cell Biol. 17*, 107–117.

100 Miyauchi, A., Hruska, K.A., Greenfield, E.M., Duncan, R., Alvarez, J., Barattolo, R., Colucci, S., Zambonin-Zallone, A., Teitelbaum, S.L., and Teti, A. (1990). Osteoclast cytosolic calcium, regulated by

voltage-gated calcium channels and extracellular calcium, controls podosome assembly and bone resorption. *J. Cell Biol. 111*, 2543–2552.

101 Yamaguchi, H., and Condeelis, J. (2006). Regulation of the actin cytoskeleton in cancer cell migration and invasion. *Biochim. Biophys. Acta 1773*, 642–652.

102 Linder, S., Wintergerst, U., Bender-Gotze, C., Schwarz, K., Pannicke, U., and Aepfelbacher, M. (2003). Macrophages of patients with X-linked thrombocytopenia display an attenuated Wiskott-Aldrich syndrome phenotype. *Immunol. Cell Biol. 81*, 130–136.

103 Linder, S., Nelson, D., Weiss, M., and Aepfelbacher, M. (1999). Wiskott–Aldrich syndrome protein regulates podosomes in primary human macrophages. *Proc. Natl. Acad. Sci. USA 96*, 9648–9653.

104 Gimona, M., and Buccione, R. (2006). Adhesions that mediate invasion. *Int. J. Biochem. Cell Biol. 38*, 1875–1892.

105 Ayala, I., Baldassarre, M., Caldieri, G., and Buccione, R. (2006). Invadopodia: a guided tour. *Eur. J. Cell Biol. 85*, 159–164.

106 Buccione, R., Orth, J.D., and McNiven, M.A. (2004). Foot and mouth: podosomes, invadopodia and circular dorsal ruffles. *Nat. Rev. Mol. Cell. Biol. 5*, 647–657.

107 Gavazzi, I., Nermut, M.V., and Marchisio, P.C. (1989). Ultrastructure and gold-immunolabelling of cell-substratum adhesions (podosomes) in RSV-transformed BHK cells. *J. Cell Sci. 94* (Pt 1), 85–99.

108 Mueller, S.C., Yeh, Y., and Chen, W.T. (1992). Tyrosine phosphorylation of membrane proteins mediates cellular invasion by transformed cells. *J. Cell Biol. 119*, 1309–1325.

109 Marchisio, P.C., Bergui, L., Corbascio, G.C., Cremona, O., D'Urso, N., Schena, M., Tesio, L., and Caligaris-Cappio, F. (1988). Vinculin, talin, and integrins are localized at specific adhesion sites of malignant B lymphocytes. *Blood 72*, 830–833.

110 Teti, A., Colucci, S., Grano, M., Argentino, L., and Zambonin Zallone, A. (1992). Protein kinase C affects microfilaments, bone resorption, and [Ca2+] sensing in cultured osteoclasts. *Am. J. Physiol. 263*, C130–C139.

111 Bruzzaniti, A., Neff, L., Sanjay, A., Horne, W.C., De Camilli, P., and Baron, R. (2005). Dynamin forms a Src kinase-sensitive complex with Cbl and regulates podosomes and osteoclast activity. *Mol. Biol. Cell 16*, 3301–3313.

112 Brabek, J., Constancio, S.S., Siesser, P.F., Shin, N.Y., Pozzi, A., and Hanks, S.K. (2005). Crk-associated substrate tyrosine phosphorylation sites are critical for invasion and metastasis of SRC-transformed cells. *Mol. Cancer Res. 3*, 307–315.

113 Chellaiah, M.A. (2006). Regulation of podosomes by integrin alphavbeta3 and Rho GTPase-facilitated phosphoinositide signaling. *Eur. J. Cell Biol. 85*, 311–317.

114 Luxenburg, C., Parsons, J.T., Addadi, L., and Geiger, B. (2006). Involvement of the Src-cortactin pathway in podosome formation and turnover during polarization of cultured osteoclasts. *J. Cell Sci. 119*, 4878–4888.

115 Gatesman, A., Walker, V.G., Baisden, J.M., Weed, S.A., and Flynn, D.C. (2004). Protein kinase Calpha activates c-Src and induces podosome formation via AFAP-110. *Mol. Cell. Biol. 24*, 7578–7597.

116 Tehrani, S., Faccio, R., Chandrasekar, I., Ross, F.P., and Cooper, J.A. (2006). Cortactin has an essential and specific role in osteoclast actin assembly. *Mol. Biol. Cell 17*, 2882–2895.

117 Webb, B.A., Eves, R., and Mak, A.S. (2006). Cortactin regulates podosome formation: roles of the protein interaction domains. *Exp. Cell Res. 312*, 760–769.

118 Chou, H.C., Anton, I.M., Holt, M.R., Curcio, C., Lanzardo, S., Worth, A., Burns, S., Thrasher, A.J., Jones, G.E., and Calle, Y. (2006). WIP regulates the stability and localization of WASP to podosomes in migrating dendritic cells. *Curr. Biol. 16*, 2337–2344.

119 Berdeaux, R.L., Diaz, B., Kim, L., and Martin, G.S. (2004). Active Rho is localized to podosomes induced by oncogenic Src and is required for their assembly and function. *J. Cell Biol. 166*, 317–323.

120 Chellaiah, M.A., Soga, N., Swanson, S., McAllister, S., Alvarez, U., Wang, D., Dowdy, S.F., and Hruska, K.A. (2000). Rho-A is critical for osteoclast podosome organization, motility, and bone resorption. *J. Biol. Chem. 275*, 11993–12002.

121 Yamaguchi, H., Lorenz, M., Kempiak, S., Sarmiento, C., Coniglio, S., Symons, M., Segall, J., Eddy, R., Miki, H., Takenawa, T., and Condeelis, J. (2005). Molecular mechanisms of invadopodium formation: the role of the N-WASP-Arp2/3 complex pathway and cofilin. *J. Cell Biol. 168*, 441–452.

122 Hiratsuka, S., Watanabe, A., Aburatani, H., and Maru, Y. (2006). Tumour-mediated upregulation of chemoattractants and recruitment of myeloid cells predetermines lung metastasis. *Nat. Cell Biol. 8*, 1369–1375.

123 Kaverina, I., Stradal, T.E., and Gimona, M. (2003). Podosome formation in cultured A7r5 vascular smooth muscle cells requires Arp2/3-dependent de-novo actin polymerization at discrete microdomains. *J. Cell Sci. 116*, 4915–4924.

124 Mizutani, K., Miki, H., He, H., Maruta, H., and Takenawa, T. (2002). Essential role of neural Wiskott-Aldrich syndrome protein in podosome formation and degradation of extracellular matrix in src-transformed fibroblasts. *Cancer Res. 62*, 669–674.

125 Olivier, A., Jeanson-Leh, L., Bouma, G., Compagno, D., Blondeau, J., Seye, K., Charrier, S., Burns, S., Thrasher, A.J., Danos, O., Vainchenker, W., and Galy, A. (2006). A partial down-regulation of WASP is sufficient to inhibit podosome formation in dendritic cells. *Mol. Ther. 13*, 729–737.

126 Tsuboi, S. (2007). Requirement for a complex of Wiskott-Aldrich Syndrome Protein (WASP) with WASP interacting protein in podosome formation in macrophages. *J. Immunol. 178*, 2987–2995.

127 Webb, B.A., Jia, L., Eves, R., and Mak, A.S. (2007). Dissecting the functional domain requirements of cortactin in invadopodia formation. *Eur. J. Cell Biol. 86*, 189–206.

128 Bokoch, G.M., and Der, C.J. (1993). Emerging concepts in the Ras superfamily of GTP-binding proteins. *FASEB J. 7*, 750–759.

129 Linder, S., Hufner, K., Wintergerst, U., and Aepfelbacher, M. (2000). Microtubule-dependent formation of podosomal adhesion structures in primary human macrophages. *J. Cell Sci. 113* (Pt 23), 4165–4176.

130 Ory, S., Munari-Silem, Y., Fort, P., and Jurdic, P. (2000). Rho and Rac exert antagonistic functions on spreading of macrophage-derived multinucleated cells and are not required for actin fiber formation. *J. Cell Sci. 113* (Pt 7), 1177–1188.

131 Burns, S., Thrasher, A.J., Blundell, M.P., Machesky, L., and Jones, G.E. (2001). Configuration of human dendritic cell cytoskeleton by Rho GTPases, the WAS protein, and differentiation. *Blood 98*, 1142–1149.

132 Suzuki, T., Shoji, S., Yamamoto, K., Nada, S., Okada, M., Yamamoto, T., and Honda, Z. (1998). Essential roles of Lyn in fibronectin-mediated filamentous actin assembly and cell motility in mast cells. *J. Immunol. 161*, 3694–3701.

133 Hauck, C.R., Hsia, D.A., Ilic, D., and Schlaepfer, D.D. (2002). v-Src SH3-enhanced interaction with focal adhesion kinase at beta 1 integrin-containing invadopodia promotes cell invasion. *J. Biol. Chem. 277*, 12487–12490.

134 van Helden, S.F., Krooshoop, D.J., Broers, K.C., Raymakers, R.A., Figdor, C.G., and Van Leeuwen, F.N. (2006). A critical role for prostaglandin E2 in podosome dissolution and induction of high-speed migration during dendritic cell maturation. *J. Immunol. 177*, 1567–1574.

135 Calle, Y., Carragher, N.O., Thrasher, A.J., and Jones, G.E. (2006). Inhibition of calpain stabilises podosomes and impairs dendritic cell motility. *J. Cell Sci. 119*, 2375–2385.

136 Marzia, M., Chiusaroli, R., Neff, L., Kim, N.Y., Chishti, A.H., Baron, R., and Horne, W.C. (2006). Calpain is required for normal osteoclast function and is down-regulated by calcitonin. *J. Biol. Chem. 281*, 9745–9754.

137 Shyu, J.F., Shih, C., Tseng, C.Y., Lin, C.H., Sun, D.T., Liu, H.T., Tsung, H.C., Chen, T.H., and Lu, R.B. (2007). Calcitonin induces podosome disassembly and detachment of osteoclasts by modulating Pyk2 and Src activities. *Bone 40*, 1329–1342.

5
Integrin Trafficking

Susan E. LaFlamme, Feng Shi, and Jane Sottile

5.1
Introduction

Integrins are adhesion and signaling receptors that play central roles in many normal and pathophysiological processes [1–3]. Integrin trafficking is emerging as a critical regulator of cell adhesion, cell migration, and extracellular matrix (ECM) remodeling. In addition, several microbial pathogens enter cells by mechanisms requiring integrin internalization. For these reasons, in recent years significant effort has been made to identify mechanisms that regulate integrin endocytosis and recycling. In this chapter, we will discuss the current understanding of the physiological importance of integrin trafficking, and the molecular mechanisms that govern this process.

5.2
Historical Perspective

Integrins form a relatively large family of α/β heterodimers. Some of the early studies on integrin trafficking could not distinguish which integrins were being endocytosed, since the tracking of integrins relied on antibodies that recognized a common β subunit, which is shared among several integrins [4–6]. Other studies showed that some integrins, including α5β1, α6β4, αMβ2 and αIIbβ3, are constitutively internalized from the cell surface and subsequently recycled [7–9], whereas others, such as α3β1, α4β1 and αLβ2, are either not endocytosed, or endocytosed at a much slower rate [7]. The differences in the trafficking of individual heterodimers, however, may be cell type- or stimulus-specific, as later studies demonstrated that αLβ2 is constitutively endocytosed and recycled when expressed in Chinese hamster ovary (CHO) cells [10], and that the trafficking of αLβ2 can be triggered in neutrophils by addition of chemotactic agents [11].

In early studies, surface integrins were labeled with anti-integrin antibodies to monitor trafficking events. This raised the issue of whether antibody crosslinking promoted the endocytosis of integrins. Iodine-labeling of cell-surface proteins demonstrated that antibody-mediated crosslinking is not responsible for triggering endocytosis, but that endocytosis is a constitutive property of integrins [7, 8, 12]. Studies using biotin-labeling of cell-surface proteins confirmed that antibody-mediated crosslinking is not required for integrin endocytosis [10, 11, 13, 14]. It is now well accepted that internalization and recycling is a common property of integrin receptors. However, antibody crosslinking may regulate the rate of integrin internalization, at least in some instances [15].

The regulation of the kinetics of integrin trafficking is clearly important physiologically. The first measurement of the rate of integrin trafficking was made by Mark Bretscher in 1989, who determined that the cycling time for α5β1 in CHO cells incubated in suspension in the presence of fetal calf serum was approximately 80 min [12]. Subsequently, the rate of integrin internalization and/or recycling has been measured in response to the addition of specific growth factors or after inhibiting or activating individual intracellular signaling molecules. These studies have indicated that the mechanisms which regulate the rate of integrin trafficking can differ depending upon the specific integrin heterodimer, the cell type, and the extracellular stimulus (as discussed below).

5.3
Clathrin versus Lipid Rafts

The endocytosis of cell-surface receptors can occur both by clathrin-dependent mechanisms and by clathrin-independent mechanisms involving lipid rafts or caveolae [16]. Early studies of integrin trafficking did not distinguish the route by which integrins were internalized. Rather, it was assumed that the integrins were endocytosed via a clathrin-mediated process, because this was the major endocytic route known at the time, and because integrins possess an NPXY motif, which at least partially conforms to a consensus [F/Y]XNPXY sequence known to be important for clathrin-mediated endocytosis [17, 18]. Evidence for the involvement of clathrin in integrin endocytosis was obtained from electron microscopy studies, which showed that β1-containing integrins could be found within clathrin-coated vesicles [6]. In a later study, β1 integrins were found to colocalize with FITC-transferrin [19], a marker for clathrin-dependent endocytosis [20]. The αvβ5 integrin has also been localized to clathrin-coated membrane domains [21]. Interestingly, potassium depletion, which is known to prevent clathrin- but not caveolae-mediated endocytosis [22], inhibited the internalization of αVβ5 [23], but not α5β1 [4]. These data suggested that integrins may be internalized by both clathrin-dependent and -independent mechanisms.

These has been much debate in the literature as to whether caveolae are sufficiently dynamic to serve as true endocytic entities. Some data have shown that caveolae are relatively stable structures, and that there is little movement of caveolae from the cell surface into the cell [24]. However, other evidence has shown that certain agents, including okadaic acid, epidermal growth factor (EGF) and SV40, can increase the rate of caveolae endocytosis [14, 25, 26]. In addition, recent studies have shown that caveolae can mediate rapid protein endocytosis or transcytosis in lung tissue [27]. In addition, numerous studies have shown that the disruption of caveolae by dominant-negative caveolin mutants, cholesterol-depleting drugs, certain kinase inhibitors, or the genetic deletion of caveolin, can inhibit the entry of various proteins or pathogens into cells [28–32]. The results of recent studies have also shown that endocytosis of both α2β1 and αLβ2 can occur via a caveolae/lipid raft pathway [14, 33]. Furthermore, glycosphingolipids have been shown to stimulate β1 integrin endocytosis via caveolae [15, 34]. The addition of glycosphingolipids induces clustering of integrins on the cell surface in lipid raft-enriched areas of the membrane, and promotes β1 endocytosis [15]. Certain integrins including α5β1, α2β1, α3β1, and αVβ3 can associate with caveolin, as shown by co-immunoprecipitation experiments [35]. For a subset of these integrins, caveolin-1 has also been shown to be important for integrin-dependent signaling, although whether this association regulates integrin internalization has not been addressed [35].

5.4
Internalization of Occupied versus Unoccupied Integrins

The question arises as to whether internalized integrins are ligand-occupied or -unoccupied, and whether the state of ligand occupancy changes the kinetics or pathways of integrin trafficking. The αLβ2 integrin can be endocytosed in a ligand-unoccupied form, since endocytosis can occur in the absence of the αLβ2 ligand, ICAM [14]; however, the ability of ICAM to alter αLβ2 endocytosis was not examined. Internalized integrins have been shown to colocalize with some ECM proteins, including vitronectin and fibronectin, suggesting that ECM ligands may be co-endocytosed with integrins [19, 23, 36]. Co-endocytosis of ECM proteins and integrins is also suggested by the ability of function blocking anti-integrin antibodies to inhibit uptake and/or degradation of certain ECM ligands [36–38]. Monoclonal antibodies have also been used to try to distinguish between occupied and unoccupied integrins. Monoclonal antibodies (mAbs) 12G10 and 13 recognize different β1 integrin conformers [39, 40]. mAb 12G10 binds to activated/ligand-bound β1 integrins, whereas mAb13 recognizes unoccupied β1 integrins. Integrins bound to either mAb 13 or mAb 12G10 are both internalized; however, localization of the endocytosed mAb 13 or 12G10 only partially overlaps [19], which suggests that the trafficking of ligand-occupied and -unoccupied integrins may be different. It will be interesting to determine whether the state of integrin activation alters the rate of integrin endocytosis or recycling. The importance of ECM ligand occupancy on integrin endocytosis has been difficult to assess, as most cells constitutively express ECM proteins, and/or are cultured in serum-containing medium, which contains both fibronectin and vitronectin. Recently, it was shown that β1 integrins are endocytosed in fibronectin null cells in the absence or presence of fibronectin and fibronectin matrix [36]. Hence, at least some β1 integrins can be endocytosed in the absence of ligand [14, 36]. Most internalized integrins are recycled to the cell surface [7, 8, 12]. In contrast, ECM proteins such as fibronectin and vitronectin are degraded by the lysosomes following internalization [37, 38, 41–43]. As with many receptor–ligand systems, it is likely that the ECM proteins and integrins are sorted in an endosomal compartment following internalization-such sorting can occur following either clathrin- or caveolae-mediated endocytosis [44].

5.5
Regulation of Integrin Trafficking

Although integrins appear to be constitutively endocytosed and recycled to the cell surface, there is growing evidence that integrin trafficking is regulated by cell signaling (Figure 5.1). Activation of the EGF receptor stimulates the internalization of α2β1 via lipid rafts [14], and the recycling of integrins from endosomal recycling compartments [45]. Treatment of human fibroblasts with bradykinin stimulates the endocytosis and recycling of β1 integrins [46]. Signaling by the platelet-derived growth factor (PDGF) receptor promotes the recycling of αVβ3 in fibroblasts [13], and the recycling of α5β1 increases following treatment of endothelial cells with tumor necrosis factor [47]. The molecular mechanisms regulating these events are rapidly being identified.

Members of the Rab family of GTPases are central regulators of vesicular trafficking [48, 49]. Rab5 regulates receptor endocytosis and trafficking to early endosomes; Rab4 and Rab11 regulate receptor recycling. At present, the role of Rab5

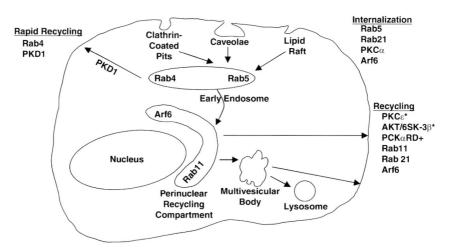

Figure 5.1 Endocytosis and recycling pathways used by integrins and extracellular matrix (ECM) proteins. Depicted are the cellular structures involved in the trafficking of integrins. Signaling proteins known to regulate the internalization, recycling and rapid recycling of integrins are listed. Fibronectin (FN) and vitronectin (VN) are trafficked to the lysosome for degradation.

in regulating integrin trafficking is not fully understood. Interestingly, dominant-negative Rab5 was found to inhibit the internalization of transferrin, but not lipid raft markers [50], thus suggesting that Rab5 may not play a role in the internalization of integrins by caveolae or lipid rafts. Consistent with this idea, the expression of a dominant-negative Rab5 mutant does not inhibit the internalization of αLβ2 [11]. However, Rab5 can co-immunoprecipitate with β1 integrins, and the overexpression of wild-type Rab5 stimulates β1 internalization [51]. It is possible that Rab5 may regulate integrin trafficking in a heterodimer- and/or stimulus-specific manner.

Several studies have demonstrated that Rab4 and Rab11 regulate integrin recycling. Depending upon the heterodimer, integrins can recycle back to the plasma membrane through a "short loop" involving Rab4 endosomes, or through a "long loop" via Rab11 recycling endosomes [13, 52]. Activation of the PDGF receptor stimulates the rapid recycling of integrin αVβ3 via a Rab4-dependent pathway by a mechanism involving PKD1 [13, 53]. To date, αvβ3 is the only integrin found to use the short loop [13, 53], which is likely to be due at least in part to the specific association of PKD1 with the β3 cytoplasmic tail [53]. The αVβ3 integrin can also use the long loop requiring Rab11 activity; this is the route used by the α5β1 integrin [13]. The Rab11-dependent recycling of integrins is stimulated by EGF or serum [45] and is regulated by the Akt-dependent inactivation of GSK-3β [54].

Other signaling proteins have also been shown to regulate integrin trafficking. PKCε promotes the recycling of β1 integrins, but only in mesenchymal and tumor cells that express the intermediate filament protein vimentin [55, 56]. The inhibition of PKC traps β1 integrins inside the cell, and PKCε activity is required for the recycling of β1 integrins to the cell surface [55]. PKCε and β1 integrins colocalize on vesicles that associate with vimentin filaments. The phosphorylation of vimentin by PKCε releases the vesicles from vimentin filaments, allowing the β1 integrins to be recycled to the cell surface [56].

PKCα can regulate the surface expression and internalization of β1 integrins [19]. The regulatory domain of PKCα associates with β1 integrins. The overexpression of PKCα or its regulatory domain is sufficient to increase the cell-surface expression of integrins, whereas the activation of PKCα promotes β1 integrin internalization [19].

Arf6 is a small GTP-binding protein that regulates vesicular trafficking [57], and Arf6 activity is required for the recycling of β1 integrins from Rab11/Arf6-positive recycling endosomes [45]. The depletion of Arf6 by siRNA inhibits integrin recycling [58]. Surprisingly, the depletion of BRAG2, a guanine nucleotide exchange factor (GEF) for Arf6, leads to the accumulation of Arf6 and β1 integrins at the cell surface, thus indicating that Arf6 also regulates integrin internalization [58]. Hence, Arf6 appears to be differentially activated at specific subcellular locations to promote the endocytosis and recycling of β1 integrins [58].

Rab21 is a member of the Rab family that associates with β1 integrins [51]. Time-lapse microscopy shows GFP-Rab21-associated vesicles moving toward and away from the cell surface, which suggests that Rab21 may regulate both integrin endocytosis and recycling. The overexpression of GFP-Rab21 increased β1 internalization and recycling of β1 integrins to the plasma membrane. Interestingly, expression of a Rab21 mutant unable to bind GTP induced the formation of large β1-containing focal adhesions.

It is not clear whether some of these signaling proteins target clathrin- or caveolea/lipid raft-dependent integrin internalization. PKCα is known to promote the endocytosis of caveolae, and Rab21 associates with α2β1 [51], which can be endocytosed via caveolae [14, 33]. However, β1 integrins internalized downstream of PKCα have been colocalized with transferrin [19], which is known to be endocytosed via clathrin-coated pits. Thus, it will be interesting to determine whether PKCα and Rab21 regulate the internalization of β1 integrins by caveolae/lipid raft-dependent or clathrin-dependent mechanisms. There is also evidence that Arf6 may regulate clathrin-dependent and -independent trafficking. Arf6 can activate phosphatidylinositol 4-phosphate 5-kinase (PIP5K) to promote the assembly of clathrin and AP2 adaptor complexes at the plasma membrane [59, 60]. Arf6 can also increase PIP2 levels to regulate membrane traffic that uses a clathrin-independent endocytic pathway [61]. Thus, it will also be important to determine whether Arf6 can promote β1 internalization by both of these mechanisms.

5.6
The Role of Integrin Cytoplasmic Domains

Most constitutively internalized receptors that enter cells via clathrin-coated pits contain tyrosine sorting motifs in their cytoplasmic domains [17]. Coated pit localization signals include the [F/Y]XNPXY motif found in the LDL receptor and the YXRΦ found in the transferrin receptor [17, 18]. The β2 and β7 integrin cytoplasmic domains contain a YXRΦ motif [62] which is required for the internalization and/or recycling of recombinant αLβ2, depending upon the cell type [10, 63]. As noted above, when cells are stimulated by fMLF, αLβ2 becomes internalized via lipid rafts and is recycled by a Rab11-dependent mechanism that requires the YXRΦ motif [11]. Most integrin β subunit cytoplasmic domains (β tails) contain NPXY motifs with the exception of β2, and β8 tails [62]. These motifs are not in the context of [F/Y]XNPXY known to be required for the internalization of the LDL receptor, and mutation of the NPXY motifs in the β1 or β3 subunit cytoplasmic domain does not inhibit the constitutive endocytosis of β1 and β3 integrins [9, 64].

Integrin β subunit cytoplasmic tails provide specificity to the regulation of integrin trafficking. PKCα binds to the β1 tail to regulate the trafficking of β1 integrins in breast carcinoma cells [65]. Such binding requires both NPXY motifs in the β1 tail, which suggests that PKCα regulates β1 trafficking by a mechanism that differs from the constitutive endocytosis and recycling of β1 described previously [64]. The αVβ3 integrin is the only integrin known to be rapidly recycled in response to growth factor signaling via a Rab4-, PKD1-dependent mechanism. This may be due to the specific association of PKD1 with the carboxy-terminal 14 amino acids of the β3 tail, which are not conserved in other β tails [53]. As mentioned above, αVβ5 is endocytosed by a clathrin-dependent mechanism [23], and the β5 cytoplasmic tail may regulate this process by associating with clathrin. A peptide SRARYEMASNPLYRKPIST modeled from the β5 tail has been shown to bind to clathrin in affinity chromatography experiments, and also to inhibit the membrane localization of clathrin in microinjection experiments. This association appears to be specific for the β5 tail, since a peptide corresponding to the analogous region of the β1 tail did not bind clathrin [21].

Recently, a role has been demonstrated for the α cytoplasmic domain in the regulation of integrin trafficking. The cytoplasmic domains of several α subunits known to heterodimerize with the integrin β1 subunit can associate with Rab21 [51]. This interaction was first identified using the α2 tail as bait in a yeast two-hybrid screen. Rab21 co-immunoprecipitates with integrins containing α1, α2, α5, or α6 tails. Interestingly, this association requires the conserved membrane-proximal amino acids, GFFKRK. Mutation of this motif to GFFAAK in the α2 tail inhibits the interaction of Rab21, but not the adhesive function of the α2β1 integrin. In fact, this mutation is likely to constitutively activate the α2β1 integrin [66]. Rab5 also associates with β1 integrins, and the overexpression of Rab5 leads to a rapid internalization of β1 integrins [51]. In the future it will be important to demonstrate whether Rab5 and Rab21 function in the same or distinct pathways.

5.7
Integrin-Dependent Endocytosis of Microbial Pathogens

There is a large body of literature describing the role of integrins as endocytic receptors for viruses and bacteria. In many cases, endocytosis of viruses and bacteria by integrins is clathrin-mediated [67–69], an example being the uptake of adenoviruses which occurs following adenovirus binding to the CAR receptor. The internalization of virions requires integrins αVβ3 or αVβ5 [70], while the release of virions within the cell depends upon αvβ5 [71]. The integrin-mediated endocytosis of adenoviruses is clathrin-mediated, since agents that block clathrin-mediated endocytosis also block adenoviral entry [68, 72]. Similarly, the entry of foot-and-mouth disease virus relies on integrin αvβ6 and is also clathrin-mediated [67, 73]. The uptake of bacteria by integrins can also be mediated by clathrin [69]. For example, β1 integrin-mediated endocytosis of *Staphylococcus aureus* is both AP2- and clathrin-dependent, and requires an intact NPXY motif in the β1 subunit cytoplasmic tail [69]. Although the integrin-mediated uptake of many pathogens occurs by a clathrin-mediated process, certain pathogens can gain entry by a caveolae/lipid raft-dependent pathway. For example, α2β1-mediated echovirus-1 endocytosis is lipid raft-dependent [74], which is consistent with the recent identification of lipid raft-dependent endocytosis of this integrin [33]. Some viruses that are endocytosed by an integrin-dependent pathway have been localized to caveolae and caveosomes (SV40) [75], or to clathrin-coated pits and coated vesicles (adenovirus) [72], indicating that their uptake is via receptor-mediated endocytosis. The integrin-mediated uptake of bacteria is likely to involve a phagocytic mechanism [76–78].

5.8
Regulation of Cell Adhesion and Migration by Integrin Trafficking

Accumulating evidence links integrin trafficking with cell migration. Indeed, many studies have demonstrated that the activation of signaling pathways which promote integrin trafficking increase cell migration, whereas the inhibition of integrin trafficking negatively regulates cell migration. For example, perturbing Arf6 function in MDA-MB-231 breast carcinoma cells inhibits β1 integrin recycling and cell migration [45]. Likewise, the expression of catalytically inactive PKCε in mesenchymal cells inhibits both integrin trafficking and cell migration [55]. Suppressing PKD1 expression by RNAi inhibits both PDGF-induced recycling of αvβ3 and cell migration [53]. Inhibition of the expression of Rab21 impairs cell migration, whereas overexpression of Rab21 stimulates migration. It is important to note that the aforementioned signaling proteins have multiple cellular targets; thus, the effects of inhibiting their function on cell migration may not be due to changes in integrin trafficking alone. A more direct association between integrin trafficking and cell migration comes from studies of mutant integrins. Mutations in the α and β subunit cytoplasmic tails have been identified that inhibit integrin trafficking but not integrin-mediated adhesion. Expression of these mutant integrins is sufficient to inhibit migration, thus demonstrating the importance of integrin trafficking for cell migration [51, 63]. Integrin trafficking is thought to promote cell migration by the delivery of integrins to the leading edge, although there is little direct evidence for this notion except for studies with neutrophils [79]. Interestingly, modulating the activities of PKD1 and Rab21 also affects the assembly/disassembly of focal adhesions, which suggests that integrin trafficking may impact cell migration by multiple mechanisms [51, 53]. Not surprisingly, changes in integrin trafficking can also impact cell adhesion and spreading by altering the surface expression of integrins, and this is best illustrated by studies altering the activities of Rab4 and Arf6. At present, two dominant-negative mutants of Rab4 have been identified: (i) S22NRab4, which significantly reduces αVβ3 recycling and inhibits cell spreading; and (ii) N121IRab4, which completely abolishes αVβ3 recycling and also inhibits cell adhesion [13]. Conversely, inhibiting the internalization of β1 integrins by the depletion of the Arf6 GEF BRAG2 promotes adhesion and spreading on fibronectin, which is a β1 integrin ligand [58]. Changes in the adhesiveness of cells due to changes in the surface expression of integrins may also impact cell migration either directly or indirectly by altering adhesion signaling events.

5.9
The Regulation of Integrin Trafficking by Cell Adhesion

There is growing evidence that integrin signaling can regulate integrin trafficking. The loss of integrin-mediated adhesion can trigger the degradation of internalized integrins in some cell types. For example, when normal rat kidney (NRK) cells are incubated in suspension for 8h, more than 50% of the cell-surface integrins are internalized and degraded [80]. In addition, during suspension-induced keratinocyte differentiation, endocytosed integrins are degraded [5], although it is not clear whether degradation is coupled to differentiation or to a loss of adhesion. The results of other studies have indicated that internalized integrins can be recycled to the cell surface in cells deprived of adhesive contacts. In fact, some of the early studies to characterize integrin recycling were performed using CHO cells incubated in suspension [7, 8, 12]. In these studies, recycling, rather than degradation, was the major fate of the endocytosed integrins. It has been found that the expression of activated Cdc42 in human fibroblasts increases the amount of internalized $\beta1$ integrins and decreases their surface expression when these cells are incubated in suspension for 30 min. In contrast, the surface expression of $\beta1$ integrins is not decreased when control cells are incubated in suspension, which indicates that the effects on integrin trafficking in suspended cells was due to the presence of activated Cdc42. Interestingly, the re-adhesion of cells expressing activated Cdc42 restores $\beta1$ surface expression, which suggests that integrin signaling can modulate the regulation of integrin trafficking by activated Cdc42 [46]. Hence, it is unclear why the loss of adhesion triggers the degradation of internalized integrins in some instances, but not others. Such variation may be due to cell type differences in the regulation of integrin trafficking by adhesion, or to differences in experimental design. Integrin degradation was observed when cells were incubated in suspension for 4h or longer [5, 80], whereas integrin recycling was the major fate of internalized integrins in studies where cells were incubated in suspension for less than 2h [7, 12].

The regulation of integrin trafficking by cell adhesion may be linked to the regulation of trafficking of lipid rafts. As discussed above, some integrins are internalized via lipid rafts, and several studies have shown that cell adhesion regulates the surface expression of lipid raft markers, such as ganglioside GM1. Depriving cells of adhesion leads to the internalization of GM1, while the re-adhesion of suspended cells results in the re-expression of GM1 at the cell surface [81–83]. If integrins were present in these lipid rafts, their surface expression would be similarly regulated. Recently, it was found that bradykinin, a known activator of Cdc42, triggers the internalization of $\beta1$ integrins and the colocalization of internalized integrins with Cdc42 and the lipid raft marker, GM1 [46].

The relationship between integrins and the trafficking of lipid rafts may be complex. The addition of divalent $\beta1$ integrin antibodies increases the caveolae-dependent uptake of lipid raft markers [15]. Furthermore, the downregulation of $\beta1$ integrin expression decreases the endocytosis of lipid raft markers and proteins known to be endocytosed via caveolae [34]. These studies were performed with adherent cells, and the results appear to contradict those of the studies described above, which showed that loss of adhesion (and presumably loss of integrin signaling) promotes the loss of lipid raft domains from the cell surface [81, 82]. It is possible that cells in suspension have different endocytic regulatory mechanisms compared with adherent cells. Alternatively, the recycling of lipid rafts to the cell surface may be adhesion-dependent, whereas integrin signaling may enhance –

but may not be strictly required for – lipid raft internalization. Clearly, much remains to be learned about the role of adhesion signaling in the regulation of integrin trafficking.

5.10
Integrin Trafficking and ECM Remodeling

The ECM undergoes constant remodeling which involves the deposition, reorganization and degradation of ECM components. ECM remodeling is especially active during angiogenesis and tissue repair [84–86], and abnormal ECM remodeling may contribute to the pathogenesis of many diseases, including fibrosis, arthritis, hypertension and cancer [87–93].

ECM degradation and removal are complicated by the ability of many ECM macromolecules to form supramolecular complexes. For example, collagens I, II, III, and fibronectin form large fibrillar structures, while collagen IV and laminin form sheet-like structures within the basement membrane [94–98]. Many ECM components can also be covalently crosslinked, which may affect their half-lives [99]. The stabilities of the ECM components vary widely; for example, in rat liver the turnover of fibronectin matrix occurs within 24 h [100], whereas in rat aorta the half-life of collagen is estimated to be 60–70 days. In contrast, under hypertensive conditions, the collagen half-life is reduced to 17 days [101]. Although the ECM can be degraded extracellularly by matrix metalloproteases (MMPs), plasmin and other proteases [102, 103], cellular uptake followed by lysosomal degradation is also an important mechanism for ECM turnover. Endocytosis and lysosomal degradation have been documented for many ECM components, including collagen I, fibronectin, vitronectin, thrombospondin, and proteoglycans [23, 38, 41, 42, 104–108]. Much recent evidence indicates that protein fragments derived from ECM proteins can possess functions distinct from the intact proteins [109–112], and can affect diverse cell functions including cell migration, growth, and survival. Therefore, endocytosis and intracellular degradation may provide cells with a mechanism to control the production and availability of bioactive ECM fragments.

The ability of integrins to undergo continuous endocytosis and recycling suggests that integrin trafficking may play an important role in regulating ECM endocytosis and hence, ECM remodeling. However, few studies of integrin trafficking have investigated the fate of ECM ligands that are bound to endocytosed integrins. Vitronectin is a transient component of the ECM, and a vitronectin–thrombin–antithrombin complex can be found in the ECM following tissue injury [113]. Vitronectin can be removed from the ECM through integrin-mediated endocytosis and intracellular degradation [23, 38, 41]. In fact, it has been speculated that vitronectin endocytosis may provide an important means of removing thrombin–serpin complexes from tissues [41]. Inhibitory antibodies to $\alpha v\beta 5$ or $\alpha v\beta 3$ do not block the binding of vitronectin to the cell surface, but significantly reduce vitronectin degradation [114]. These data suggest that integrins are functionally required for vitronectin endocytosis. It is assumed that αvβ3 or αvβ5 integrins are internalized in a vitronectin-bound form; subsequently, the integrin is most likely recycled back to the cell surface, while vitronectin is degraded in the lysosomes.

Additional evidence for the involvement of integrins in ECM remodeling has been derived from studies of fibrinogen endocytosis. Immobilized fibrinogen can be endocytosed in A549 alveolar epithelial cells by an αvβ3-mediated process [115]. Integrin trafficking is also involved in the turnover of fibronectin [36]; recently, the turnover of ECM fibronectin was shown to occur via a β1 integrin- [36] and caveolin-1-dependent process [36, 42]. Evidence for the involvement of β1 integrins in fibronectin endocytosis include the extensive colocalization of β1 integrins with internalized fibronectin (Figure 5.2), and the ability of β1 integrin-blocking antibodies to inhibit fibronectin endoctyosis [36]. The downregulation of caveolin-1 by siRNA results in impaired β1 integrin endocytosis and an inhibition of fibronectin matrix turnover [36]. These data indicate that caveolin-1-dependent fibronectin matrix turnover is due, at least in part, to its regulation of β1 integrin endocytosis.

Evidence that integrins are involved in collagen endocytosis has been obtained from studies in which the ability of cells to phagocytose collagen-coated beads has been examined. For example, monoclonal antibodies to α2 integrin significantly reduce the phagocytosis of collagen-coated beads, indicating that α2β1 is required for collagen phagocytosis [116]. However, the relevance of collagen bead phagocytosis to the endoctyosis of ECM collagen is not clear. Recent data have shown that a non-integrin receptor, Endo180, plays an important *in-vitro* and *in-vivo* role in regulating collagen endocytosis and degradation [104, 106, 117]. Hence, the importance of integrins in removal of ECM collagen has not been clearly established.

Integrin trafficking plays an important role in regulating cell migration [11, 52, 79, 118, 119]. In addition, it has been shown that ECM remodeling may be an important consequence of integrin trafficking [36, 38, 114]. Since ECM remodeling is also a key event that occurs during cell migration, it is possible that it is part of the mechanism by which integrin trafficking regulates cell migration. It will be interesting to test this hypothesis directly.

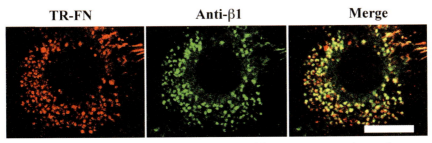

Figure 5.2 Internalized fibronectin colocalizes with β1 integrins. Rat aortic smooth muscle cells (SMC) were incubated with 10μg mL^{-1} Texas Red-conjugated fibronectin (TR-FN) and 50μM chloroquine for 12h (left). Cells were stained using an anti-β1 integrin antibody, followed by a FITC-conjugated secondary antibody (center). Right: merge of TR-FN and Anti-β1. All images are optical sections collected from a confocal microscope. Scale bar = 20μm.

While integrin trafficking is functionally important in regulating ECM remodeling, few studies have addressed whether such remodeling events can reciprocally regulate integrin trafficking. Binding of ligands, such as antibodies and pathogens, may trigger integrin endocytosis [15, 75]. It has also been suggested that the conformation of ECM ligands may be an important determinant for their clearance from the matrix. For example, denatured vitronectin is preferentially endocytosed and degraded in comparison to native vitronectin [120]. Immobilized fibrinogen can also be endocytosed [115], but fibrinogen is not endocytosed when added in soluble form to cells [115]. By using antibodies that preferentially recognize occupied and unoccupied integrins, it was shown that ligand-occupied and unoccupied integrins only partially colocalize in intracellular vesicles [19], which suggests that ECM-bound integrins may traffic in a distinctly different way from unoccupied integrins. Recent studies on fibronectin trafficking have provided evidence that ECM remodeling may affect integrin recycling [36]. The data obtained show that cell-surface levels of the α5 integrin increase approximately 1.5- to 2.3-fold under conditions which induce fibronectin matrix turnover and increase fibronectin endocytosis. However, this increase in cell-surface integrin levels is not due to increased integrin synthesis, which suggests that the turnover of matrix fibronectin can promote integrin recycling. Integrins may modulate the activities and localization of some MMPs, the primary function of which is to degrade ECM proteins. For example, αVβ3 integrins serve as cell-surface receptors for MMP2 and MT1-MMP [121, 122]. Disruption of the αVβ3/MMP2 complex can inhibit angiogenesis *in vivo* [123]. MT1-MMP is recruited to the leading edge by αVβ3 integrins during endothelial cell migration [122]. It has also been shown recently that the clustering of β1 integrins can polarize MT1-MMP on the cell surface via a Rab8-regulated exocytic pathway [124]. Rab8 is known to be involved in the polarized transport of proteins to the plasma membrane [125, 126]. Therefore it is possible that integrins are involved in two aspects of ECM remodeling: one involving localized extracellular ECM degradation; and one involving ECM endocytosis and intracellular degradation.

5.11
Conclusions

Integrins play important roles in regulating cell adhesion, spreading, migration, growth, differentiation, and survival. Many studies have focused on the role of integrin signaling in regulating these processes. Integrin trafficking is emerging as another critical control point that regulates integrin activity, and hence, cell function. Integrin trafficking influences integrin-dependent functions by a variety of mechanisms, including the control of integrin cell-surface levels, integrin localization on the cell surface, ECM turnover, and lipid raft dynamics. In the future, it will be interesting to determine whether integrin endocytosis serves to attenuate integrin signaling, or whether integrin signaling can persist in endosomal compartments, and may be important for the duration or localization of specific intracellular signals or signaling components. It will also be interesting to determine whether integrin activation influences the rate of integrin endocytosis and recycling. Further insights into how integrin endocytosis affects ECM remodeling, and how the ECM and its remodeling affect integrin trafficking will also be important for understanding the complex interplay between cells and their extracellular environment.

Acknowledgments

Two of the authors (J.S. and F.S.) were supported by NIH grants HL070261 and GM069729. S.L. was supported by an Established Investigator Award from the American Heart Association.

References

1. Sheppard, D. (2000) In vivo functions of integrins: lessons from null mutations in mice. *Matrix Biol. 19*, 203–209.
2. van der Flier, A., Sonnenberg, A. (2001) Function and interactions of integrins. *Cell Tissue Res. 305*, 285–298.
3. Hynes, R.O. (2002) Integrins: bidirectional, allosteric signaling machines. *Cell 110*, 673–687.
4. Altankov, G., Grinnell, F. (1995) Fibronectin receptor internalization and AP-2 complex reorganization in potassium-depleted fibroblasts. *Exp. Cell Res. 216*, 299–309.
5. Hotchin, N.A., Gandarillas, A., Watt, F.M. (1995) Regulation of cell surface beta 1 integrin levels during keratinocyte terminal differentiation. *J. Cell Biol. 128*, 1209–1219.
6. Raub, T.J., Kuentzel, S.L. (1989) Kinetic and morphological evidence for endocytosis of mammalian cell integrin receptors by using an anti-fibronectin receptor beta subunit monoclonal antibody. *Exp. Cell Res. 184*, 407–426.
7. Bretscher, M.S. (1992) Circulating integrins: alpha 5 beta 1, alpha 6 beta 4 and Mac-1, but not alpha 3 beta 1, alpha 4 beta 1 or LFA-1. *EMBO J. 11*, 405–410.
8. Sczekan, M.M., Juliano, R.L. (1990) Internalization of the fibronectin receptor is a constitutive process. *J. Cell Physiol. 142*, 574–580.
9. Ylanne, J., Huuskonen, J., O'Toole, T.E., Ginsberg, M.H., Virtanen, I., Gahmberg, C.G. (1995) Mutation of the cytoplasmic domain of the integrin beta 3 subunit. Differential effects on cell spreading, recruitment to adhesion plaques, endocytosis, and phagocytosis. *J. Biol. Chem. 270*, 9550–9557.
10. Fabbri, M., Fumagalli, L., Bossi, G., Bianchi, E., Bender, J.R., Pardi, R. (1999) A tyrosine-based sorting signal in the beta2 integrin cytoplasmic domain mediates its recycling to the plasma membrane and is required for ligand-supported migration. *EMBO J. 18*, 4915–4925.
11. Fabbri, M., Di Meglio, S., Gagliani, M.C., Consonni, E., Molteni, R., Bender, J.R., Tacchetti, C., Pardi, R. (2005) Dynamic partitioning into lipid rafts controls the endo-exocytic cycle of the alphaL/beta2 integrin, LFA-1, during leukocyte chemotaxis. *Mol. Biol. Cell 16*, 5793–5803.
12. Bretscher, M.S. (1989) Endocytosis and recycling of the fibronectin receptor in CHO cells. *EMBO J. 8*, 1341–1348.
13. Roberts, M., Barry, S., Woods, A., van der Sluijs, P., Norman, J. (2001) PDGF-regulated rab4-dependent recycling of alphaVbeta3 integrin from early

endosomes is necessary for cell adhesion and spreading. *Curr. Biol.* 11, 1392–1402.

14 Ning, Y., Buranda, T., Hudson, L.G. (2007) Activated epidermal growth factor receptor induces integrin alpha2 internalization via caveolae/raft-dependent endocytic pathway. *J. Biol. Chem.* 282, 6380–6387.

15 Sharma, D.K., Brown, J.C., Cheng, Z., Holicky, E.L., Marks, D.L., Pagano, R.E. (2005) The glycosphingolipid, lactosylceramide, regulates beta1-integrin clustering and endocytosis. *Cancer Res.* 65, 8233–8241.

16 Nichols, B.J., Lippincott-Schwartz, J. (2001) Endocytosis without clathrin coats. *Trends Cell Biol.* 11, 406–412.

17 Bonifacino, J.S., Traub, L.M. (2003) Signals for sorting of transmembrane proteins to endosomes and lysosomes. *Annu. Rev. Biochem.* 72, 395–447.

18 Mellman, I. (1996) Endocytosis and molecular sorting. *Annu. Rev. Cell Dev. Biol.* 12, 575–625.

19 Ng, T., Shima, D., Squire, A., Bastiaens, P.I., Gschmeissner, S., Humphries, M.J., Parker, P.J. (1999) PKCalpha regulates beta1 integrin-dependent cell motility through association and control of integrin traffic. *EMBO J.* 18, 3909–3923.

20 Harding, C., Heuser, J., Stahl, P. (1983) Receptor-mediated endocytosis of transferrin and recycling of the transferrin receptor in rat reticulocytes. *J. Cell Biol.* 97, 329–339.

21 De Deyne, P.G., O'Neill, A., Resneck, W.G., Dmytrenko, G.M., Pumplin, D.W., Bloch, R.J. (1998) The vitronectin receptor associates with clathrin-coated membrane domains via the cytoplasmic domain of its beta5 subunit. *J. Cell Sci.* 111, 2729–2740.

22 Marks, D.L., Singh, R.D., Choudhury, A., Wheatley, C.L., Pagano, R.E. (2005) Use of fluorescent sphingolipid analogs to study lipid transport along the endocytic pathway. *Methods* 36, 186–195.

23 Memmo, L.M., McKeown-Longo, P. (1998) The alphaVbeta5 integrin functions as an endocytic receptor for vitronectin. *J. Cell Sci.* 111, 425–433.

24 Thomsen, P., Roepstorff, K., Stahlhut, M., van Deurs, B. (2002) Caveolae are highly immobile plasma membrane microdomains, which are not involved in constitutive endocytic trafficking. *Mol. Biol. Cell* 13, 238–250.

25 Parton, R.G., Joggerst, B., Simons, K. (1994) Regulated internalization of caveolae. *J. Cell Biol.* 127, 1199–1215.

26 Tagawa, A., Mezzacasa, A., Hayer, A., Longatti, A., Pelkmans, L., Helenius, A. (2005) Assembly and trafficking of caveolar domains in the cell: caveolae as stable, cargo-triggered, vesicular transporters. *J. Cell Biol.* 170, 769–779.

27 Oh, P., Borgstrom, P., Witkiewicz, H., Li, Y., Borgstrom, B.J., Chrastina, A., Iwata, K., Zinn, K.R., Baldwin, R., Testa, J.E., Schnitzer, J.E. (2007) Live dynamic imaging of caveolae pumping targeted antibody rapidly and specifically across endothelium in the lung. *Nat. Biotechnol.* 25, 327–337.

28 Anderson, H.A., Chen, Y., Norkin, L.C. (1996) Bound simian virus 40 translocates to caveolin-enriched membrane domains, and its entry is inhibited by drugs that selectively disrupt caveolae. *Mol. Biol. Cell* 7, 1825–1834.

29 Pelkmans, L., Fava, E., Grabner, H., Hannus, M., Habermann, B., Krausz, E., Zerial, M. (2005) Genome-wide analysis of human kinases in clathrin- and caveolae/raft-mediated endocytosis. *Nature* 436, 78–86.

30 Pelkmans, L., Puntener, D., Helenius, A. (2002) Local actin polymerization and dynamin recruitment in SV40-induced internalization of caveolae. *Science* 296, 535–539.

31 Roy, S., Luetterforst, R., Harding, A., Apolloni, A., Etheridge, M., Stang, E., Rolls, B., Hancock, J.F., Parton, R.G. (1999) Dominant-negative caveolin inhibits H-Ras function by disrupting cholesterol-rich plasma membrane domains. *Nat. Cell Biol.* 1, 98–105.

32 Schubert, W., Frank, P.G., Razani, B., Park, D.S., Chow, C.W., Lisanti, M.P. (2001) Caveolae-deficient endothelial cells show defects in the uptake and transport

of albumin in vivo. *J. Biol. Chem.* 276, 48619–48622.
33. Upla, P., Marjomaki, V., Kankaanpaa, P., Ivaska, J., Hyypia, T., Van Der Goot, F.G., Heino, J. (2004) Clustering induces a lateral redistribution of alpha2 beta1 integrin from membrane rafts to caveolae and subsequent protein kinase C-dependent internalization. *Mol. Biol. Cell* 15, 625–636.
34. Singh, R.D., Holicky, E.L., Cheng, Z.J., Kim, S.Y., Wheatley, C.L., Marks, D.L., Bittman, R., Pagano, R.E. (2007) Inhibition of caveolar uptake, SV40 infection, and beta1-integrin signaling by a nonnatural glycosphingolipid stereoisomer. *J. Cell Biol.* 176, 895–901.
35. Wary, K.K., Mariotti, A., Zurzolo, C., Giancotti, F.G. (1998) A requirement for caveolin-1 and associated kinase Fyn in integrin signaling and anchorage-dependent cell growth. *Cell* 94, 625–634.
36. Shi, F., Sottile, J. (2007) Caveolin-1 dependent integrin endocytosis is a critical regulator of fibronectin turnover (submitted).
37. Panetti, T.S., Wilcox, S.A., Horzempa, C., McKeown-Longo, P.J. (1995) AlphaVbeta5 integrin receptor-mediated endocytosis of vitronectin is protein kinase C-dependent. *J. Biol. Chem.* 270, 18593–18597.
38. Pijuan-Thompson, V., Gladson, C.L. (1997) Ligation of integrin alpha5beta1 is required for internalization of vitronectin by integrin alphavbeta3. *J. Biol. Chem.* 272, 2736–2743.
39. Mould, A.P., Akiyama, S.K., Humphries, M.J. (1996) The inhibitory anti-beta1 integrin monoclonal antibody 13 recognizes an epitope that is attenuated by ligand occupancy. Evidence for allosteric inhibition of integrin function. *J. Biol. Chem.* 271, 20365–20374.
40. Mould, A.P., Garratt, A.N., Askari, J.A., Akiyama, S.K., Humphries, M.J. (1995) Identification of a novel anti-integrin monoclonal antibody that recognises a ligand-induced binding site epitope on the beta 1 subunit. *FEBS Lett.* 363, 118–122.
41. McKeown-Longo, P.J., Panetti, T.S. (1993) Receptor mediated endocytosis of vitronectin by fibroblast monolayers. In: K. T. Preissner, S. Rosenblatt, C. Kost, J. Wegerhoff, D. F. Mosher (Eds.), *Biology of vitronectins and their receptors*, Elsevier Science Publishers, pp. 111–118.
42. Sottile, J., Chandler, J. (2005) Fibronectin matrix turnover occurs through a caveolin-1-dependent process. *Mol. Biol. Cell* 16, 757–768.
43. Sottile, J., Hocking, D.C. (2002) Fibronectin polymerization regulates the composition and stability of extracellular matrix fibrils and cell-matrix adhesions. *Mol. Biol. Cell* 13, 3546–3559.
44. Le Roy, C., Wrana, J.L. (2005) Clathrin- and non-clathrin-mediated endocytic regulation of cell signalling. *Nat. Rev. Mol. Cell Biol.* 6, 112–126.
45. Powelka, A.M., Sun, J., Li, J., Gao, M., Shaw, L.M., Sonnenberg, A., Hsu, V.W. (2004) Stimulation-dependent recycling of integrin beta1 regulated by ARF6 and Rab11. *Traffic* 5, 20–36.
46. Starinski, J.S., Mastrangelo, A.M., R., M., LaFlamme, S.E. (2008) Cdc42 and cell adhesion regulates the trafficking of beta1 integrins. (submitted).
47. Gao, B., Curtis, T.M., Blumenstock, F.A., Minnear, F.L., Saba, T.M. (2000) Increased recycling of (alpha)5(beta)1 integrins by lung endothelial cells in response to tumor necrosis factor. *J. Cell Sci.* 113 (Pt 2), 247–257.
48. Miaczynska, M., Zerial, M. (2002) Mosaic organization of the endocytic pathway. *Exp. Cell Res.* 272, 8–14.
49. Zerial, M., McBride, H. (2001) Rab proteins as membrane organizers. *Nat. Rev. Mol. Cell Biol.* 2, 107–117.
50. Sharma, D.K., Choudhury, A., Singh, R.D., Wheatley, C.L., Marks, D.L., Pagano, R.E. (2003) Glycosphingolipids internalized via caveolar-related endocytosis rapidly merge with the clathrin pathway in early endosomes and form microdomains for recycling. *J. Biol. Chem.* 278, 7564–7572.
51. Pellinen, T., Arjonen, A., Vuoriluoto, K., Kallio, K., Fransen, J.A., Ivaska, J. (2006) Small GTPase Rab21 regulates cell adhesion and controls endosomal traffic

of beta1-integrins. *J. Cell Biol.* **173**, 767–780.

52. Caswell, P.T., Norman, J.C. (2006) Integrin trafficking and the control of cell migration. *Traffic* **7**, 14–21.
53. Woods, A.J., White, D.P., Caswell, P.T., Norman, J.C. (2004) PKD1/PKCmu promotes alphaVbeta3 integrin recycling and delivery to nascent focal adhesions. *EMBO J.* **23**, 2531–2543.
54. Roberts, M.S., Woods, A.J., Dale, T.C., Van Der Sluijs, P., Norman, J.C. (2004) Protein kinase B/Akt acts via glycogen synthase kinase 3 to regulate recycling of alpha v beta 3 and alpha 5 beta 1 integrins. *Mol. Cell. Biol.* **24**, 1505–1515.
55. Ivaska, J., Whelan, R.D., Watson, R., Parker, P.J. (2002) PKC epsilon controls the traffic of beta1 integrins in motile cells. *EMBO J.* **21**, 3608–3619.
56. Ivaska, J., Vuoriluoto, K., Huovinen, T., Izawa, I., Inagaki, M., Parker, P.J. (2005) PKCepsilon-mediated phosphorylation of vimentin controls integrin recycling and motility. *EMBO J.* **24**, 3834–3845.
57. D'Souza-Schorey, C., Chavrier, P. (2006) ARF proteins: roles in membrane traffic and beyond. *Nat. Rev. Mol. Cell Biol.* **7**, 347–358.
58. Dunphy, J.L., Moravec, R., Ly, K., Lasell, T.K., Melancon, P., Casanova, J.E. (2006) The Arf6 GEF GEP100/BRAG2 regulates cell adhesion by controlling endocytosis of beta1 integrins. *Curr. Biol.* **16**, 315–320.
59. Krauss, M., Kinuta, M., Wenk, M.R., De Camilli, P., Takei, K., Haucke, V. (2003) ARF6 stimulates clathrin/AP-2 recruitment to synaptic membranes by activating phosphatidylinositol phosphate kinase type Igamma. *J. Cell Biol.* **162**, 113–124.
60. Padron, D., Wang, Y.J., Yamamoto, M., Yin, H., Roth, M.G. (2003) Phosphatidylinositol phosphate 5-kinase Ibeta recruits AP-2 to the plasma membrane and regulates rates of constitutive endocytosis. *J. Cell Biol.* **162**, 693–701.
61. Brown, F.D., Rozelle, A.L., Yin, H.L., Balla, T., Donaldson, J.G. (2001) Phosphatidylinositol 4,5-bisphosphate and Arf6-regulated membrane traffic. *J. Cell Biol.* **154**, 1007–1017.
62. LaFlamme, S.E., Homan, S.M., Bodeau, A.L., Mastrangelo, A.M. (1997) Integrin cytoplasmic domains as connectors to the cell's signal transduction apparatus. *Matrix Biol.* **16**, 153–163.
63. Tohyama, Y., Katagiri, K., Pardi, R., Lu, C., Springer, T.A., Kinashi, T. (2003) The critical cytoplasmic regions of the alphaL/beta2 integrin in Rap1-induced adhesion and migration. *Mol. Biol. Cell* **14**, 2570–2582.
64. Vignoud, L., Usson, Y., Balzac, F., Tarone, G., Block, M.R. (1994) Internalization of the alpha5beta1 integrin does not depend on 'NPXY' signals. *Biochem. Biophys. Res. Commun.* **199**, 603–611.
65. Parsons, M., Keppler, M.D., Kline, A., Messent, A., Humphries, M.J., Gilchrist, R., Hart, I.R., Quittau-Prevostel, C., Hughes, W.E., Parker, P.J., Ng, T. (2002) Site-directed perturbation of protein kinase C-integrin interaction blocks carcinoma cell chemotaxis. *Mol. Cell. Biol.* **22**, 5897–5911.
66. Hughes, P.E., Diaz-Gonzalez, F., Leong, L., Wu, C., McDonald, J.A., Shattil, S.J., Ginsberg, M.H. (1996) Breaking the integrin hinge. A defined structural constraint regulates integrin signaling. *J. Biol. Chem.* **271**, 6571–654.
67. Berryman, S., Clark, S., Monaghan, P., Jackson, T. (2005) Early events in integrin alphaVbeta6-mediated cell entry of foot-and-mouth disease virus. *J. Virol.* **79**, 8519–8534.
68. Nemerow, G.R., Stewart, P.L. (1999) Role of alpha(V) integrins in adenovirus cell entry and gene delivery. *Microbiol. Mol. Biol. Rev.* **63**, 725–734.
69. Van Nhieu, G.T., Krukonis, E.S., Reszka, A.A., Horwitz, A.F., Isberg, R.R. (1996) Mutations in the cytoplasmic domain of the integrin beta1 chain indicate a role for endocytosis factors in bacterial internalization. *J. Biol. Chem.* **271**, 7665–7672.
70. Wickham, T.J., Mathias, P., Cheresh, D.A., Nemerow, G.R. (1993) Integrins alphaVbeta3 and alphaVbeta5 promote adenovirus internalization but not virus attachment. *Cell* **73**, 309–319.

71. Wickham, T.J., Filardo, E.J., Cheresh, D.A., Nemerow, G.R. (1994) Integrin alphaVbeta5 selectively promotes adenovirus mediated cell membrane permeabilization. *J. Cell Biol.* 127, 257–264.
72. Meier, O., Greber, U.F. (2004) Adenovirus endocytosis. *J. Gene Med.* 6 (Suppl 1), S152–S163.
73. O'Donnell, V., LaRocco, M., Duque, H., Baxt, B. (2005) Analysis of foot-and-mouth disease virus internalization events in cultured cells. *J. Virol.* 79, 8506–8518.
74. Pietiainen, V., Marjomaki, V., Upla, P., Pelkmans, L., Helenius, A., Hyypia, T. (2004) Echovirus 1 endocytosis into caveosomes requires lipid rafts, dynamin II, and signaling events. *Mol. Biol. Cell* 15, 4911–4925.
75. Pelkmans, L., Kartenbeck, J., Helenius, A. (2001) Caveolar endocytosis of simian virus 40 reveals a new two-step vesicular-transport pathway to the ER. *Nat. Cell Biol.* 3, 473–483.
76. Pelkmans, L., Burli, T., Zerial, M., Helenius, A. (2004) Caveolin-stabilized membrane domains as multifunctional transport and sorting devices in endocytic membrane traffic. *Cell* 118, 767–780.
77. Veiga, E., Cossart, P. (2006) The role of clathrin-dependent endocytosis in bacterial internalization. *Trends Cell Biol.* 16, 499–504.
78. Wong, K.W., Isberg, R.R. (2005) Emerging views on integrin signaling via Rac1 during invasin-promoted bacterial uptake. *Curr. Opin. Microbiol.* 8, 4–9.
79. Lawson, M.A., Maxfield, F.R. (1995) Ca(2+)- and calcineurin-dependent recycling of an integrin to the front of migrating neutrophils. *Nature* 377, 75–79.
80. Dalton, S.L., Scharf, E., Briesewitz, R., Marcantonio, E.E., Assoian, R.K. (1995) Cell adhesion to extracellular matrix regulates the life cycle of integrins. *Mol. Biol. Cell* 6, 1781–1791.
81. del Pozo, M.A., Alderson, N.B., Kiosses, W.B., Chiang, H.H., Anderson, R.G., Schwartz, M.A. (2004) Integrins regulate Rac targeting by internalization of membrane domains. *Science* 303, 839–842.
82. del Pozo, M.A., Balasubramanian, N., Alderson, N.B., Kiosses, W.B., Grande-Garcia, A., Anderson, R.G., Schwartz, M.A. (2005) Phospho-caveolin-1 mediates integrin-regulated membrane domain internalization. *Nat. Cell Biol.* 7, 901–908.
83. Palazzo, A.F., Eng, C.H., Schlaepfer, D.D., Marcantonio, E.E., Gundersen, G.G. (2004) Localized stabilization of microtubules by integrin- and FAK-facilitated Rho signaling. *Science* 303, 836–839.
84. Clark, R.A.F. (1996) Wound repair: Overview and general considerations. In: R. A. F. Clark (Ed.), *The molecular and cellular biology of wound repair*. Plenum Press, New York, pp. 3–50.
85. Davis, G.E., Senger, D.R. (2005) Endothelial extracellular matrix: biosynthesis, remodeling, and functions during vascular morphogenesis and neovessel stabilization. *Circ. Res.* 97, 1093–1107.
86. Ingber, D.E. (2002) Mechanical signaling and the cellular response to extracellular matrix in angiogenesis and cardiovascular physiology. *Circ. Res.* 91, 877–887.
87. Hotary, K.B., Allen, E.D., Brooks, P.C., Datta, N.S., Long, M.W., Weiss, S.J. (2003) Membrane type I matrix metalloproteinase usurps tumor growth control imposed by the three-dimensional extracellular matrix. *Cell* 114, 33–45.
88. Intengan, H.D., Deng, L.Y., Li, J.S., Schiffrin, E.L. (1999) Mechanics and composition of human subcutaneous resistance arteries in essential hypertension. *Hypertension* 33, 569–574.
89. Kurban, G., Hudon, V., Duplan, E., Ohh, M., Pause, A. (2006) Characterization of a von Hippel Lindau pathway involved in extracellular matrix remodeling, cell invasion, and angiogenesis. *Cancer Res.* 66, 1313–1319.
90. Liotta, L.A., Kohn, E.C. (2001) The microenvironment of the tumour-host interface. *Nature* 411, 375–379.
91. Mutsaers, S.E., Bishop, J.E., McGrouther, G., Laurent, G.J. (1997) Mechanisms of tissue repair: from wound healing to fibrosis. *Int. J. Biochem. Cell Biol.* 29, 5–17.

92 Poole, A.R., Nelson, F., Dahlberg, L., Tchetina, E., Kobayashi, M., Yasuda, T., Laverty, S., Squires, G., Kojima, T., Wu, W., Billinghurst, R.C. (2003) Proteolysis of the collagen fibril in osteoarthritis. *Biochem. Soc. Symp.* 115–123.

93 Prescott, M.F., Sawyer, W.K., Von Linden-Reed, J., Jeune, M., Chou, M., Caplan, S.L., Jeng, A.Y. (1999) Effect of matrix metalloproteinase inhibition on progression of atherosclerosis and aneurysm in LDL receptor-deficient mice overexpressing MMP-3, MMP-12, and MMP-13 and on restenosis in rats after balloon injury. *Ann. N. Y. Acad. Sci.* 878, 179–190.

94 Kuivaniemi, H., Tromp, G., Prockop, D.J. (1991) Mutations in collagen genes: causes of rare and some common diseases in humans. *FASEB J.* 5, 2052–2060.

95 McKeown-Longo, P.J., Mosher, D.F. (1989) The assembly of fibronectin matrix in cultured human fibroblast cells. In: D. F. Mosher (Ed.), *Fibronectin*. Academic Press, New York, pp. 162–179.

96 Sasaki, T., Fassler, R., Hohenester, E. (2004) Laminin: the crux of basement membrane assembly. *J. Cell Biol.* 164, 959–963.

97 Schwarzbauer, J.E., Sechler, J.L. (1999) Fibronectin fibrillogenesis: a paradigm for extracellular matrix assembly. *Curr. Opin. Cell Biol.* 11, 622–627.

98 van der Rest, M., Garrone, R. (1991) Collagen family of proteins. *FASEB J.* 5, 2814–2823.

99 Zhou, X., Jamil, A., Nash, A., Chan, J., Trim, N., Iredale, J.P., Benyon, R.C. (2006) Impaired proteolysis of collagen I inhibits proliferation of hepatic stellate cells: implications for regulation of liver fibrosis. *J. Biol. Chem.* 281, 39757–39765.

100 Rotundo, R.F., Vincent, P.A., McKeown-Longo, P.J., Blumenstock, F.A., Saba, T.M. (1999) Hepatic fibronectin matrix turnover in rats: involvement of the asialoglycoprotein receptor. *Am. J. Physiol.* 277, G1189–G1199.

101 Nissen, R., Cardinale, G.J., Udenfriend, S. (1978) Increased turnover of arterial collagen in hypertensive rats. *Proc. Natl. Acad. Sci. USA* 75, 451–453.

102 Marchina, E., Barlati, S. (1996) Degradation of human plasma and extracellular matrix fibronectin by tissue type plasminogen activator and urokinase. *Int. J. Biochem. Cell Biol.* 28, 1141–1150.

103 Shapiro, S.D. (1998) Matrix metalloproteinase degradation of extracellular matrix: biological consequences. *Curr. Opin. Cell Biol.* 10, 602–608.

104 Curino, A.C., Engelholm, L.H., Yamada, S.S., Holmbeck, K., Lund, L.R., Molinolo, A.A., Behrendt, N., Nielsen, B.S., Bugge, T.H. (2005) Intracellular collagen degradation mediated by uPARAP/Endo180 is a major pathway of extracellular matrix turnover during malignancy. *J. Cell Biol.* 169, 977–985.

105 East, L., McCarthy, A., Wienke, D., Sturge, J., Ashworth, A., Isacke, C.M. (2003) A targeted deletion in the endocytic receptor gene Endo180 results in a defect in collagen uptake. *EMBO Rep.* 4, 710–716.

106 Engelholm, L.H., List, K., Netzel-Arnett, S., Cukierman, E., Mitola, D.J., Aaronson, H., Kjoller, L., Larsen, J.K., Yamada, K.M., Strickland, D.K., Holmbeck, K., Dano, K., Birkedal-Hansen, H., Behrendt, N., Bugge, T.H. (2003) uPARAP/Endo180 is essential for cellular uptake of collagen and promotes fibroblast collagen adhesion. *J. Cell Biol.* 160, 1009–1015.

107 Godyna, S., Liau, G., Popa, I., Stefansson, S., Argraves, W.S. (1995) Identification of the low density lipoprotein receptor-related protein (LRP) as an endocytic receptor for thrombospondin-1. *J. Cell Biol.* 129, 1403–1410.

108 Tkachenko, E., Rhodes, J.M., Simons, M. (2005) Syndecans: new kids on the signaling block. *Circ. Res.* 96, 488–500.

109 Giannelli, G., Falk-Marzillier, J., Schiraldi, O., Stetler-Stevenson, W.G., Quaranta, V. (1997) Induction of cell migration by matrix metalloprotease-2 cleavage of laminin-5. *Science* 277, 225–228.

110 O'Reilly, M.S., Boehm, T., Shing, Y., Fukai, N., Vasios, G., Lane, W.S., Flynn,

E., Birkhead, J.R., Olsen, B.R., Folkman, J. (1997) Endostatin: an endogenous inhibitor of angiogenesis and tumor growth. *Cell* 88, 277–285.
111. Sasaki, T., Larsson, H., Tisi, D., Claesson-Welsh, L., Hohenester, E., Timpl, R. (2000) Endostatins derived from collagens XV and XVIII differ in structural and binding properties, tissue distribution and anti-angiogenic activity. *J. Mol. Biol.* 301, 1179–1190.
112. Xu, J., Rodriguez, D., Petitclerc, E., Kim, J.J., Hangai, M., Moon, Y.S., Davis, G.E., Brooks, P.C. (2001) Proteolytic exposure of a cryptic site within collagen type IV is required for angiogenesis and tumor growth in vivo. *J. Cell Biol.* 154, 1069–1079.
113. de Boer, H.C., Preissner, K.T., Bouma, B.N., de Groot, P.G. (1995) Internalization of vitronectin-thrombin-antithrombin complex by endothelial cells leads to deposition of the complex into the subendothelial matrix. *J. Biol. Chem.* 270, 30733–30740.
114. Panetti, T.S., McKeown-Longo, P.J. (1993) The alphaVbeta5 integrin receptor regulates receptor-mediated endocytosis of vitronectin. *J. Biol. Chem.* 268, 11492–11495.
115. Odrljin, T.M., Haidaris, C.G., Lerner, N.B., Simpson-Haidaris, P.J. (2001) Integrin alphaVbeta3-mediated endocytosis of immobilized fibrinogen by A549 lung alveolar epithelial cells. Am. *J. Respir. Cell. Mol. Biol.* 24, 12–21.
116. Lee, W., Sodek, J., McCulloch, C.A. (1996) Role of integrins in regulation of collagen phagocytosis by human fibroblasts. *J. Cell Physiol.* 168, 695–704.
117. Sheikh, H., Yarwood, H., Ashworth, A., Isacke, C.M. (2000) Endo180, an endocytic recycling glycoprotein related to the macrophage mannose receptor is expressed on fibroblasts, endothelial cells and macrophages and functions as a lectin receptor. *J. Cell Sci.* 113 (Pt 6), 1021–1032.
118. Pierini, L.M., Lawson, M.A., Eddy, R.J., Hendey, B., Maxfield, F.R. (2000) Oriented endocytic recycling of alpha5beta1 in motile neutrophils. *Blood* 95, 2471–2480.
119. Proux-Gillardeaux, V., Gavard, J., Irinopoulou, T., Mege, R.M., Galli, T. (2005) Tetanus neurotoxin-mediated cleavage of cellubrevin impairs epithelial cell migration and integrin-dependent cell adhesion. *Proc. Natl. Acad. Sci. USA* 102, 6362–6367.
120. Panetti, T.S., McKeown-Longo, P.J. (1993) Receptor-mediated endocytosis of vitronectin is regulated by its conformational state. *J. Biol. Chem.* 268, 11988–11993.
121. Brooks, P.C., Stromblad, S., Sanders, L.C., von Schalscha, T.L., Aimes, R.T., Stetler-Stevenson, W.G., Quigley, J.P., Cheresh, D.A. (1996) Localization of matrix metalloproteinase MMP-2 to the surface of invasive cells by interaction with integrin alphaVbeta 3. *Cell* 85, 683–693.
122. Galvez, B.G., Matias-Roman, S., Yanez-Mo, M., Sanchez-Madrid, F., Arroyo, A.G. (2002) ECM regulates MT1-MMP localization with beta1 or alphavbeta3 integrins at distinct cell compartments modulating its internalization and activity on human endothelial cells. *J. Cell Biol.* 159, 509–521.
123. Brooks, P.C., Silletti, S., von Schalscha, T.L., Friedlander, M., Cheresh, D.A. (1998) Disruption of angiogenesis by PEX, a noncatalytic metalloproteinase fragment with integrin binding activity. *Cell* 92, 391–400.
124. Bravo-Cordero, J.J., Marrero-Diaz, R., Megias, D., Genis, L., Garcia-Grande, A., Garcia, M.A., Arroyo, A.G., Montoya, M.C. (2007) MT1-MMP proinvasive activity is regulated by a novel Rab8-dependent exocytic pathway. *EMBO J.* 26, 1499–1510.
125. Hattula, K., Furuhjelm, J., Tikkanen, J., Tanhuanpaa, K., Laakkonen, P., Peranen, J. (2006) Characterization of the Rab8-specific membrane traffic route linked to protrusion formation. *J. Cell Sci.* 119, 4866–4877.
126. Sato, T., Mushiake, S., Kato, Y., Sato, K., Sato, M., Takeda, N., Ozono, K., Miki, K., Kubo, Y., Tsuji, A., Harada, R., Harada, A. (2007) The Rab8 GTPase regulates apical protein localization in intestinal cells. *Nature* 448, 366–369.

6
Hemidesmosomes and their Components: Adhesion versus Signaling in Health and Disease

Kristina Kligys, Kevin Hamill, and Jonathan C. R. Jones

6.1
Introduction

Hemidesmosomes are important adhesive devices that link the keratin cytoskeleton to laminins in the extracellular matrix (ECM). They are found primarily in mammalian epithelial tissues which are subject to mechanical stress, such as the skin, bladder, lungs and cornea, as well as in various glandular tissues such as the breast and prostate gland [1, 2]. Hemidesmosomes were first defined by electron microscopy and, because of their ultrastructural appearance, have been likened to spot welds [3] (Figure 6.1A).

Each hemidesmosome is a triangular, multilayered electron-dense region where epithelial cells abut the basement membrane (Figure 6.1A). In the cytoplasm, keratin-type intermediate filaments bind to the innermost region of a membrane-associated plaque. Along the extracellular face of each hemidesmosome, a sub-basal dense plate parallels the plasma membrane, while anchoring filaments traverse the lamina lucida and connect the hemidesmosome to the lamina densa of the basement membrane zone [1, 2] (Figure 6.1A). Over the past 20 years, at least nine proteins that constitute the hemidesmosome have been identified and characterized (Figure 6.1B). Specifically, two plakin family member proteins in the inner cytoplasmic plaque, BP230 (BPAG1) and plectin, bind keratin [1, 4, 5]. Spanning the plasma membrane are the collagen-related molecule BP180 (BPAG2, type XVII collagen), the tetraspanin CD151, and the heterodimer $\alpha 6 \beta 4$ integrin [1, 2, 5–8]. The cytoplasmic tails of $\alpha 6 \beta 4$ integrin and the head of BP180 contribute to the structure of the outer plaque, whereas their extracellular domains are components of either the sub-basal dense plate and/or anchoring filaments [1, 2, 9]. The extracellular domains of BP180 and $\alpha 6 \beta 4$ integrin interact with the matrix molecule laminin-332 (formerly known as laminin-5) [1, 2, 10–13]. Although many of these molecules are the targets of autoimmune and/or genetic diseases, certain hemidesmosome proteins also play important roles in the signaling and migration of cells in normal tissues undergoing remodeling, and in tumor cells.

K. K. and K. H. contributed equally to this chapter.

Cell Junctions. Adhesion, Development, and Disease.
Edited by Susan E. LaFlamme and Andrew Kowalczyk
Copyright © 2008 WILEY-VCH Verlag GmbH & Co. KGaA, Weinheim
ISBN: 978-3-527-31882-7

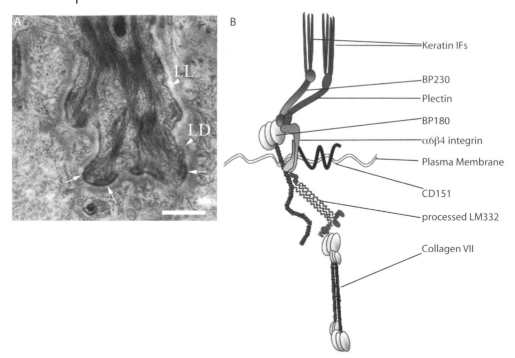

Figure 6.1 (A) An electron micrograph of the region of interaction between human epidermis and the dermis. The arrows indicate numerous hemidesmosomes along the epidermal–dermal border. LL, lamina lucida; LD, lamina densa. Scale bar = 0.5 μm. (B) Schematic of the molecular components of the hemidesmosome.

This chapter will include a discussion of hemidesmosome structure and its role in disease, as well as some details of recent studies which were conducted to define the molecular mechanisms via which hemidesmosome proteins – primarily when outside the confines of a hemidesmosome (as defined by ultrastructure) – regulate signal pathways and cell motility.

6.2
The Structural Components of the Hemidesmosome

6.2.1
The Plaque Proteins

6.2.1.1 Plectin

Plectin, also referred to as HD1 and IFAP300, is a high-molecular-weight (>600 kDa) member of the plakin family of proteins, and serves as one of the two linkers that mediate keratin interaction with the plaque of the hemidesmosome [14–18]. Plectin is thought to homodimerize, and has been described as a "dumb-bell"-shaped

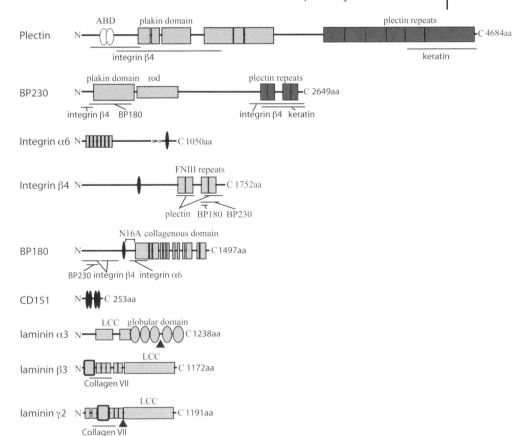

Figure 6.2 Stick diagrams of the hemidesmosome components. The black ovals in α6, β4, BP180 and CD151 depict transmembrane domains. Mapped interactions between hemidesmosome protein components and domain structures are indicated. ABD, actin-binding domain; FN, fibronectin type; LCC, laminin coiled-coil.

molecule with globular domains at each of its ends joined by a coiled-coil rod structure [19]. An actin-binding domain is found at the amino terminus, whereas the carboxy terminus contains an intermediate filament-binding domain [19–21] (Figure 6.2). Numerous splice variants of plectin have been described, each with a different expression pattern [22]. For example, plectin1a and 1c are expressed in mouse and human keratinocytes, but only plectin 1a has been localized to hemidesmosomes [23, 24]. The notion that plectin is an essential component of hemidesmosomes derives, in part, from analyses of mice lacking plectin. The basal epidermal cells in the skin of these animals fail to assemble hemidesmosomes, resulting in epithelial cell–connective tissue detachment and death within 2–3 days after birth [25].

At the site of the hemidesmosome, plectin anchors to the cell surface by binding to the cytoplasmic tail of β4 integrin. The actin-binding domain of plectin binds to the second fibronectin type III repeat as well as to amino acid residues 1329–1359 within the connecting segment of β4 integrin, while its plakin domain interacts with a second region within the connecting segment and a region following the fourth fibronectin type III repeat [26, 27] (Figure 6.2). Post-translational modifications of the cytoplasmic tail of β4 integrin, especially phosphorylation of serine residues 1356, 1360, and 1364, disrupt the interaction between β4 integrin and plectin [28]. This leads to hemidesmosome disassembly, underscoring the importance of plectin and α6β4 association in the maintenance of hemidesmosome integrity [28].

Plectin also binds BP180, although this weak interaction must occur in the presence of α6β4 integrin [29] (Figure 6.2). It has been suggested that the binding of plectin to β4 integrin may alter the conformational state of the latter such that it allows recruitment of BP180 to the hemidesmosome [30].

6.2.1.2 BP230

A second plakin protein, BP230 (BPAG1), is also involved in the anchorage of keratin to the hemidesmosomal plaque. BP230 was first identified as a 230–240 kDa protein that was recognized by autoantibodies from patients with bullous pemphigoid (BP) [31]. Four main splice isoforms of BP230 have been identified, of which only the epithelial-specific form (BPAG1-e), formed from an alternative promoter and first exon, has been localized to the inner cytoplasmic plaque of hemidesmosomes [5, 32, 33].

BP230 is highly homologous to plectin and desmoplakin and, as in the case of plectin, when purified it appears rod-shaped and is thought to dimerize [34]. BP230 contains globular domains at the ends of the protein that are separated by a coiled-coil rod domain (Figure 6.2). In addition, as does plectin, BP230 binds directly to keratin intermediate filaments via its plectin repeats located at the carboxy terminus [33, 35, 36] (Figure 6.2).

The amino terminus of BP230 can interact with BP180 and β4 integrin (Figure 6.2); specifically, via its plakin domain, BP230 binds the amino terminus of BP180 [37]. In addition, the first 56 residues of BP230 interact with the third and fourth fibronectin type III repeats of β4 integrin and a region of the connecting segment [29]. However, the carboxy terminus of BP230 has also been shown to bind the second pair of fibronectin type III repeats of β4 integrin [37] (Figure 6.2).

The skin of mice lacking BP230 expression exhibits defective hemidesmosomes that lack a keratin cytoskeleton linkage [35] (Figure 6.3). This reveals that plectin is unable to compensate for the loss of BP230 in maintaining cytoskeletal connection at the site of the hemidesmosome.

6.2.2
Membrane Elements

6.2.2.1 α6β4 Integrin
α6β4 integrin, which is also referred to as TSP180 [38], resides at the core of each hemidesmosome, and is a receptor for laminin-332 [12, 39]. α6β4 integrin also has

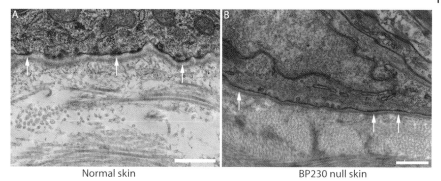

Figure 6.3 (A) An electron micrograph of normal mouse skin showing hemidesmosomes (arrows) along the region where epidermal cells abut the basement membrane zone. (B) In BP230-null mouse skin, only rudimentary hemidesmosome are observed (arrows). Scale bar = 0.5 µm.

been suggested to be a receptor for laminin-111 (previously laminin-1) and laminin-511/-521 (previously laminin-10/11), although neither of these laminin isoforms has been localized to the hemidesmosome proper [40, 41].

The β4 integrin subunit of α6β4 integrin is atypical in that its cytoplasmic tail is unusually long, consisting of approximately 1000 amino acids organized into duplicate sets of fibronectin type III repeats divided by a connecting segment [42] (see Figure 6.2). As mentioned above, this domain interacts with plakin molecules and, hence, indirectly binds keratin to the cell surface. Five different variants of the cytoplasmic domain of β4 integrin exist (β4A–E), due to alternatively spliced pre-mRNA [43, 44]. β4 integrin can undergo proteolytic processing in a tissue-specific manner; such cleavage most likely results in the disruption of hemidesmosomes, alters β4 integrin–cytoskeletal interactions, and may also occur as a stimulus for cell migration [45, 46].

All β integrin subunits contain specificity-determining loops (SDL) within their extracellular domains. The SDL of β4 integrin, however, is distinct from the equivalent domain of other β integrin subunits [47, 48]. Nevertheless, the SDL of β4 integrin is an important regulator of α6β4 integrin-ligand binding. In particular, a point mutation in the SDL of β4 integrin inhibits its binding to laminin-332, and perturbs hemidesmosome assembly and the co-localization of α6β4 integrin with other hemidesmosome proteins namely plectin, BP180, and BP230 [47].

The α6 integrin is composed of an amino-terminal heavy chain and a carboxy-terminal light chain, which are produced by post-translational cleavage of the pre-protein [43, 49] (see Figure 6.2). The heavy chain contains the region that interacts with the β integrin subunit [43]. The gene for α6 integrin, like β4 integrin, encodes several splice variants, α6A and α6B. The isoform α6p is another variant of α6 integrin, but is a proteolytic cleavage product of the α6A isoform [50, 51]. However, the α6A isoform is the most abundant and is the variant found in mature hemidesmosomes [50].

The first indication that α6β4 integrin was involved in hemidesmosome formation and was essential for maintenance of hemidesmosome structural integrity was derived from studies using 804G rat bladder cells which assemble numerous hemidesmosomes when maintained *in vitro* [52]. Treatment of these cells with an antibody against β4 integrin inhibits hemidesmosome formation [6]. Furthermore, hemidesmosome assembly is also perturbed in 804G cells induced to express β4 integrin mutants that lack most of its cytoplasmic tail [53]. The notion that α6β4 integrin is crucial for hemidesmosome assembly gained further credence following the development of mice engineered to lack expression of either α6 or β4 integrin. Mature hemidesmosomes were not formed in the epidermis of these mice [54–56]; moreover, the mice displayed extensive skin blistering and died shortly after birth, indicating that hemidesmosomes are important in maintaining epidermal cell integrity with the underlying dermis [54–56]. Intriguingly, the expression of α6β4 integrin does not appear to play a role in skin morphogenesis, as the stratified epidermis in mice lacking either α6 or β4 integrin appears normal before any blisters are formed [57].

α6β4 integrin is also expressed in some simple epithelia where others have suggested that, when ligated by laminin-332, it assembles into a hemidesmosome variant or a type II hemidesmosome [58]. Type II hemidesmosomes lack either of the BP antigens but contain plectin. Expression of the α6β4 integrin has also been reported in endothelial cells, immature T cells and Schwann cells [59–62], although whether any of these cell types assemble adhesive devices akin to the hemidesmosome is unclear. It remains possible that α6β4 integrin in these non-epithelial cell types may play an important role in signaling rather than adhesion. The subject of α6β4 integrin-mediated signal transduction will be revisited later in the chapter.

6.2.2.2 BP180

The 180 kDa transmembrane protein BP180 (BPAG2, type XVII collagen) was first identified as a protein that reacted with autoantibodies from BP patients [63, 64]. Like BP230 autoantibodies, BP180 autoantibodies localize to the hemidesmosome [5]. BP180 is a type II membrane protein that assembles at the cell surface as a homotrimer and has the ability to bind almost all of the proteins within the hemidesmosome [29, 65, 66] (see Figure 6.2). Specifically, amino acids 506–519 of BP180 bind the extracellular domain of α6 integrin, the cytoplasmic domain of BP180 binds the third fibronectin type III repeat and a region of the connecting segment of β4 integrin, and the amino terminus of BP180 can interact with both the amino and carboxy termini of BP230 [29, 30, 37, 67, 68] (Figure 6.2). The binding sites for BP180 on β4 integrin are distinct from those for plectin on β4 integrin; hence, α6β4 integrin, plectin, and BP180 can interact simultaneously. Although it has been suggested that the binding between BP180 and BP230 occurs on the same residues as that of BP180 and plectin, it appears that the interaction between BP180 and BP230 is favored over that between BP180 and plectin [29]. Thus, it appears that the major role for BP180 is to recruit BP230 into the hemidesmosome and, thereby, to anchor keratin to the cytoplasmic hemidesmosome plaque [37].

6.2.2.3 CD151

The tetraspanin CD151 has recently been shown to localize to hemidesmosomes [8] (see Figure 6.1B), it being recruited to the hemidesmosome through an interaction with the α6 subunit of the α6β4 integrin [8]. Tetraspanins, which are characterized by their four transmembrane domains, interact with various proteins to generate multimolecular complexes referred to as "tetraspanin webs" [69, 70]. Although it was originally suggested that CD151 plays a role in hemidesmosome stability [8], recent data have challenged this hypothesis as hemidesmosomes in the skin of mice lacking CD151 expression appear normal [71]. Yet, it should be noted that mice also express a tetraspanin (BAB22942) that bears 57% amino acid sequence homology to CD151. It remains possible that this CD151 relative might compensate for a lack of CD151 in mouse cells and tissues [69, 71]. Thus, the role of CD151 in hemidesmosome formation requires further investigation.

6.3 Laminin-332

Laminin-332 (laminin-5, nicein, epiligrin, kalinin) is a major component of the basement membrane of many epithelial tissues [10, 11, 72]. It is expressed in stratified epithelium in the basal layer of cells that interact with the basement membrane zone. These cells assemble type I hemidesmosomes [1]. Laminin-332 is also found in simple epithelial cells, such as those in the gut, which assemble type II hemidesmosome structures [58].

Laminin-332 is comprised of α3, β3 and γ2 subunits that associate intracellularly via their coiled-coil domains to form a cruciform-shaped heterotrimer [11, 73, 74] (see Figure 6.2). These three chains are shorter than conventional laminin subunits, and therefore, lack the domains required for self-polymerization [75]. A higher-order network formation may require interaction with laminin-311 (previously laminin-6), and laminin-321 (previously laminin-7), although cultured keratinocytes assemble extensive arrays of laminin-332 with no apparent laminin-311/-321 [76, 77]. Laminin-332 directly interacts with collagen type VII, the major protein of anchoring fibrils, through the amino terminus of the β3 laminin subunit and, to a lesser extent, the γ2 laminin subunit [78, 79] (see Figure 6.1B). This presumably strengthens the epidermal anchorage to the connective tissue.

The processing of the laminin-332 subunits is an important regulator of its function. In human cells maintained *in vitro*, laminin-332 is secreted initially as a 460 kDa species [73]. This "precursor" laminin-332 incorporates into the matrix, and it has been reported that precursor laminin-332 supports the migration of cells [80, 81]. Subsequently, *in vitro*, laminin-332 is processed to a predominant 440 kDa form in low calcium concentrations and to a predominant 400 kDa form in high calcium concentrations [73]. These size shifts are due to processing of the α3 chain from 190/200 kDa to 165 kDa, and processing of the γ2 chain from 155 kDa to 105 kDa [80, 82–87].

The identity of the specific proteases involved in processing the subunits of laminin-332 is controversial. Cleavage of the α3 laminin subunit *in vitro* can be mediated by the protease plasmin, and occurs within the globular (G) domain of the molecule [80]. In addition to plasmin, all of the bone morphogenetic protein-1 isoenzymes (BMP, Mammalian Tolloid; mTLD, Mammalian tolloid-like1 and -2; mTLL1, mTLL2) are capable of processing α3 laminin *in vitro* [86, 87]. The cleavage within the G domain has been localized to the linker between the LG3–LG4 sub-domains [88] (see Figure 6.2). However, this linker sequence is not a consensus for either plasmin or BMP-1, but for matrix metalloproteinases (MMP), the ADAM family, or astacin family of proteases. Thus, these enzymes are more likely to be responsible for α3 laminin cleavage *in vivo*. There is some indication that the cleaved LG4–LG5 module may support adhesion, independent of the remainder of the laminin-332 molecule, although the physiological relevance of this remains unclear [89]. Regardless, the processed α3 laminin subunit is capable of supporting stable adhesion by nucleating hemidesmosome assembly in an α6β4 integrin-dependent fashion [80, 90] (see below).

BMPs are likely to play important roles in processing the γ2 subunit of laminin-332, *in vivo*, as there is a dramatic reduction of γ2 laminin processing in the skin of mTLD/BMP1-deficient mice [86]. The hemidesmosomes in skin from mTLD/BMP1-null mice appear rudimentary, indicating the importance of correct processing of laminin-332 in hemidesmosome formation and integrity [86, 87]. *In vitro*, several proteases have been reported to be capable of cleaving γ2 laminin; these include the matrix metalloproteinases MMP-2, MT1-MMP (MMP-14), MMP-19 and BMPs [84, 85, 91].

Both, *in-vitro* and *in-vivo* analyses have shown that laminin-332 is crucial to the nucleation of hemidesmosome assembly and the maintenance of their structural integrity [74, 83, 90, 92]. In particular, the skin of mice lacking expression of α3 laminin exhibits abnormal hemidesmosomes, without discernable sub-basal dense plates [92]. In addition, some regions of basal cells completely lack hemidesmosomes [92]. Moreover, even though the mutant animals are normal at birth, they develop progressive blistering of the forepaws, limbs and oral mucosa, caused by separation at the dermal–epidermal junction. This reveals that laminin-332 is an important regulator of epidermal–dermal adhesion as well as hemidesmosome formation [92]. In tumor cells and in tissues undergoing remodeling, the processing of γ2 laminin to a 80 kDa species may be responsible for the dissolution of hemidesmosomes and the release of cells from the constraints of the basement membrane [84, 93]. In addition, the amino-terminal fragment of the cleaved γ2 laminin subunit has epidermal growth factor (EGF)-like properties and may stimulate the migration of cells by activating the EGF receptor [93].

6.4
Hemidesmosome Assembly and Disassembly

As discussed previously, both *in-vitro* and *in-vivo* studies have confirmed the importance of α6β4 integrin and its ligand laminin-332 in hemidesmosome assembly [6, 54–56, 74, 90, 92]. Thus, the question to be asked is how does laminin-332 induce hemidesmosome assembly? Despite extensive studies, this area remains controversial. It is assumed that hemidesmosome formation involves laminin-332-induced clustering of α6β4 integrin which, *in vitro*, leads to the dephosphorylation of α6 integrin, which in turn enhances the recruitment of keratin filaments to the cell surface via plectin [94]. In addition, ligand-induced clustering of α6β4 integrin results in the phosphorylation of tyrosine residues within the cytoplasmic tail of β4 integrin in a region termed as a putative tyrosine activation motif (TAM). This has been suggested to trigger hemidesmosome assembly [95, 96]. However, it should also be noted that some data indicate that the phosphorylation of tyrosine residues within the β4 integrin cytoplasmic tail may promote the exact opposite – that is, hemidesmosome disassembly [97, 98].

Sonnenberg and coworkers have proposed that pre-hemidesmosomal clusters, consisting of CD151 and α3β1 integrin, are formed at the basal cell surface of epithelial cells in contact with laminin-332 [2]. Pre-hemidesmosome clusters are thought to serve as docking sites for α6β4 integrin, which initiates hemidesmosome assembly by interacting with plectin. This initial interaction subsequently recruits BP180 and BP230 to the developing hemidesmosome structure. Although CD151 remains in the mature hemidesmosome, α3β1 integrin does not, and relocates to either cell–cell contacts or focal contacts. However, analyses of knockout mice lacking either α3β1 integrin or CD151 indicate that neither is necessary for hemidesmosome assembly [71, 99].

During wound healing, and also during the early stages of tumorigenesis, migrating cells lose their hemidesmosomes [1, 2, 100, 101]. Previously, it was reported that hemidesmosome complexes are endocytosed as cells are released from the basement membrane [102], but recent data have shown that γ2 laminin undergoes proteolytic processing during tissue reorganization and that β4 integrin is cleaved by caspases upon the initiation of cell migration [45, 84, 103]. Both of these events presumably result in the disassembly of hemidesmosomes. EGF is also known to trigger the dissociation of the α6β4 integrin/laminin-332 complex from other proteins in the hemidesmosome by inducing the phosphorylation of specific β4 integrin tyrosine residues [28, 97, 104–106]. This allows the α6β4 integrin to relocate to lipid rafts or lamellipodia, where it becomes competent to activate various signaling pathways, leading to an enhancement of cell migration, survival and proliferation (see below) [104, 107]. The residues on which EGF acts in this context are reported to be distinct from tyrosine residues phosphorylated in response to ligand binding and initiation of hemidesmosome assembly [95, 98, 106].

The exposure of cells to EGF also leads to the phosphorylation of serine residues within the connecting segment of the β4 integrin cytoplasmic tail, through the activity of protein kinase C alpha or delta (PKCα or PKCδ). This phosphorylation also results in hemidesmosome disassembly [28, 104, 105]. PKCα, along with the growth factor, macrophage-stimulating protein (MSP), have been shown to mediate the phosphorylation of three evolutionarily conserved serine residues within the tail of β4 integrin that also serve as docking sites for 14-3-3 proteins [28, 108]. A complex between Ron (the receptor for MSP) and α6β4 integrin forms in response to 14-3-3 protein binding, leading to the displacement of α6β4 integrin from the hemidesmosome and its transfer to the lamellipodia, where it plays a role in regulating cell migration [108].

6.5
Hemidesmosomes and Disease

6.5.1
Inherited Skin Diseases Involving Hemidesmosome Proteins

Mutations in the genes encoding hemidesmosomal components are now known to lead to a variety of skin blistering diseases. Generally, these diseases fall into the epidermolysis bullosa group of bullous disorders, and include junctional epidermolysis bullosa (JEB), generalized atrophic epidermolysis bullosa (GABEB), and muscular dystrophy with associated epidermolysis bullosa simplex (MD-EBS) [109–112]. The study of these rare disorders has greatly aided in the understanding of the basic functions of hemidesmosomes and their protein components.

6.5.1.1 Disease Involving α6β4 Integrin

Recessive mutations in either of the genes encoding α6 or β4 integrin (*ITGA6* or *ITGB4*) have been identified as pathogenic in JEB accompanied by pyloric atresia (PA-JEB) [113–116]. Clinically, PA-JEB is characterized by mucocutaneous fragility with intestinal abnormalities. α6 or β4 integrin is either absent or attenuated in PA-JEB skin [117, 118], and ultrastructurally there are reduced numbers of hemidesmosomes in the basal epidermal layers. Those hemidesmosomes that are detected appear rudimentary, lacking their sub-basal dense plates and frequently an inner cytoplasmic plaque, while retaining keratin attachment [118]. Blister cleavage planes are most frequently seen through the lamina lucida, but they may also occur at the level of the plasma membrane [118].

The most severe phenotypes of blistering diseases involving hemidesmosomal integrins result from mutations that lead to the generation of a premature termination codon (PTC), either directly through nucleotide substitution, or indirectly through out-of-frame insertions or deletions. Total loss of protein expression is due to mRNA instability, yet mutations affecting protein stability also give rise to severe phenotypes. Two such examples that result in lethality are an 11-amino acid deletion in β4 integrin (amino acids 59–69) and a serine to lysine substitution at residue 47 (S47L) in α6 integrin [119, 120]. These mutations target the altered β4 or α6 integrin proteins for degradation via proteasomal or lysosomal pathways, respectively [119, 120]. Severe phenotypes also occur where critical residues for α6β4 interactions or protein folding are changed, for example β4 C61Y [121], C245G [122], Q425P [123] and C562R [121].

Milder phenotypes are associated with mutations in less critically conserved residues. For example, a 50-amino acid deletion (amino acids 1450–1499) affecting the third FNIII repeat of β4 integrin which effectively removes the BP180 interaction site, leads to only a mild blistering phenotype, with cleavage occurring at the intradermal level and without associated pyloric atresia [124]. A loss of plectin/β4 integrin interaction ability, through homozygous inheritance of R1281W or compound heterozygous inheritance of R1281W/R1225H, R1821W/R252C affecting the second FNIII repeat of β4 integrin, also leads to milder blistering [121, 125].

6.5.1.2 Disease Involving Plectin

Muscular dystrophy-EBS (MD-EBS) is an autosomal recessive disorder characterized by the dual phenotype of skin blistering and late-onset neuromuscular disease [109, 126]. Blistering in MD-EBS patients occurs very close to the basal plasma membrane, with a loss of staining for plectin and a concomitant reduction in hemidesmosome number [17, 109]. Pathogenic mutations in the plectin gene were subsequently identified in MD-EBS patients [21, 109].

Almost all of the mutations in plectin that have been identified to date lead to the generation of a PTC and protein truncation. Truncation of the carboxy terminus of plectin effectively removes the intermediate filament binding region, thus inhibiting keratin attachment to the plaque of the hemidesmosome [127, 128].

6.5.1.3 Disease Involving BP180

Some patients with GABEB, recently termed non-Herlitz JEB (non-HJEB), harbor mutations in the gene *COL17A1* encoding BP180. The majority of pathogenic mutations in BP180, in patients with severe symptoms, lead to a PTC and subsequent nonsense-mediated mRNA decay of the mutant transcripts [129–132]. Missense or in-frame deletions that impact BP180 trimer formation result in protein instability and subsequent protein degradation [133, 134]. Complete deficiency of BP180 is phenotypically characterized by congenital generalized blistering with mild atrophic scarring, mild mucosal involvement, rudimentary nails, incomplete universal atrophic alopecia, pigmentary changes and enamel hypoplasia [116].

In GABEB skin, α6β4 integrin, BP230, and laminin-332 are distributed normally, but the hemidesmosomes are either absent or lack well-developed cytoplasmic plaques [111, 135, 136]. Keratinocytes isolated from GABEB patients are able to form hemidesmosome-like structures, *in vitro*, containing α6β4 integrin and plectin, but not BP230, thereby supporting the consideration of a BP230/BP180 interaction [37, 137]. Furthermore, the restoration of BP180 expression is able to relocate BP230 to the hemidesmosome [137]. These data, combined with the less severe blistering in patients lacking BP180 (relative to those lacking α6β4 integrin) suggest that BP180, although important, is less critical for hemidesmosome formation.

Interestingly, in some patients, a second-site mutation in the coding or regulatory sequences of the BP180 gene, or skipping of PTC-containing exons, leads to local expression of a functional form of the protein and therefore, localized protection from blistering [138]. Another example of this mosaicism was reported where partial rescue of the BP180 transcript occurred through outsplicing of the mutated exon, leading to expression of a truncated peptide at about 15% of normal BP180 levels [139]. The blistering in this patient was much milder than would be predicted from this genetic mutation and was confined to the palms, the backs of the hands and the lower legs [139].

6.5.1.4 Disease Involving CD151

Although studies from *CD151* knockout mice have suggested that the protein is not necessary for hemidesmosome formation, mutations in CD151 in humans appear to lead to an instability of the dermal–epidermal junction [140]. The skin phenotype of the human mutation may be more severe than the mouse equivalent because the human genome lacks the potential compensatory CD151-related protein seen in mouse [71]. Patients studied to date are homozygous for a single nucleotide insertion leading to a frameshift, and generation of a PTC. Any protein translated from this gene would be truncated between transmembrane domains three and four and, therefore, would lack the ability to interact with α6β4 integrin and associate with the plasma membrane.

6.5.1.5 Disease Involving Laminin-332

JEB patients whose disease is the result of mutations in the genes encoding the subunits of laminin-332 (*LAMA3*, *LAMB3*, *LAMC2*) can be divided into two primary subtypes: Herlitz JEB (HJEB), and non-HJEB. Patients with HJEB have generalized blistering from birth, whereas non-HJEB patients have a less severe generalized blistering that improves with age [141]. In most cases, HJEB is associated with homozygous or compound heterozygous PTC mutations (nonsense, frameshift or out-of-frame splicing mutations) in the sequences encoding laminin-332 subunits. These mutations normally lead to mRNA instability and a complete loss of expression of one of the subunits [141–143]. The absence of any of the laminin-332 subunits prevents heterotrimer assembly and secretion, and leads to a total absence of laminin-332 in the basement membrane zone. Mature hemidesmosome assembly, therefore, is inhibited [142–144] (Figure 6.4). The resulting instability leads to a disruption of the epidermal–dermal interaction along the plane of the lamina lucida of the basement membrane (Figure 6.4).

In non-HJEB, usually one allele carries a PTC and the other allele carries a missense or small in-frame deletion leading to the production of a reduced quantity of laminin-332, containing one defective but partially functional subunit. The phenotype of such patients is relatively mild [142, 143, 145].

Revertant mosaicism has been described in patients with JEB [138]. In one patient, the mutated *LAMB3* mRNA was rescued through skipping of the PTC-containing exon; this led to the rescue of that transcript and a milder phenotype [146]. In a second case, mRNA rescue occurred later in life through outsplicing of the mutated exon, and the previously blistered skin became phenotypically normal

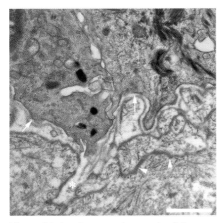

Figure 6.4 An electron micrograph of the region of interaction between human epidermis and dermis in the skin of a junctional epidermolysis bullosa patient. Note the absence of hemidesmosomes along the basement membrane zone (arrows) and the apparent duplication of the lamina densa in the connective tissue (arrow heads). Also, note the widening in some areas of the lamina lucida (asterisk). Scale bar = 1 μm.

[147]. A third patient who initially suffered from generalized non-HJEB due to heterozygous missense and nonsense mutations experienced localized reversion in a previously blistered area [138].

6.5.2
Autoimmmune Diseases Associated with Hemidesmosomes

6.5.2.1 BP and Related Diseases

BP is the most frequent acquired skin disease of the epidermal–dermal anchoring complex [148]. It is most often associated with older patients and is phenotypically characterized by tense blisters with severe itching. Immunohistologically, BP features include epidermal–dermal separation with inflammatory cell infiltrate in the upper dermis and antibodies, present in the circulation, bound to the epidermal side of the blistered skin [148]. In addition to antibodies, complement activation and neutrophil infiltration are prerequisites for disease induction. Various inflammatory cells and their secreted mediators are thought to be involved in blister formation of BP. For example, early in BP lesion formation, mast cells degranulate close to the basement membrane zone [149]. As the lesion evolves, eosinophils influx and proinflammatory cytokines are released by T cells and macrophages, which increases the inflammatory reaction. This contributes to an influx of granulocytes and this, through their subsequent release of proteinases and cytotoxic agents, ultimately causes epidermal–dermal separation [148].

The majority of BP patients harbor IgG antibodies that recognize a specific epitope on BP180, NC16A, which lies in the non-collagenous domain just extracellular to the transmembrane domain [9, 150] (see Figure 6.2). The importance of these BP180 autoantibodies in the pathogenesis of disease has been established using animal models [151]. BP230 IgG and IgE antibodies are also frequently present in BP sera, but are more closely correlated with a focal rather than general disease presentation, and a better prognosis [152, 153]. To date, there is no strong evidence that BP230 autoantibodies cause disease.

Anti-BP180 and anti-BP230 antibodies have also been isolated from other autoimmune disorders. For example, in patients with pemphigoid gestationis (PG, also known as herpes gestationis), IgG antibodies against BP230, and more commonly the NC16A domain of BP180, were isolated [9, 64, 154]. PG most commonly occurs during the second and third trimesters of pregnancy, and presents with a range of symptoms including papules, urticarial plaques, eczematous lesion, erythema multiform-like changes, and blisters [148]. Additionally, circulating antibodies to BP180 and BP230 have been identified in patients with lichen planus pemphigoides (LPP), a rare variant of BP where bullous lesions arise on lichen planus papules or on clinically uninvolved skin [155, 156]. LPP generally has a lower age of onset with a less severe clinical course compared to BP [148]. These antibodies against the NC16A domain of BP180 contain a distinct epitope from that recognized by BP sera [148].

6.5.2.2 Cicatricial Pemphigoid

Cicatricial pemphigoid (CP) is an acquired autoimmune blistering disease with circulating IgG antibodies that bind to the dermal side of split skin at the lamina lucida/lamina densa. Patients present with blistering, erosive, and scarring lesions of mucous membranes, particularly of the oral and ocular mucosa [157]. Antibodies in the sera of a majority of CP patients are against BP180 [148]. However, in about 5% of patients, autoantibodies are primarily against $\alpha 3$ laminin [148, 157–160]. *In vitro*, antibodies against $\alpha 3$ laminin block keratinocyte adhesion and cause the detachment of epithelia in organotypic cultures. *In vivo*, these antibodies can induce blistering in mouse skin graft models, suggesting that the human autoantibody equivalent is likely to be pathogenic in CP [11, 159].

6.5.2.3 Autoimmune Diseases Involving $\alpha 6 \beta 4$ Integrin

Circulating autoantibodies to $\beta 4$ and $\alpha 6$ integrins have been identified in ocular cicatricial pemphigoid (OCP) and oral pemphigoid (OP), respectively [161]. OCP is a subset of mucous membrane pemphigoid where conjunctival involvement is more pronounced than in other mucous membranes [162]. Patients present with progressive subepithelial fibrosis following chronic conjunctivitis, ultimately leading to ocular keratinization, corneal scarring, and blindness. Autoantibodies isolated from patients recognize the intracellular portion of $\beta 4$ integrin and are capable of causing lysis of a corneal explant model [162–164].

OP is a subset of CP in which the affected tissue is limited to the oral cavity. Antibodies from OP patients recognize the extracellular domain of $\alpha 6$ integrin, specifically amino acids 292–305. Rabbit antibodies raised to this region are capable of causing oral mucosa basement membrane separation in organ culture [165]. Interestingly, circulating autoantibodies to this integrin fragment are present in 52% of BP patients, suggesting that α6 integrin, in addition to BP180 and BP230, may be targeted in classical BP [166].

6.6
α6β4 Integrin and Laminin-332 Expression in Cancer

It is clear from the study of diseases such as JEB and BP that a major function for hemidesmosomes and their protein components is to mediate the stable adherence of epithelial cells to the matrix. Yet, a hallmark of certain aggressive cancers is the upregulation of α6β4 integrin and/or the subunits of its ligand laminin-332, despite the absence of hemidesmosomes in tumor cells. In fact, β4 integrin was first identified as a tumor antigen [38]. Upregulation in α6β4 integrin expression has been documented in the lung, squamous cells of the skin, oral cavity, cervix, and in the breast [167–173]. Moreover, an increase in α6β4 integrin expression has been shown to correlate with the invasive potential of certain tumor types [171, 174]. In skin tumors, α6β4 integrin expression is mislocalized to the suprabasal layer and is only found upregulated in squamous cell carcinomas but not basal cell carcinomas [175, 176]. Although α6β4 integrin expression is downregulated in prostate adenocarcinomas, highly invasive prostate tumors that have lost androgen sensitivity, show a subsequent increase in α6β4 integrin expression [177–179].

Increased laminin-332 expression is often localized to invasive areas of carcinomas. For example, colorectal and gastric cancers measure laminin-332 expression as a marker for invasiveness [180, 181]. In thyroid carcinoma, thyroid atypical adenoma, and in lung squamous and adenocarcinoma, expression of laminin-332 is evident at the epithelial–stromal interface [180, 182].

Taken together, the data from tumor analyses provides indirect, correlative evidence that α6β4 integrin and laminin-332 may have an impact on cancer cell survival, proliferation or metastasis, in addition to their role as structural components of hemidesmosomes in normal tissue. Recent studies in a mouse model of skin cancer have underscored the importance of laminin-332 and its receptor α6β4 integrin in the formation of tumor cells [183]. There is now considerable evidence that α6β4 integrin modulates cancer cell behavior via its ability to regulate multiple signaling pathways. The signaling ability of α6β4 integrin is not likely to be restricted to tumor cells, but may play an important role in epithelial cells during wound healing or development [97].

6.7
Signaling by the α6β4 Integrin in Pathological States and Wound Healing

It has become increasingly clear that signaling mediated by hemidesmosome proteins, most notably α6β4 integrin, is most likely activated when these proteins are outside the confines of the hemidesmosome, as defined by electron microscopy. Intriguingly, α6β4 integrin appears to be able to induce signaling both in a ligated and non-ligated state [184].

The main signaling cascade that is activated upon ligation of α6β4 integrin is the phosphoinositide 3-OH kinase (PI3K) pathway [185], the activation of which promotes the invasion, proliferation, and survival of carcinoma cells, *in vitro* [185–188]. PI3K-mediated invasion, in response to α6β4 integrin ligation, requires Rac1 activity [185], and this activity is involved in the formation of the lamellipodium at the leading edge of the cell [189]. The EGF receptor family member ErbB-2 functionally interacts with α6β4 integrin and phosphorylates β4 integrin on tyrosine residues within its cytoplasmic tail; both of these events promote the PI3K-mediated invasion of tumor cells [187, 190]. A recent report has extended these results to suggest that activation of the PI3K pathway by α6β4 integrin also regulates the translation of ErbB-3, which then heterodimerizes with ErbB-2, further promoting activation of PI3K signaling [191].

The activation of PI3K signaling by α6β4 integrin also plays a role in cell proliferation [107]. The palmitoylation of cysteine residues within the cytoplasmic tail of β4 integrin leads to the incorporation of a portion of α6β4 integrin into lipid rafts, which are signaling compartments within the cell [107]. Once in the lipid raft, α6β4 integrin associates with Src family kinases to mediate EGF-induced proliferation of epithelial cells, which is both PI3K- and ERK-dependent [107].

The ligation of α6β4 integrin also activates a signaling cascade from PI3K to AKT that promotes the survival of carcinoma cells by inhibiting apoptosis [186]. In addition, α6β4 integrin ligation to laminin-332 inhibits the apoptosis of breast cancer cells by regulating the translation of vascular endothelial growth factor (VEGF) [188]. This is accomplished through α6β4 integrin-dependent activation of PI3K signaling to mTOR, which leads to the phosphorylation and inactivation of 4E-BP1. This, in turn, allows for eIF-4E-dependent VEGF translation [188]. The inhibition of PI3K signaling or downregulation of VEGF negates the α6β4 integrin-mediated inhibition of apoptosis [188]. These results can be replicated *in vivo*, which suggests a role for α6β4 integrin in breast carcinoma cell survival [192]. The p53 status of the tumor cells plays a role in α6β4 integrin-dependent cell survival [193, 194]. For example, in cells lacking p53 activity, α6β4 integrin-dependent PI3K signaling promotes cell survival [193]. In contrast, α6β4 integrin signaling promotes apoptosis in cells containing wild-type p53 activity [194], due partly to an increase in p53 activity in response to α6β4 integrin signaling [194], and partly to the p53-dependent upregulation of caspase-3, which cleaves and inactivates AKT, a downstream target of PI3K [193]. Thus, PI3K signaling in response to α6β4 integrin ligation to laminin-332 promotes a host of cellular functions.

Since α6β4 integrin and the regulatory subunit of PI3K, p85, do not physically interact, it is unclear how α6β4 integrin activates the PI3K signaling. One report has suggested that the cytoplasmic adaptor proteins insulin receptor substrate-1 and -2 (IRS-1) and (IRS-2) bridge α6β4 integrin to PI3K [195]. Specifically, the ligation of α6β4 integrin by laminin-332 leads to the phosphorylation of IRS-1 and IRS-2, and binding of these proteins to PI3K leads to the subsequent activation of the PI3K signaling cascade [195]. The compartmentalization of α6β4 integrin into lipid rafts has also been suggested as a mode for activating a PI3K signaling pathway [107]. This hypothesis is supported by data that demonstrate efficient PI3K activation by H-Ras, which is also localized in lipid rafts [196]. Lastly, it has been suggested that α6β4 integrin ligation by laminin-332 can activate PI3K signaling through ErbB-2/ErbB-3 directly. As these EGFR family members have been shown to bind PI3K directly and to phosphorylate it, activation of the PI3K signaling pathway may be through the cooperation of α6β4 integrin and ErbB-2 [187, 197, 198]. It is important to note that PI3K activation by α6β4 integrin ligation is most likely due to a combination of factors that act in concert rather than separately.

In addition to PI3K signaling, α6β4 integrin ligation by laminin-332 induces mitogen-activated protein kinase (MAPK) and c-jun N-terminal kinase (JNK) signaling [199, 200]. These signaling cascades promote cellular proliferation, migration, and tumor cell invasion. As stated earlier, α6β4 integrin-mediated adhesion in the presence of EGF mediates Shc binding to the β4 integrin cytoplasmic tail and downstream MAPK activation [199]. MAPK signaling leads to the nuclear translocation of phospho-ERK and phospho-JNK, and, in turn, to the transcription of early response genes that mediate epithelial cell proliferation [199, 200].

The invasion of tumor cells is mediated by the interaction of α6β4 integrin with the growth factor receptors Met and Ron [108, 184, 201]. α6β4 integrin activation of PI3K signaling occurs when the Met receptor is activated by HGF, an event which leads to the downstream activation of Ras and additional PI3K signaling through the Met receptor. Therefore, the interaction between α6β4 integrin and the Met receptor amplifies PI3K signaling to promote the invasion of tumor cells [184]. In addition, the interaction of α6β4 integrin with the receptor Ron supports epithelial cell migration [108]. The interaction of Ron with its ligand MSP leads to phosphorylation of tyrosine residues within the β4 integrin cytoplasmic tail, and subsequent p38MAPK and NF-κB signaling, which play a role in the migration of keratinocytes over a wound bed [108, 202].

Similar signaling cascades that are activated by ligated α6β4 integrin are also activated when α6β4 integrin is not bound to laminin-332 ligand – that is, the PI3K pathway and the Ras/ERK pathway [184, 203–205]. Anchorage-independent growth of tumor cells has been shown to be mediated by Shp2 binding to phosphorylated tyrosine residues within the β4 integrin cytoplasmic tail, which occurs in response to Met receptor activity [205]. Shp2 binding ultimately activates the MAPK/ERK pathway in these tumor cells [205]. PI3K signaling is important in inhibiting transforming growth factor β (TGFβ) activity in tumor cells with suprabasal expression of α6β4 integrin [204]. The altered localization of α6β4 integrin activates PI3K activity, which promotes cell proliferation by preventing TGFβ-mediated translocation of SMAD2/3 heterodimers to the nucleus [204].

6.8
Laminin-332, α6β4 Integrin and Migration

The studies of skin disease and animal models have provided strong evidence that laminin-332 and its receptor α6β4 integrin are crucial for the stable adhesion of a variety of epithelial cell types. However, the same proteins have been implicated as regulators of epithelial cell migration, both *in vivo* and *in vitro* [77, 183, 185, 200, 206]. For example, the migration of many normal and tumor epithelial cells occurs over matrices composed of laminin-332 [10, 77, 207, 208]. As mentioned above, laminin-332 is initially secreted with a full-length α3 subunit [80, 207], and this unprocessed laminin-332 supports cell migration via binding to α3β1 integrin [80, 207]. The presence of α6β4 integrin in actin-rich filipodia in migrating cells, and the ability of α6β4 integrin to activate PI3K suggest that α6β4 integrin also promotes invasion by directly mediating cell movement [185, 209, 210]. In this regard, Santoro and others have shown that when MSP activates PI3K, both the MSP receptor Ron and β4 integrin become phosphorylated, leading to the disassembly of hemidesmosomes and the recruitment of a Ron/α6β4 integrin complex to lamellipodia [108]. However, evidence that the latter complex plays a direct role in migration is lacking. Rather, Santoro and others have hypothesized that MSP-induced migration of keratinocytes on laminin-332 is mediated by α3β1 as opposed to α6β4 integrin, with the latter having a "transdominating" inhibitory impact on motility mediated by α3β1 integrin [108, 211, 212]. Some recently published results suggest that α6β4 integrin promotes epithelial cell migration by regulating laminin-332 matrix assembly into tracks along which cells move [77]. α6β4 integrin does so by regulating Rac1 activity which, in turn, regulates the activity of the actin-severing protein cofilin [77]. Based on these data, a model has been proposed in which cofilin modulates the dynamic properties of α6β4 integrin in the plane of the membrane. The model is premised on the concept that α6β4 integrin, in some way, physically organizes laminin-332 in the ECM in an actin-dependent manner [77]. This has profound implications for the mechanism regulating the directed migration of cells during wound healing and tumorigenesis [77, 200, 208].

6.9
Conclusions

The hemidesmosome is an important adhesive structure in various epithelial tissue types. Today, its major structural components have been defined, and the interactions of these proteins have been well documented. Intriguingly, some hemidesmosome proteins, when not incorporated into a hemidesmosome, play important signaling roles in migrating cells and tumors. Recently, this duality of function of hemidesmosome components has begun to be elucidated, and the information obtained from such studies is providing invaluable new insights into the survival, proliferation and invasion of cancer cells and the migration of cells during wound healing and tissue remodeling.

References

1 Jones, J. C., Hopkinson, S. B., and Goldfinger, L. E., *BioEssays* 1998, 20, 488–494.

2 Litjens, S. H., de Pereda, J. M., and Sonnenberg, A., *Trends Cell Biol.* 2006, 16, 376–383.

3 Staehelin, L. A., *Int. Rev. Cytol.* 1974, 39, 191–283.
4 Jefferson, J. J., Leung, C. L., and Liem, R. K., *Nat. Rev. Mol. Cell. Biol.* 2004, 5, 542–553.
5 Klatte, D. H., Kurpakus, M. A., Grelling, K. A., and Jones, J. C., *J. Cell Biol.* 1989, 109, 3377–3390.
6 Jones, J. C., Kurpakus, M. A., Cooper, H. M., and Quaranta, V., *Cell Regul.* 1991, 2, 427–438.
7 Stepp, M. A., Spurr-Michaud, S., Tisdale, A., Elwell, J., and Gipson, I. K., *Proc. Natl. Acad. Sci. USA* 1990, 87, 8970–8974.
8 Sterk, L. M., Geuijen, C. A., Oomen, L. C., Calafat, J., Janssen, H., and Sonnenberg, A., *J. Cell Biol.* 2000, 149, 969–982.
9 Giudice, G. J., Emery, D. J., Zelickson, B. D., Anhalt, G. J., Liu, Z., and Diaz, L. A., *J. Immunol.* 1993, 151, 5742–5750.
10 Carter, W. G., Ryan, M. C., and Gahr, P. J., *Cell* 1991, 65, 599–610.
11 Rousselle, P., Lunstrum, G. P., Keene, D. R., and Burgeson, R. E., *J. Cell Biol.* 1991, 114, 567–576.
12 Aumailley, M., Bruckner-Tuderman, L., Carter, W. G., Deutzmann, R., Edgar, D., Ekblom, P., Engel, J., Engvall, E., Hohenester, E., Jones, J. C., Kleinman, H. K., Marinkovich, M. P., Martin, G. R., Mayer, U., Meneguzzi, G., Miner, J. H., Miyazaki, K., Patarroyo, M., Paulsson, M., Quaranta, V., Sanes, J. R., Sasaki, T., Sekiguchi, K., Sorokin, L. M., Talts, J. F., Tryggvason, K., Uitto, J., Virtanen, I., von Der Mark, K., Wewer, U. M., Yamada, Y., and Yurchenco, P. D., *Matrix Biol.* 2005, 24, 326–232.
13 Tasanen, K., Tunggal, L., Chometon, G., Bruckner-Tuderman, L., and Aumailley, M., *Am. J. Pathol.* 2004, 164, 2027–2038.
14 Herrmann, H. and Wiche, G., *J. Biol. Chem.* 1987, 262, 1320–1325.
15 Hieda, Y., Nishizawa, Y., Uematsu, J., and Owaribe, K., *J. Cell Biol.* 1992, 116, 1497–1506.
16 Skalli, O., Jones, J. C., Gagescu, R., and Goldman, R. D., *J. Cell Biol.* 1994, 125, 159–170.
17 Gache, Y., Chavanas, S., Lacour, J. P., Wiche, G., Owaribe, K., Meneguzzi, G., and Ortonne, J. P., *J. Clin. Invest.* 1996, 97, 2289–2298.
18 Nikolic, B., MacNulty, E., Mir, B., and Wiche, G., *J. Cell Biol.* 1996, 134, 1455–67.
19 Wiche, G., *J. Cell Sci.* 1998, 111, 2477–2486.
20 Liu, C. G., Maercker, C., Castanon, M. J., Hauptmann, R., and Wiche, G., *Proc. Natl. Acad. Sci. USA* 1996, 93, 4278–4283.
21 McLean, W. H., Pulkkinen, L., Smith, F. J., Rugg, E. L., Lane, E. B., Bullrich, F., Burgeson, R. E., Amano, S., Hudson, D. L., Owaribe, K., McGrath, J. A., McMillan, J. R., Eady, R. A., Leigh, I. M., Christiano, A. M., and Uitto, J., *Genes Dev.* 1996, 10, 1724–1735.
22 Fuchs, P., Zorer, M., Rezniczek, G. A., Spazierer, D., Oehler, S., Castanon, M. J., Hauptmann, R., and Wiche, G., *Hum. Mol. Genet.* 1999, 8, 2461–2472.
23 Andra, K., Kornacker, I., Jorgl, A., Zorer, M., Spazierer, D., Fuchs, P., Fischer, I., and Wiche, G., *J. Invest. Dermatol.* 2003, 120, 189–197.
24 Rezniczek, G. A., Abrahamsberg, C., Fuchs, P., Spazierer, D., and Wiche, G., *Hum. Mol. Genet.* 2003, 12, 3181–3194.
25 Andra, K., Lassmann, H., Bittner, R., Shorny, S., Fassler, R., Propst, F., and Wiche, G., *Genes Dev.* 1997, 11, 3143–3156.
26 Niessen, C. M., Hulsman, E. H., Oomen, L. C., Kuikman, I., and Sonnenberg, A., *J. Cell Sci.* 1997, 110, 1705–1716.
27 Rezniczek, G. A., de Pereda, J. M., Reipert, S., and Wiche, G., *J. Cell Biol.* 1998, 141, 209–225.
28 Rabinovitz, I., Tsomo, L., and Mercurio, A. M., *Mol. Cell. Biol.* 2004, 24, 4351–4360.
29 Koster, J., Geerts, D., Favre, B., Borradori, L., and Sonnenberg, A., *J. Cell Sci.* 2003, 116, 387–399.
30 Schaapveld, R. Q., Borradori, L., Geerts, D., van Leusden, M. R., Kuikman, I., Nievers, M. G., Niessen, C. M., Steenbergen, R. D., Snijders, P. J., and Sonnenberg, A., *J. Cell Biol.* 1998, 142, 271–284.

31 Stanley, J. R., *Adv. Immunol.* 1993, *53*, 291–325.
32 Tamai, K., Sawamura, D., Do, H. C., Tamai, Y., Li, K., and Uitto, J., *J. Clin. Invest.* 1993, *92*, 814 822.
33 Leung, C. L., Liem, R. K., Parry, D. A., and Green, K. J., *J. Cell Sci.* 2001, *114*, 3409–3410.
34 Klatte, D. H. and Jones, J. C., *J. Invest. Dermatol.* 1994, *102*, 39–44.
35 Guo, L., Degenstein, L., Dowling, J., Yu, Q. C., Wollmann, R., Perman, B., and Fuchs, E., *Cell* 1995, *81*, 233–243.
36 Fontao, L., Favre, B., Riou, S., Geerts, D., Jaunin, F., Saurat, J. H., Green, K. J., Sonnenberg, A., and Borradori, L., *Mol. Biol. Cell* 2003, *14*, 1978–1992.
37 Hopkinson, S. B. and Jones, J. C., *Mol. Biol. Cell* 2000, *11*, 277–286.
38 Kennel, S. J., Foote, L. J., Falcioni, R., Sonnenberg, A., Stringer, C. D., Crouse, C., and Hemler, M. E., *J. Biol. Chem.* 1989, *264*, 15515–15521.
39 Niessen, C. M., Hogervorst, F., Jaspars, L. H., de Melker, A. A., Delwel, G. O., Hulsman, E. H., Kuikman, I., and Sonnenberg, A., *Exp. Cell Res.* 1994, *211*, 360–367.
40 Lee, E. C., Lotz, M. M., Steele, G. D., Jr., and Mercurio, A. M., *J. Cell Biol.* 1992, *117*, 671–678.
41 Kikkawa, Y., Sanzen, N., Fujiwara, H., Sonnenberg, A., and Sekiguchi, K., *J. Cell Sci.* 2000, *113*, 869–876.
42 Hogervorst, F., Kuikman, I., von dem Borne, A. E., and Sonnenberg, A., *EMBO J.* 1990, *9*, 765–770.
43 Tamura, R. N., Rozzo, C., Starr, L., Chambers, J., Reichardt, L. F., Cooper, H. M., and Quaranta, V., *J. Cell Biol.* 1990, *111*, 1593–1604.
44 Clarke, A. S., Lotz, M. M., and Mercurio, A. M., *Cell Adhes. Commun.* 1994, *2*, 1–6.
45 Werner, M. E., Chen, F., Moyano, J. V., Yehiely, F., Jones, J. C., and Cryns, V. L., *J. Biol. Chem.* 2007, *282*, 5560–5569.
46 Giancotti, F. G., Stepp, M. A., Suzuki, S., Engvall, E., and Ruoslahti, E., *J. Cell Biol.* 1992, *118*, 951–959.
47 Tsuruta, D., Hopkinson, S. B., Lane, K. D., Werner, M. E., Cryns, V. L., and Jones, J. C., *J. Biol. Chem.* 2003, *278*, 38707–38714.
48 Takagi, J., DeBottis, D. P., Erickson, H. P., and Springer, T. A., *Biochemistry* 2002, *41*, 4339–4347.
49 Seidah, N. G., Day, R., Marcinkiewicz, M., and Chretien, M., *Ann. N. Y. Acad. Sci.* 1998, *839*, 9–24.
50 Hogervorst, F., Admiraal, L. G., Niessen, C., Kuikman, I., Janssen, H., Daams, H., and Sonnenberg, A., *J. Cell Biol.* 1993, *121*, 179–191.
51 Demetriou, M. C., Pennington, M. E., Nagle, R. B., and Cress, A. E., *Exp. Cell Res.* 2004, *294*, 550–558.
52 Riddelle, K. S., Green, K. J., and Jones, J. C., *J. Cell Biol.* 1991, *112*, 159–168.
53 Spinardi, L., Einheber, S., Cullen, T., Milner, T. A., and Giancotti, F. G., *J. Cell Biol.* 1995, *129*, 473–487.
54 Georges-Labouesse, E., Messaddeq, N., Yehia, G., Cadalbert, L., Dierich, A., and Le Meur, M., *Nat. Genet.* 1996, *13*, 370–373.
55 van der Neut, R., Krimpenfort, P., Calafat, J., Niessen, C. M., and Sonnenberg, A., *Nat. Genet.* 1996, *13*, 366–369.
56 Dowling, J., Yu, Q. C., and Fuchs, E., *J. Cell Biol.* 1996, *134*, 559–572.
57 DiPersio, C. M., van der, N. R., Georges-Labouesse, E., Kreidberg, J. A., Sonnenberg, A., and Hynes, R. O., *J. Cell Sci.* 2000, *113*, 3051–3062.
58 Uematsu, J., Nishizawa, Y., Sonnenberg, A., and Owaribe, K., *J. Biochem. (Tokyo)* 1994, *115*, 469 476.
59 Sepp, N. T., Cornelius, L. A., Romani, N., Li, L. J., Caughman, S. W., Lawley, T. J., and Swerlick, R. A., *J. Invest. Dermatol.* 1995, *104*, 266–270.
60 Watt, S. M., Thomas, J. A., Edwards, A. J., Murdoch, S. J., and Horton, M. A., *Exp. Hematol.* 1992, *20*, 1101–1111.
61 Niessen, C. M., Cremona, O., Daams, H., Ferraresi, S., Sonnenberg, A., and Marchisio, P. C., *J. Cell Sci.* 1994, *107*, 543–552.
62 Hiran, T. S., Mazurkiewicz, J. E., Kreienberg, P., Rice, F. L., and LaFlamme, S. E., *J. Cell Sci.* 2003, *116*, 3771–3781.
63 Giudice, G. J., Emery, D. J., and Diaz, L. A., *J. Invest. Dermatol.* 1992, *99*, 243–50.
64 Diaz, L. A., Ratrie, H.3rd, , Saunders, W. S., Futamura, S., Squiquera, H. L.,

Anhalt, G. J., and Giudice, G. J., *J. Clin. Invest.* 1990, *86*, 1088–1094.

65 Hopkinson, S. B., Riddelle, K. S., and Jones, J. C., *J. Invest. Dermatol.* 1992, *99*, 264–270.

66 Hirako, Y., Usukura, J., Nishizawa, Y., and Owaribe, K., *J. Biol. Chem.* 1996, *271*, 13739–13745.

67 Borradori, L., Koch, P. J., Niessen, C. M., Erkeland, S., van Leusden, M. R., and Sonnenberg, A., *J. Cell Biol.* 1997, *136*, 1333–1347.

68 Hopkinson, S. B., Findlay, K., deHart, G. W., and Jones, J. C., *J. Invest. Dermatol.* 1998, *111*, 1015–1022.

69 Hemler, M. E., *J. Cell Biol.* 2001, *155*, 1103–1107.

70 Boucheix, C. and Rubinstein, E., *Cell. Mol. Life Sci.* 2001, *58*, 1189–1205.

71 Wright, M. D., Geary, S. M., Fitter, S., Moseley, G. W., Lau, L. M., Sheng, K. C., Apostolopoulos, V., Stanley, E. G., Jackson, D. E., and Ashman, L. K., *Mol. Cell. Biol.* 2004, *24*, 5978–5988.

72 Verrando, P., Partouche, O., Pisani, A., and Ortonne, J. P., *Exp. Dermatol.* 1992, *1*, 52–58.

73 Marinkovich, M. P., Lunstrum, G. P., and Burgeson, R. E., *J. Biol. Chem.* 1992, *267*, 17900–17906.

74 Baker, S. E., Hopkinson, S. B., Fitchmun, M., Andreason, G. L., Frasier, F., Plopper, G., Quaranta, V., and Jones, J. C., *J. Cell Sci.* 1996, *109*, 2509–2520.

75 Cheng, Y. S., Champliaud, M. F., Burgeson, R. E., Marinkovich, M. P., and Yurchenco, P. D., *J. Biol. Chem.* 1997, *272*, 31525–31532.

76 Champliaud, M. F., Lunstrum, G. P., Rousselle, P., Nishiyama, T., Keene, D. R., and Burgeson, R. E., *J. Cell Biol.* 1996, *132*, 1189–1198.

77 Sehgal, B. U., DeBiase, P. J., Matzno, S., Chew, T. L., Claiborne, J. N., Hopkinson, S. B., Russell, A., Marinkovich, M. P., and Jones, J. C., *J. Biol. Chem.* 2006, *281*, 35487–35498.

78 Chen, M., Marinkovich, M. P., Veis, A., Cai, X., Rao, C. N., O'Toole, E. A., and Woodley, D. T., *J. Biol Chem.* 1997, *272*, 14516–14522.

79 Chen, M., Marinkovich, M. P., Jones, J. C., O'Toole, E. A., Li, Y. Y., and Woodley, D. T., *J. Invest. Dermatol.* 1999, *112*, 177–183.

80 Goldfinger, L. E., Stack, M. S., and Jones, J. C., *J. Cell Biol.* 1998, *141*, 255–265.

81 Gagnoux-Palacios, L., Allegra, M., Spirito, F., Pommeret, O., Romero, C., Ortonne, J. P., and Meneguzzi, G., *J. Cell Biol.* 2001, *153*, 835–850.

82 Nguyen, B. P., Gil, S. G., and Carter, W. G., *J. Biol. Chem.* 2000, *275*, 31896–31907.

83 Baudoin, C., Fantin, L., and Meneguzzi, G., *J. Invest. Dermatol.* 2005, *125*, 883–888.

84 Giannelli, G., Falk-Marzillier, J., Schiraldi, O., Stetler-Stevenson, W. G., and Quaranta, V., *Science* 1997, *277*, 225–228.

85 Sadowski, T., Dietrich, S., Koschinsky, F., Ludwig, A., Proksch, E., Titz, B., and Sedlacek, R., *Cell. Mol. Life Sci.* 2005, *62*, 870–880.

86 Veitch, D. P., Nokelainen, P., McGowan, K. A., Nguyen, T. T., Nguyen, N. E., Stephenson, R., Pappano, W. N., Keene, D. R., Spong, S. M., Greenspan, D. S., Findell, P. R., and Marinkovich, M. P., *J. Biol. Chem.* 2003, *278*, 15661–15668.

87 Amano, S., Scott, I. C., Takahara, K., Koch, M., Champliaud, M. F., Gerecke, D. R., Keene, D. R., Hudson, D. L., Nishiyama, T., Lee, S., Greenspan, D. S., and Burgeson, R. E., *J. Biol. Chem.* 2000, *275*, 22728–22735.

88 Tsubota, Y., Mizushima, H., Hirosaki, T., Higashi, S., Yasumitsu, H., and Miyazaki, K., *Biochem. Biophys. Res. Commun.* 2000, *278*, 614–620.

89 Hirosaki, T., Tsubota, Y., Kariya, Y., Moriyama, K., Mizushima, H., and Miyazaki, K., *J. Biol. Chem.* 2002, *277*, 49287–49295.

90 Langhofer, M., Hopkinson, S. B., and Jones, J. C., *J. Cell Sci.* 1993, *105*, 753–764.

91 Koshikawa, N., Schenk, S., Moeckel, G., Sharabi, A., Miyazaki, K., Gardner, H., Zent, R., and Quaranta, V., *FASEB J.* 2004, *18*, 364–366.

92 Ryan, M. C., Lee, K., Miyashita, Y., and Carter, W. G., *J. Cell Biol.* 1999, *145*, 1309–1323.

93 Schenk, S., Hintermann, E., Bilban, M., Koshikawa, N., Hojilla, C., Khokha, R.,

and Quaranta, V., *J. Cell Biol.* 2003, *161*, 197–209.
94 Baker, S. E., Skalli, O., Goldman, R. D., and Jones, J. C., *Cell Motil. Cytoskeleton* 1997, *37*, 271–286.
95 Mainiero, F., Pepe, A., Wary, K. K., Spinardi, L., Mohammadi, M., Schlessinger, J., and Giancotti, F. G., *EMBO J.* 1995, *14*, 4470–4481.
96 Dellambra, E., Prislei, S., Salvati, A. L., Madeddu, M. L., Golisano, O., Siviero, E., Bondanza, S., Cicuzza, S., Orecchia, A., Giancotti, F. G., Zambruno, G., and De Luca, M., *J. Biol. Chem.* 2001, *276*, 41336–41342.
97 Mainiero, F., Pepe, A., Yeon, M., Ren, Y., and Giancotti, F. G., *J. Cell Biol.* 1996, *134*, 241–253.
98 Dans, M., Gagnoux-Palacios, L., Blaikie, P., Klein, S., Mariotti, A., and Giancotti, F. G., *J. Biol. Chem.* 2001, *276*, 1494–1502.
99 DiPersio, C. M., Hodivala-Dilke, K. M., Jaenisch, R., Kreidberg, J. A., and Hynes, R. O., *J. Cell Biol.* 1997, *137*, 729–742.
100 Borradori, L. and Sonnenberg, A., *J. Invest. Dermatol.* 1999, *112*, 411–418.
101 Kurpakus, M. A., Quaranta, V., and Jones, J. C., *J. Cell Biol.* 1991, *115*, 1737–1750.
102 Takahashi, Y., Mutasim, D. F., Patel, H. P., Anhalt, G. J., Labib, R. S., and Diaz, L. A., *J. Invest. Dermatol.* 1985, *85*, 309–313.
103 Hintermann, E. and Quaranta, V., *Matrix Biol.* 2004, *23*, 75–85.
104 Rabinovitz, I., Toker, A., and Mercurio, A. M., *J. Cell Biol.* 1999, *146*, 1147–1160.
105 Alt, A., Ohba, M., Li, L., Gartsbein, M., Belanger, A., Denning, M. F., Kuroki, T., Yuspa, S. H., and Tennenbaum, T., *Cancer Res.* 2001, *61*, 4591–4598.
106 Mariotti, A., Kedeshian, P. A., Dans, M., Curatola, A. M., Gagnoux-Palacios, L., and Giancotti, F. G., *J. Cell Biol.* 2001, *155*, 447–458.
107 Gagnoux-Palacios, L., Dans, M., van't Hof, W., Mariotti, A., Pepe, A., Meneguzzi, G., Resh, M. D., and Giancotti, F. G., *J. Cell Biol.* 2003, *162*, 1189–1196.
108 Santoro, M. M., Gaudino, G., and Marchisio, P. C., *Dev. Cell.* 2003, *5*, 257–271.
109 Smith, F. J., Eady, R. A., Leigh, I. M., McMillan, J. R., Rugg, E. L., Kelsell, D. P., Bryant, S. P., Spurr, N. K., Geddes, J. F., Kirtschig, G., Milana, G., De Bono, A. G., Owaribe, K., Wiche, G., Pulkkinen, L., Uitto, J., McLean, W. H., and Lane, E. B., *Nat. Genet.* 1996, *13*, 450–457.
110 Kivirikko, S., McGrath, J. A., Baudoin, C., Aberdam, D., Ciatti, S., Dunnill, M. G., McMillan, J. R., Eady, R. A., Ortonne, J. P., Meneguzzi, G., et al., *Hum. Mol. Genet.* 1995, *4*, 959–962.
111 Jonkman, M. F., de Jong, M. C., Heeres, K., Pas, H. H., van der Meer, J. B., Owaribe, K., Martinez de Velasco, A. M., Niessen, C. M., and Sonnenberg, A., *J. Clin. Invest.* 1995, *95*, 1345–1352.
112 Uitto, J., *J. Invest. Dermatol.* 2004, *123*, xii–xiii.
113 Ruzzi, L., Gagnoux-Palacios, L., Pinola, M., Belli, S., Meneguzzi, G., D'Alessio, M., and Zambruno, G., *J. Clin. Invest.* 1997, *99*, 2826–2831.
114 Pulkkinen, L., Kurtz, K., Xu, Y., Bruckner-Tuderman, L., and Uitto, J., *Lab. Invest.* 1997, *76*, 823–833.
115 Vidal, F., Aberdam, D., Miquel, C., Christiano, A. M., Pulkkinen, L., Uitto, J., Ortonne, J. P., and Meneguzzi, G., *Nat. Genet.* 1995, *10*, 229–234.
116 Jonkman, M. F., *J. Dermatol. Sci.* 1999, *20*, 103–121.
117 Gil, S. G., Brown, T. A., Ryan, M. C., and Carter, W. G., *J. Invest. Dermatol.* 1994, *103*, 31S–38S.
118 Niessen, C. M., van der Raaij-Helmer, M. H., Hulsman, E. H., van der Neut, R., Jonkman, M. F., and Sonnenberg, A., *J. Cell Sci.* 1996, *109*, 1695–1706.
119 Micheloni, A., De Luca, N., Tadini, G., Zambruno, G., and D'Alessio, M., *Br. J. Dermatol.* 2004, *151*, 796–802.
120 Allegra, M., Gagnoux-Palacios, L., Gache, Y., Roques, S., Lestringant, G., Ortonne, J. P., and Meneguzzi, G., *J. Invest. Dermatol.* 2003, *121*, 1336–1343.
121 Pulkkinen, L., Rouan, F., Bruckner-Tuderman, L., Wallerstein, R., Garzon, M., Brown, T., Smith, L., Carter, W., and Uitto, J., *Am. J. Hum. Genet.* 1998, *63*, 1376–1387.

122 Pulkkinen, L., Kim, D. U., and Uitto, J., *Am. J. Pathol.* 1998, *152*, 157–166.

123 Masunaga, T., Ishiko, A., Takizawa, Y., Kim, S. C., Lee, J. S., Nishikawa, T., and Shimizu, H., *Exp. Dermatol.* 2004, *13*, 61–64.

124 Jonkman, M. F., Pas, H. H., Nijenhuis, M., Kloosterhuis, G., and Steege, G., *J. Invest. Dermatol.* 2002, *119*, 1275–1281.

125 Koster, J., Kuikman, I., Kreft, M., and Sonnenberg, A., *J. Invest. Dermatol.* 2001, *117*, 1405–1411.

126 Shimizu, H., Takizawa, Y., Pulkkinen, L., Murata, S., Kawai, M., Hachisuka, H., Udono, M., Uitto, J., and Nishikawa, T., *J. Am. Acad. Dermatol.* 1999, *41*, 950–956.

127 Schroder, R., Kunz, W. S., Rouan, F., Pfendner, E., Tolksdorf, K., Kappes-Horn, K., Altenschmidt-Mehring, M., Knoblich, R., van der Ven, P. F., Reimann, J., Furst, D. O., Blumcke, I., Vielhaber, S., Zillikens, D., Eming, S., Klockgether, T., Uitto, J., Wiche, G., and Rolfs, A., *J. Neuropathol. Exp. Neurol.* 2002, *61*, 520–530.

128 Chavanas, S., Pulkkinen, L., Gache, Y., Smith, F. J., McLean, W. H., Uitto, J., Ortonne, J. P., and Meneguzzi, G., *J. Clin. Invest.* 1996, *98*, 2196–2200.

129 McGrath, J. A., Gatalica, B., Christiano, A. M., Li, K., Owaribe, K., McMillan, J. R., Eady, R. A., and Uitto, J., *Nat. Genet.* 1995, *11*, 83–86.

130 Gatalica, B., Pulkkinen, L., Li, K., Kuokkanen, K., Ryynanen, M., McGrath, J. A., and Uitto, J., *Am. J. Hum. Genet.* 1997, *60*, 352–365.

131 Darling, T. N., McGrath, J. A., Yee, C., Gatalica, B., Hametner, R., Bauer, J. W., Pohla-Gubo, G., Christiano, A. M., Uitto, J., Hintner, H., and Yancey, K. B., *J. Invest. Dermatol.* 1997, *108*, 463–468.

132 Huber, A., Yee, C., Darling, T. N., and Yancey, K. B., *Exp. Dermatol.* 2002, *11*, 75–81.

133 Wu, Y., Li, G., and Zhu, X., *J. Dermatol. Sci.* 2002, *28*, 181–186.

134 Chavanas, S., Gache, Y., Vailly, J., Kanitakis, J., Pulkkinen, L., Uitto, J., Ortonne, J., and Meneguzzi, G., *Hum. Mol. Genet.* 1999, *8*, 2097–2105.

135 Pohla-Gubo, G., Lazarova, Z., Giudice, G. J., Liebert, M., Grassegger, A., Hintner, H., and Yancey, K. B., *Exp. Dermatol.* 1995, *4*, 199–206.

136 Matsumura, Y., Horiguchi, Y., Toda, K., Fujii, H., Kore-Eda, S., Tachibana, T., Ohta, K., Okamoto, H., and Imamura, S., *Br. J. Dermatol.* 1997, *136*, 757–761.

137 Borradori, L., Chavanas, S., Schaapveld, R. Q., Gagnoux-Palacios, L., Calafat, J., Meneguzzi, G., and Sonnenberg, A., *Exp. Cell Res.* 1998, *239*, 463–476.

138 Jonkman, M. F., Castellanos Nuijts, M., and van Essen, A. J., *Clin. Exp. Dermatol.* 2003, *28*, 625–631.

139 Pasmooij, A. M., van Zalen, S., Nijenhuis, A. M., Kloosterhuis, A. J., Zuiderveen, J., Jonkman, M. F., and Pas, H. H., *Exp. Dermatol.* 2004, *13*, 125–128.

140 Karamatic Crew, V., Burton, N., Kagan, A., Green, C. A., Levene, C., Flinter, F., Brady, R. L., Daniels, G., and Anstee, D. J., *Blood* 2004, *104*, 2217–2223.

141 Nakano, A., Chao, S. C., Pulkkinen, L., Murrell, D., Bruckner-Tuderman, L., Pfendner, E., and Uitto, J., *Hum. Genet.* 2002, *110*, 41–51.

142 Matsui, C., Pereira, P., Wang, C. K., Nelson, C. F., Kutzkey, T., Lanigan, C., Woodley, D., Morohashi, M., Welsh, E. A., and Hoeffler, W. K., *J. Exp. Med.* 1998, *187*, 1273–1283.

143 Pulkkinen, L. and Uitto, J., *Matrix Biol.* 1999, *18*, 29–42.

144 Matsui, C., Wang, C. K., Nelson, C. F., Bauer, E. A., and Hoeffler, W. K., *J. Biol. Chem.* 1995, *270*, 23496–23503.

145 Scaturro, M., Posteraro, P., Mastrogiacomo, A., Zaccaria, M. L., De Luca, N., Mazzanti, C., Zambruno, G., and Castiglia, D., *Biochem. Biophys. Res. Commun.* 2003, *309*, 96–103.

146 McGrath, J. A., Ashton, G. H., Mellerio, J. E., Salas-Alanis, J. C., Swensson, O., McMillan, J. R., and Eady, R. A., *J. Invest. Dermatol.* 1999, *113*, 314–321.

147 Gache, Y., Allegra, M., Bodemer, C., Pisani-Spadafora, A., de Prost, Y., Ortonne, J. P., and Meneguzzi, G., *Hum. Mol. Genet.* 2001, *10*, 2453–2461.

148 Zillikens, D., *J. Dermatol. Sci.* 1999, *20*, 134–154.

149 Wintroub, B. U., Mihm, M. C., Jr., Goetzl, E. J., Soter, N. A., and Austen, K. F., *N. Engl. J. Med.* 1978, *298*, 417–421.

150 Matsumura, K., Amagai, M., Nishikawa, T., and Hashimoto, T., *Arch. Dermatol. Res.* 1996, *288*, 507–509.

151 Liu, Z., Giudice, G. J., Swartz, S. J., Fairley, J. A., Till, G. O., Troy, J. L., and Diaz, L. A., *J. Clin. Invest.* 1995, *95*, 1539–1544.

152 Thoma-Uszynski, S., Uter, W., Schwietzke, S., Hofmann, S. C., Hunziker, T., Bernard, P., Treudler, R., Zouboulis, C. C., Schuler, G., Borradori, L., and Hertl, M., *J. Invest. Dermatol.* 2004, *122*, 1413–1422.

153 Ghohestani, R. F., Cozzani, E., Delaporte, E., Nicolas, J. F., Parodi, A., and Claudy, A., *J. Clin. Immunol.* 1998, *18*, 202–209.

154 Murakami, H., Amagai, M., Higashiyama, M., Hashimoto, K., Chorzelski, T. P., Bhogal, B. S., Jenkins, R. E., Black, M. M., Zillikens, D., Nishikawa, T., and Hashimoto, T., *J. Dermatol. Sci.* 1996, *13*, 112–117.

155 Swale, V. J., Black, M. M., and Bhogal, B. S., *Clin. Exp. Dermatol.* 1998, *23*, 132–135.

156 Ogg, G. S., Bhogal, B. S., Hashimoto, T., Coleman, R., and Barker, J. N., *Br. J. Dermatol.* 1997, *136*, 412–414.

157 Domloge-Hultsch, N., Gammon, W. R., Briggaman, R. A., Gil, S. G., Carter, W. G., and Yancey, K. B., *J. Clin. Invest.* 1992, *90*, 1628–1633.

158 Lazarova, Z., Hsu, R., Yee, C., and Yancey, K. B., *Br. J. Dermatol.* 1998, *139*, 791–797.

159 Lazarova, Z., Hsu, R., Yee, C., and Yancey, K. B., *J. Invest. Dermatol.* 2000, *114*, 178–184.

160 Zambruno, G. and Failla, C. M., *Eur. J. Dermatol.* 1999, *9*, 437–442.

161 Rashid, K. A., Stern, J. N., and Ahmed, A. R., *J. Immunol.* 2006, *176*, 1968–1977.

162 Chan, R. Y., Bhol, K., Tesavibul, N., Letko, E., Simmons, R. K., Foster, C. S., and Ahmed, A. R., *Invest. Ophthalmol. Vis. Sci.* 1999, *40*, 2283–2290.

163 Bhol, K. C., Dans, M. J., Simmons, R. K., Foster, C. S., Giancotti, F. G., and Ahmed, A. R., *J. Immunol.* 2000, *165*, 2824–2829.

164 Tyagi, S., Bhol, K., Natarajan, K., Livir-Rallatos, C., Foster, C. S., and Ahmed, A. R., *Proc. Natl. Acad. Sci. USA* 1996, *93*, 14714–14719.

165 Rashid, K. A., Gurcan, H. M., and Ahmed, A. R., *J. Invest. Dermatol.* 2006, *126*, 2631–2636.

166 Kiss, M., Perenyi, A., Marczinovits, I., Molnar, J., Dobozy, A., Kemeny, L., and Husz, S., *Ann. N. Y. Acad. Sci.* 2005, *1051*, 104–110.

167 Kimmel, K. A. and Carey, T. E., *Cancer Res.* 1986, *46*, 3614–3623.

168 Tennenbaum, T., Weiner, A. K., Belanger, A. J., Glick, A. B., Hennings, H., and Yuspa, S. H., *Cancer Res.* 1993, *53*, 4803–4810.

169 Carico, E., French, D., Bucci, B., Falcioni, R., Vecchione, A., and Mariani-Costantini, R., *Gynecol. Oncol.* 1993, *49*, 61–66.

170 Mariani Costantini, R., Falcioni, R., Battista, P., Zupi, G., Kennel, S. J., Colasante, A., Venturo, I., Curio, C. G., and Sacchi, A., *Cancer Res.* 1990, *50*, 6107–6112.

171 Rabinovitz, I. and Mercurio, A. M., *Biochem. Cell. Biol.* 1996, *74*, 811–821.

172 Chung, J. and Mercurio, A. M., *Mol. Cell* 2004, *17*, 203–209.

173 Diaz, L. K., Cristofanilli, M., Zhou, X., Welch, K. L., Smith, T. L., Yang, Y., Sneige, N., Sahin, A. A., and Gilcrease, M. Z., *Mod. Pathol.* 2005, *18*, 1165–1175.

174 Mercurio, A. M., *Cancer Cell* 2003, *3*, 201–202.

175 Savoia, P., Trusolino, L., Pepino, E., Cremona, O., and Marchisio, P. C., *J. Invest. Dermatol.* 1993, *101*, 352–358.

176 Rossen, K., Dahlstrom, K. K., Mercurio, A. M., and Wewer, U. M., *Acta Dermatol. Venereol.* 1994, *74*, 101–105.

177 Knox, J. D., Cress, A. E., Clark, V., Manriquez, L., Affinito, K. S., Dalkin, B. L., and Nagle, R. B., *Am. J. Pathol.* 1994, *145*, 167–174.

178 Natali, P. G., Nicotra, M. R., Botti, C., Mottolese, M., Bigotti, A., and Segatto, O., *Br. J. Cancer* 1992, *66*, 318–322.

179 Bonaccorsi, L., Carloni, V., Muratori, M., Salvadori, A., Giannini, A., Carini, M., Serio, M., Forti, G., and Baldi, E., *Endocrinology* 2000, *141*, 3172–3182.

180 Lohi, J., *Int. J. Cancer* 2001, *94*, 763–767.

181 Malina, R., Motoyama, S., Hamana, S., and Maruo, T., *Kobe J. Med. Sci.* 2004, *50*, 123–130.

182 Moriya, Y., Niki, T., Yamada, T., Matsuno, Y., Kondo, H., and Hirohashi, S., *Cancer* 2001, *91*, 1129–1141.

183 Dajee, M., Lazarov, M., Zhang, J. Y., Cai, T., Green, C. L., Russell, A. J., Marinkovich, M. P., Tao, S., Lin, Q., Kubo, Y., and Khavari, P. A., *Nature* 2003, *421*, 639–643.

184 Trusolino, L., Bertotti, A., and Comoglio, P. M., *Cell* 2001, *107*, 643–654.

185 Shaw, L. M., Rabinovitz, I., Wang, H. H., Toker, A., and Mercurio, A. M., *Cell* 1997, *91*, 949–960.

186 Tang, K., Nie, D., Cai, Y., and Honn, K. V., *Biochem. Biophys. Res. Commun.* 1999, *264*, 127–132.

187 Gambaletta, D., Marchetti, A., Benedetti, L., Mercurio, A. M., Sacchi, A., and Falcioni, R., *J. Biol. Chem.* 2000, *275*, 10604–10610.

188 Chung, J., Bachelder, R. E., Lipscomb, E. A., Shaw, L. M., and Mercurio, A. M., *J. Cell Biol.* 2002, *158*, 165–174.

189 Ridley, A. J., Paterson, H. F., Johnston, C. L., Diekmann, D., and Hall, A., *Cell* 1992, *70*, 401–410.

190 Falcioni, R., Antonini, A., Nistico, P., Di Stefano, S., Crescenzi, M., Natali, P. G., and Sacchi, A., *Exp. Cell Res.* 1997, *236*, 76–85.

191 Folgiero, V., Bachelder, R. E., Bon, G., Sacchi, A., Falcioni, R., and Mercurio, A. M., *Cancer Res.* 2007, *67*, 1645–1652.

192 Lipscomb, E. A., Simpson, K. J., Lyle, S. R., Ring, J. E., Dugan, A. S., and Mercurio, A. M., *Cancer Res.* 2005, *65*, 10970–10976.

193 Bachelder, R. E., Ribick, M. J., Marchetti, A., Falcioni, R., Soddu, S., Davis, K. R., and Mercurio, A. M., *J. Cell Biol.* 1999, *147*, 1063–1072.

194 Bachelder, R. E., Marchetti, A., Falcioni, R., Soddu, S., and Mercurio, A. M., *J. Biol. Chem.* 1999, *274*, 20733–20737.

195 Shaw, L. M., *Mol. Cell. Biol.* 2001, *21*, 5082–5093.

196 Yan, J., Roy, S., Apolloni, A., Lane, A., and Hancock, J. F., *J. Biol. Chem.* 1998, *273*, 24052–24056.

197 Ram, T. G. and Ethier, S. P., *Cell Growth Differ.* 1996, *7*, 551–561.

198 Schulze, W. X., Deng, L., and Mann, M., *Mol. Syst. Biol.* 2005, *1*, E1–E13.

199 Mainiero, F., Murgia, C., Wary, K. K., Curatola, A. M., Pepe, A., Blumemberg, M., Westwick, J. K., Der, C. J., and Giancotti, F. G., *EMBO J.* 1997, *16*, 2365–2375.

200 Nikolopoulos, S. N., Blaikie, P., Yoshioka, T., Guo, W., Puri, C., Tacchetti, C., and Giancotti, F. G., *Mol. Cell. Biol.* 2005, *25*, 6090–6102.

201 Chung, J., Yoon, S. O., Lipscomb, E. A., and Mercurio, A. M., *J. Biol. Chem.* 2004, *279*, 32287–32293.

202 Harper, E. G., Alvares, S. M., and Carter, W. G., *J. Cell Sci.* 2005, *118*, 3471–3485.

203 Gilcrease, M. Z., Zhou, X., and Welch, K., *Cancer Res.* 2004, *64*, 7395–7398.

204 Owens, D. M., Romero, M. R., Gardner, C., and Watt, F. M., *J. Cell Sci.* 2003, *116*, 3783–3791.

205 Bertotti, A., Comoglio, P. M., and Trusolino, L., *J. Cell Biol.* 2006, *175*, 993–1003.

206 Lipscomb, E. A., Dugan, A. S., Rabinovitz, I., and Mercurio, A. M., *Clin. Exp. Metastasis* 2003, *20*, 569–576.

207 Goldfinger, L. E., Hopkinson, S. B., deHart, G. W., Collawn, S., Couchman, J. R., and Jones, J. C., *J. Cell Sci.* 1999, *112*, 2615–2629.

208 Frank, D. E. and Carter, W. G., *J. Cell Sci.* 2004, *117*, 1351–1363.

209 Rabinovitz, I. and Mercurio, A. M., *J. Cell Biol.* 1997, *139*, 1873–1884.

210 Mercurio, A. M., Rabinovitz, I., and Shaw, L. M., *Curr. Opin. Cell Biol.* 2001, *13*, 541–545.

211 Hintermann, E., Bilban, M., Sharabi, A., and Quaranta, V., *J. Cell Biol.* 2001, *153*, 465–478.

212 Russell, A. J., Fincher, E. F., Millman, L., Smith, R., Vela, V., Waterman, E. A., Dey, C. N., Guide, S., Weaver, V. M., and Marinkovich, M. P., *J. Cell Sci.* 2003, *116*, 3543–3556.

7
Cell Matrix Adhesion in Three Dimensions
Patricia J. Keely

7.1
Introduction

Tissues are complex communities of specific and differing cell types organized within a specific connective tissue environment. Components of the extracellular matrix (ECM) are important regulators of cellular behavior, gene expression, and tissue architecture during normal development and tissue homeostasis. Interactions between cells and the ECM are altered during wound healing, thrombosis, the immune response, and various pathologic states, including inflammation, carcinoma progression and metastasis, and cardiovascular disease.

During recent years, much progress has been made in understanding the molecular structures by which cells interact with the ECM. It is clear that such interaction is mediated through integrins, and through non-integrin surface proteins including cell-surface proteoglycans such as the syndecans [1] and cell-surface chondroitan sulfate proteoglycans [2], and glycoproteins such as DDR1 [3]. The clustering of integrins and proteoglycans into focal adhesions nucleates signaling complexes, which include a few hundred signaling molecules [4]. The combination of exact signaling molecules is most likely the basis for specific cellular responses to the ECM. At this point the reader is referred to the other chapters in this volume, which provide excellent descriptions of the structure, function and regulation of these adhesive complexes.

Much of the present understanding of cell adhesion to ECM components has arisen from cell culture approaches, under conditions of coating plastic or glass surfaces with ECM components (either singly or in combination), and analyzing the response of cells. However, it is clear that tissue architecture and the properties of the connective tissue microenvironment are not fully recapitulated by this approach, and thus there has been a trend towards developing tissue culture models that utilize three-dimensional (3D) matrices upon which – or within which – to culture cells. In this chapter, the approaches to study cell adhesions in three dimensions are detailed, and the molecular insights that have arisen as a result are described.

Cell Junctions. Adhesion, Development, and Disease.
Edited by Susan E. LaFlamme and Andrew Kowalczyk
Copyright © 2008 WILEY-VCH Verlag GmbH & Co. KGaA, Weinheim
ISBN: 978-3-527-31882-7

7.2
Model 3D Matrices for Investigations of Cell Behavior

Several different cell types, including epithelial and endothelial cells, do not recapitulate the differentiated structures seen *in vivo* when cultured on two-dimensional (2D) surfaces *in vitro*. Thus, in- vitro 3D model systems were developed in order to study epithelial biology in a more relevant context [5, 6], and these have also been used for several other epithelial, fibroblast, neuronal, smooth muscle, and endothelial cell systems. The choice of 3D system used depends largely on the cell type under investigation, and the questions being addressed.

7.2.1
Matrigel/Reconstituted Basement Membrane (rBM)

Extracts of the Engelbreth–Holm–Swarm (EHS) sarcoma serve as a source of reconstituted basement membrane containing laminin, collagen IV, entactin, and chondroitan sulfate proteoglycan (CSPG), among several other components [7]. Both. home-made and commercial preparations of this basement membrane (which are often marketed under the name Matrigel®) are an excellent source of a relevant basement membrane environment. Matrigel/rBM forms 3D gels spontaneously, and cells can either be embedded within rBM, or cultured on a thin layer of rBM to which an overlay of rBM is then added.

The use of Matrigel/rBM allows the creation of a culture support system that recapitulates several cell behaviors. For example, the culture of breast epithelial cells in rBM normally promotes the formation of cyst-like, acinar structures [8–12], which is consistent with the finding *in vivo* that both mouse and human mammary gland have abundant laminin and collagen IV surrounding ascini [13, 14].

One difficulty with gels composed of Matrigel/rBM is that it is a complex mixture of several components, and includes growth factors and undefined components that copurify. Thus, there is often variability from lot to lot, and there is not always a clear indication of the full spectrum of stimuli for the cells. However, if such possible variation is borne in mind, then appropriately designed and controlled experiments can capitalize on the excellent environment provided by this source.

7.2.2
Collagen Gels

Gels composed of collagen I are a readily adaptable approach to providing 3D culture environments. Collagen gels are not the same composition as the basement membrane, but rather are composed of stromal ECM components, predominantly collagen I (~97%) and to a lesser extent collagen III (~3%). For experimental systems requiring basement membrane components, this may not be the appropriate environment. However, gels composed of collagen have certain advantages over those composed of rBM/Matrigel. When breast epithelial cells are cultured in floating 3D collagen gels, they differentiate and can form their own basement membrane [5, 6, 15–17]. It could be argued that this basement membrane, when surrounded by a stromal matrix, is a more accurate representation of the biochemical and physical microenvironment *in vivo* than are gels of pure Matrigel.

Collagen I is localized around breast ducts in both the mouse [13] and human [18], while laminin and collagen IV are found near alveoli and endbuds in the mouse mammary gland [13] and around the acini in human breast [14]. Thus, whilst culture in Matrigel promotes ascinar formation, culture in collagen I gels typically promotes ductal morphogenesis [13, 19].

Several experimental questions require the recapitulation of a stromal environment, and for this collagen gels are a better choice than Matrigel/rBM. Systems investigating fibroblasts [20] or mixed epithelial/fibroblast cocultures [21] often utilize 3D collagen gel culture. Moreover, for investigations of carcinoma invasion, collagen gels represent the physiologically appropriate stromal environment through which cells invade, as such invasion occurs along collagen fibers [22] and collagen is reorganized around the tumor boundary to facilitate 3D invasion *in vivo* [23]. Previously, collagen gels have provided an excellent system to investigate the 3D aspects of cell invasion [24–26].

One advantage that collagen gels have over other 3D matrices is that the collagen can be directly imaged, even in live cell cultures, through the use of multiphoton microscopy and second-harmonic generation (SHG) [23, 27–29]. In addition, collagen structures are readily observed by SHG in 3D collagen gels and fresh, unfixed tissues (Figure 7.1).

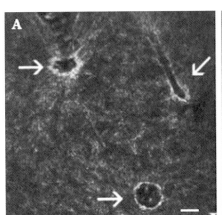
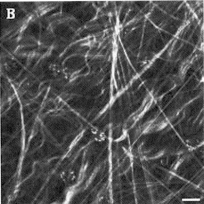

Figure 7.1 Collagen can be readily visualized using multiphoton microscopy in unfixed, unstained preparations. (A) Multiphoton/second-harmonic generation (SHG) image of a fresh, unfixed, unstained 3D collagen gel at 1.3 mg mL^{-1}, containing MCF10A cells that have formed ascinar and tubule structures. Collagen is autofluorescent by this approach, and appears as bright fibers in the matrix. Note that the cells "pull" the collagen fibers around the forming tubules (arrows). Scale bar = 50 μm. (Image courtesy of Gregory Palmer.) (B) Multiphoton/SHG image of fresh mouse mammary tissue that is unfixed and unstained, showing straight and curly collagen fibers in the connective tissue compartment. Scale bar = 50 μm. (Image courtesy of Paolo Provenzano). For both panels, multiphoton imaging was performed with 890 nm excitation as described in Ref. [23].

Cells respond not only to the biochemical composition of their microenvironment, but also to the physical properties (see below). It has been found that gels composed of collagen I may be manipulated in a manner that allows control of the mechanical properties of the culture environment. Collagen gels that are detached from the culture dish create a different mechanical environment compared to gels left attached to the culture dish [20, 30]. In addition, the mechanical properties of the matrix can be altered by varying the concentration of collagen [31, 32]. Such manipulations of Matrigel/rBM are not possible, as the medium is too pliable and the concentrations are not readily altered. As one approach, it is possible to create 3D environments of mixed collagen/rBM in an attempt to achieve the benefits of both rBM and collagen in terms of biochemical composition and physical properties. Mixed gels have not been widely used, most likely because they are less defined than either rBM or collagen gels alone.

7.2.3
Fibrin Gels

The thrombin-mediated cleavage of fibrinogen into insoluble fibrin gels forms part of the normal hemostatic process and the formation of blood clots. However, this process can also be exploited to form a 3D gel of fibrin, which is particularly useful for endothelial cell culture and the formation of microvasculature, as fibrin gel structure regulates the formation of vascular cords [33]. Fibrin gels have also found use for many cell culture systems, including the differentiation of mouse embryonic stem cells [34], cartilage [35, 36], and neurite outgrowth [37]. Moreover, the use of tissue engineering and applied magnetic forces has allowed the patterning of fibrin gels to create defined topographic 3D matrices [38].

7.2.4
Cell-Derived Fibronectin-Based Matrices

One major short-coming of gels composed of either Matrigel or collagen is that they form fibrils and polymerize spontaneously, but do so randomly and thus do not fully recapitulate the organization or even the exact fibrillar structure of a cell-derived ECM [39]. Collagen fibrillogenesis is organized on a previously formed fibronectin (FN) matrix, created by stromal fibroblasts [40–42]. Thus, a useful approach has been to culture fibroblasts, allow them to generate a cell-derived matrix containing FN among other components, and then to remove the cells by detergent lysis [39, 43]. This approach is preferred in studies of skin differentiation, as the biochemical and physical properties of dermal fibroblast-derived matrices are more similar to skin than either collagen or fibrin gels alone [44].

7.2.5
Engineered 3D Matrices and Microfluidic Chambers

Several different scaffolds and biomaterials have been exploited for tissue engineering purposes [45, 46]. Flexible polyacrylamide gels can be coated with ECM ligands, and these have proved useful for studies of mechanical forces in cell adhesion and migration [47, 48]. Likewise, synthetic hydrogels composed of various compounds can be engineered with various ECM ligands, and mixed ligands, to create combinatorial matrices of desired composition and properties. For example, 3D gels composed of poly(ethylene terephthalate) (PET) cause human mesenchymal stem cells to secrete a complex ECM of collagen I, FN, collagen IV, and laminin, and to differentiate down osteoblast and adipocyte lineages [49]. Polymers based on poly(ethylene glycol) (PEG) can be derivatized with combinations of ECM ligands and matrix metalloproteinase (MMP) substrates to create models for cell invasion [50].

Microfluidic chambers, created by the use of polydimethyl siloxand (PDMS) channels, represent an alternative for creating highly regulated 3D matrices, as they can be used to align collagen fibers or to create gradients [51–53]. These chambers have the advantage of being optically clear, and allow the visualization of cells and matrices within the support, while maintaining control of several aspects of the extracellular environment. In certain models, such as investigations into the organization and differentiation of endothelial cells, it is important to add a flow component to more accurately reflect the blood flow found *in vivo*. In particular, interstitial flow directs the organization of endothelial cells in 3D fibrin matrices [54] and causes the differentiation of myofibroblasts and alignment of collagen fibers [55]. Bioengineering approaches, such as the use of microfluidic channels, more easily allow the application of flow to 3D matrices.

7.3
Cell Behavior and Adhesive Structures Differ between 3D and 2D Matrices

Cells which are adherent to a 2D ECM-coated surface form large focal adhesions, are generally proliferative, and activate several signaling pathways. However, when cultured in a 3D matrix of identical composition these same cells often demonstrate strikingly different behaviors. For example, breast epithelial cells on a 2D surface coated with collagen I are spread, are proliferative and migratory, and activate focal adhesion kinase (FAK), whereas those cultured in 3D floating collagen I gels organize into tubule structures, have a diminished proliferation, and do not cluster FAK into focal adhesions [32]. Breast epithelial cells also differ in their apoptotic response to anti-proliferative and chemotherapeutic agents, depending on whether they are cultured in 2D or 3D matrices [56].

Other cell types which are also sensitive to differences between 2D and 3D culture conditions include fibroblasts [57] and vascular smooth muscle cells, which become proliferative and synthesize matrix when cultured under 2D compared to 3D conditions [58, 59].

Cell adhesions formed in a 3D cell-derived FN matrix differ in the composition of their adhesions when compared to 2D matrices, and are termed "matrix adhesions" [43] in order to differentiate them from the classic "focal adhesion" described for cells cultured on rigid 2D ECM-coated glass or plastic surfaces. 3D matrix adhesions are characterized by a decrease in the phosphorylation of FAK at Y397, and a switch in integrin usage [60]. Diminished FAK phosphorylated at Y397, and small to no focal complexes are also observed in breast epithelial cells cultured in 3D matrices [32].

7.4
"Compliancy" or Elastic Modulus Determines Cellular Response to ECMs

At this point, one obvious question to be asked is why does the same ligand, when immobilized to a 2D support such as glass or plastic, alter the response of cells compared to a 3D matrix? One possibility is that cells recognize whether they have ECM ligands on all of their surfaces or only on their "ventral" surface, as would be the case for 2D culture. In order to test this proposal, cells have been cultured on surfaces composed of various polymers and crosslinked polyacrylamide gels coated with ECM ligands. Several research groups have reported that cells respond differently to a compliant 2D surface compared to a rigid 2D surface, thus demonstrating that it is likely that cells respond to the physical properties of the microenvironment.

Indeed, "compliancy" is also perceived by cells in a 3D matrix, and represents a major aspect of cellular responses to 3D matrices [61]. Indeed, it was noted several years ago that breast epithelial cells could polarize into ascinar structures and differentiate only if cultured within floating 3D collagen gels, but not if the same gels remained attached to the sides and bottom of the dish [5, 6, 15]. When cultured in 3D collagen gels that are attached, the cells form matrix adhesions that contain phospho-FAK, demonstrate increased proliferation, and activate the Rho GTPase relative to cells cultured in floating 3D collagen matrices [32].

Fibroblasts also respond to "stressed" or attached collagen gels with a proliferative response, but become less proliferative and alter their behavior in "relaxed" or floating collagen gels [30, 62, 63]. Smooth muscle cell proliferation is similarly regulated in response to attached versus floating collagen gels [58], and this is most likely a response to the physical nature of attached versus floating gels, rather than being specific to collagen gels, as the rigidity of a fibrin gel regulates the formation of vascular cords [33]. Remarkably, stem cell differentiation is regulated in a manner such that the lineage matches the elasticity of particular tissues, with soft matrices which more closely resemble brain tissue inducing neurogenesis, while stiffer matrices that are more like muscle tissue stimulate myogenesis [64]. An increasing amount of data derived from several cell types highlights the important role of matrix elasticity on controlling the proliferation and differentiation of stem cells [64, 65], myotubes [66], and endothelial cells [67], as well as proliferation and apoptosis [68, 69].

"Compliancy" may also be measured as the elastic modulus of the matrix, in several different ways. For example, when measured as the modulus of compression, the tissues generally have an elastic modulus in the Pa range [61]. However, as cells also experience forces related to the pull of the cell upon the matrix, another option was to monitor the elastic modulus by tension. Consequently, this value was found to show a sigmoidal relationship to the concentration of a collagen gel in the kPa range, consistent with published results for the elastic modulus of collagen gels [31]. Thus, as the collagen concentration is increased, the elastic modulus increases and the gel becomes less compliant. Hence, it is not only the collagen density but also the population of collagen gels by fibroblasts that alters the viscoelastic properties [70].

When cells are cultured in 3D collagen gels of increased collagen concentration, or density, they respond as if in an attached collagen gel, with increases in matrix adhesion, phospho-FAK, and proliferation [32]. Moreover, the expression of a few thousand genes is altered by the density of a collagen gel in mouse mammary epithelial cells (P. Provenzano, unpublished observations), thereby demonstrating a clear alteration of cellular signaling events and cell phenotype in response to collagen matrix density. This becomes significant in light of the observation that an increased collagen density in human breast tissue is linked to a four- to six-fold increase in the incidence of breast cancer [71].

7.4.1
Cellular Contractility and the Response to Matrix Compliance

Cells clearly sense and respond to the compliance of their ECM environment [72]. In fact, although there are most likely several molecules underlying this response, the exact mechanosensor is not precisely known. One mechanism which is involved is linked to cellular-generated contractility through actin/myosin and Rho GTPase. The cells contract a compliant 3D collagen gel, but are unable to contract an attached or a dense collagen gel [20, 32]. This contraction is not simply a byproduct of cellular response to compliant 3D matrices, but rather is required for cells to respond and sense matrix compliance, as inhibitors of actin/myosin contractility (e.g., blebbistatin) inhibit not only gel contraction but also tubule formation, the control of proliferation, and gene expression in response to a compliant 3D collagen gel (Figure 7.2).

7.4.2
Rho Regulates the Cytoskeleton and Contractility

The small GTPase, Rho, is an important regulator of cellular contractility, as well as of the actin cytoskeleton, cell shape, and cell phenotype [65]. The results of several studies have supported a role for Rho in epithelial differentiation, and in the formation of adherens and tight junctions in epithelial cells [73]. Rho regulates the formation of focal adhesions and actin stress fibers [74] through the coordinated regulation of several downstream pathways regulating actin assembly and cellular contractility. Rho activates Rho kinase (ROCK), which results in the phosphorylation of myosin light chain (MLC) by both direct phosphorylation of MLC and negative regulation of myosin phosphatase [75, 76]. A second Rho effector, the formin mDia (the mammalian homologue *Drosophila* diaphanous) also

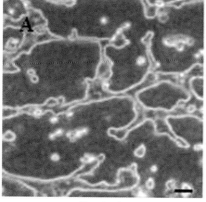 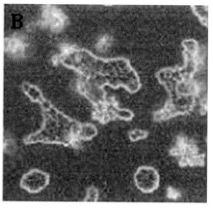

Control　　　　　　　　　　　　　Blebbistatin

Figure 7.2 Inhibition of actin–myosin contractility inhibits ductal morphogenesis in 3D collagen gels. T47D cells were cultured in floating 1.3 mg mL^{-1} collagen gels as described [32], and then treated with 100 μM blebbistatin (B) or vehicle only (A). Note that blebbistatin, an inhibitor of myosin, blocks the formation of tubules. Scale bar = 50 μm. (Figure courtesy of Michele Wozniak.)

regulates actin dynamics through effects on profilin [77]. ROCK and mDia are key regulators of focal adhesion formation in fibroblasts [78–80].

Rho is particularly relevant not only to cellular contractility, but also to mechanosensing. Rho functioning through both ROCK and mDia is necessary in order to orient the stress fibers with respect to the axis of strain [81], which regulates the elastic properties of the actin network with respect to the mechanical environment. Rho/ROCK is an important regulator of cellular responses to 3D matrices, as inhibition of this pathway blocks contraction of collagen gels and cellular responses such as tubule formation [32] and fibroblast traction forces [82].

Part of the effect of cellular contractility is most likely the result of changes in cell shape, as the cells become more rounded when they contract the ECM around them. Shape is a key regulator of the proliferation of several cells [67, 83]; for example, the culture of endothelial cells on increasingly smaller 2D islands of adhesive ligand, which restricts their spreading, regulates not only the cells' proliferation [84] but also their focal adhesion formation [85].

Indeed, Rho/ROCK has emerged as a key regulatory pathway controlling the response of cell proliferation and the differentiation of stem cells [65, 86]. Both, the tubule formation of breast epithelial cells and the branching morphogenesis of lung epithelial cells are regulated by Rho-mediated contractile events [32, 87].

7.4.3
Rho/ROCK, Contractility, and Focal Adhesion Formation

A paradigm-shifting result in the field of cell adhesion was the realization that focal adhesions are formed in part from the inside-out, and that integrins are clustered as a result of Rho-driven cellular contractility [88, 89]. ECM rigidity induces the strengthening of integrin-cytoskeletal linkages [90], and this was demonstrated by allowing fibronectin-coated beads to attach to cells, and then applying force to the beads via a laser trap. When a restraining force is applied to a bead attached to the cell surface, integrins are clustered, the components of the focal adhesion and actin cytoskeleton are recruited, and the cell increases the amount of force that the adhesion can withstand [90–94]. Thus, cells are able to tune their response to the physical compliance of the ECM. These results suggest that ligand-induced signaling events through integrins and Rho serve as a feed-forward mechanism to cause cellular contractility and the strengthening of cell adhesions.

When considering these observations, it is not surprising therefore that cells cultured in rigid or dense 3D matrices have enhanced focal adhesion formation compared to cells in compliant matrices, and that this process is dependent on Rho-mediated contractility [32]. The imaging of cells within 3D matrices demonstrates the localization of FAK, vinculin, and other molecules in a manner that differs for cells cultured in compliant versus rigid or dense collagen gels (Figure 7.3) [32]. Recently, several approaches to visualize signaling molecules within live cells cultured in 3D matrices have been developed, and have shown that multi-photon microscopy and SHG allows the simultaneous localization of vinculin-GFP relative to collagen fibers (Figure 7.3C). This approach can be readily adapted to the investigation of several signaling molecules within 3D collagen matrices.

7.4.4
Focal Adhesions as Transducers of Biophysical Signals

Integrins are poised to transduce biophysical signals related to ECM rigidity, as their cytoplasmic tails link directly to the actin cytoskeleton, and nucleate signaling complexes/focal adhesions [95]. A role for integrins as mechanosensors is suggested by the observation that force applied to integrin ligands strengthens the interaction of integrins with the actin cytoskeleton [90]. Moreover, the clustering of integrins and the formation of focal adhesions in migrating fibroblasts can be induced by local forces [47, 80, 92, 96–99], and can be stimulated in osteoblasts [100].

Several integrin signaling pathways are activated in response to mechanical forces, including the phosphorylation of FAK and ERK, and the activation of src [43, 101–104]. The phosphorylation on tyrosine 397 of FAK is a key event in response to mechanical signaling in fibroblasts and epithelial cells [32, 43, 96, 103–105]. Small GTPases are also affected, as cell tension regulates Rac [106, 108] and Rho [32].

Several molecules bind to integrin cytoplasmic tails and contribute to integrin adhesion strengthening in response to mechanical forces, including talin [93, 94], vinculin [80, 92], and filamin [94, 107, 108]. Talin is a crucial regulator of integrin function downstream of several integrin-activation pathways [109–112]. An enhanced filamin binding to integrin β cytoplasmic domains regulates integrin function [113].

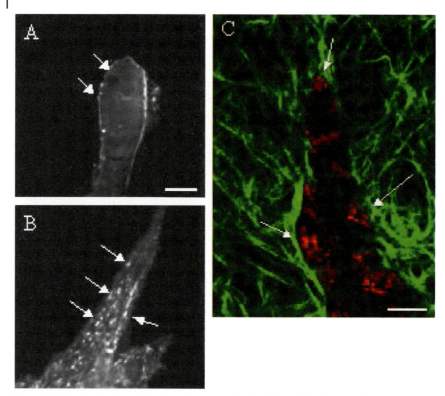

Figure 7.3 Focal adhesions differ in collagen gels of different densities. NMuMG mouse mammary epithelial cells were cultured in 3D collagen gels at (A) 3.0 mg mL^{-1} (compliant) or (B) 4.0 mg mL^{-1} (dense). Cells in collagen gels were fixed and stained with anti-vinculin antibodies (arrows) and with DAPI to visualize the nuclei. Note that vinculin is uniformly distributed around the plasma membrane for cells that are in compliant 3D gels, but organizes into discrete focal adhesions, seen as bright dots, in cells in dense 3D collagen gels. Scale bar = 20 µm. (Figure courtesy of Paolo Provenzano.) (C) Simultaneous imaging of collagen and vinculin in live 3D collagen gels. NMuMG breast carcinoma cells were stably transfected with GFP-vinculin, and cultured in 3D collagen gels of 1.3 mg mL^{-1}. Fresh, unfixed, unstained gels were imaged using multiphoton microscopy. Following excitation with a Ti-sapphire laser tuned to 890 nm, the resultant emissions were separated using a 445-nm SHG narrow band pass filter for (pseudocolored green) and a 464-nm (cut-on) MPE long pass filter (pseudocolored red). GFP-vinculin localizes to cellular adhesions (arrows) associated with collagen fibers. Scale bar = 20 µm. (Figure courtesy of Paolo Provenzano.)

7.5
A Model for Cellular Response to 3D Matrices Varying in Compliance

One model to explain the observations regarding how cells respond within 3D matrices is to consider that contractility through Rho and ROCK allows cells to "sense" the physical rigidity of the matrix (Figure 7.4). Cells encountering a rigid matrix will cluster integrins and strengthen their adhesions, consistent with the

7.5 A Model for Cellular Response to 3D Matrices Varying in Compliance

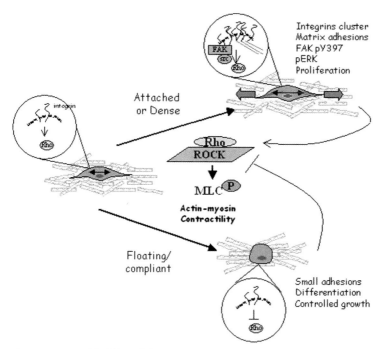

Figure 7.4 A model of the cellular response to 3D matrices differing in compliance or density. See text for details.

finding that physical forces result in adhesion strengthening [90]. It is proposed that the formation of 3D focal adhesions will scaffold several signaling molecules including FAK, and lead to proliferation, sustained activation of Rho, and changes in gene expression. In contrast, cells encountering a compliant matrix will not meet with resistance, will not cluster or strengthen their adhesions, and will not form the same signaling complexes. These cells will be more rounded and organize their actin cytoskeletons accordingly. The outcome is ultimately differentiation-specific gene expression and the control of cell proliferation, consistent with observations that rounded cells are less proliferative [83, 114].

There is an important feedback mechanism by which cells respond to 3D matrices, as Rho not only contributes to the sensing of matrix rigidity, but is itself regulated by the mechanical properties of 3D matrices. Epithelial cells cultured in attached or dense collagen gels have increased Rho-GTP, while cells cultured in compliant floating collagen matrices have a downregulated Rho activity [32, 115]. Interestingly, Rho may be localized to sites of active matrix engagement, as a reporter of active Rho, GFP-RBD, localizes to sites where cells attach to collagen fibers (P. Keely, S. Trier, and S. Ponik, unpublished results). Thus, a localized topography of 3D matrices is likely to lead to localized forces and to cause spatial regulation of Rho signaling pathways and matrix adhesion formation.

Additional approaches to investigate the temporal and spatial regulation of signaling pathways in response to the biochemical and mechanical properties of 3D extracellular matrices will yield novel insights with regard to how cells respond to their complex microenvironment, and should also help in an understanding of the complex *in-vivo* processes that govern tissue morphogenesis, remodeling, disease states, and cancer progression and metastasis.

Acknowledgments

The author wishes to acknowledge Dr. Paolo Provenzano, Dr. Michele Wozniak, Steven Trier, and Dr. Gregory Palmer for contributions of images in the figures, and Dr. Suzanne Ponik for the T47D GFP-RBD cells used in imaging. The studies described were supported by grants from NIH CA076537, the American Cancer Society RSG-00-339-04-CSM, and the Susan G. Komen Foundation to P. Keely, and by the DOD BC031277 to P. Provenzano.

References

1 D.M. Beauvais and A.C. Rapraeger, *Reprod. Biol. Endocrinol. 2* (2004) 3.
2 J. Iida, A.M. Meijne, J.R. Knutson, L.T. Furcht and J.B. McCarthy, *Semin. Cancer Biol. 7* (1996) 155–162.
3 W. Vogel, G.D. Gish, F. Alves and T. Pawson, *Mol. Cell 1* (1997) 13–23.
4 E. Zamir and B. Geiger, *J. Cell Sci. 114* (2001) 3583–3590.
5 J.T. Emerman and D.R. Pitelka, *In Vitro 13* (1977) 316–328.
6 E.Y. Lee, G. Parry and M.J. Bissell, *J. Cell Biol. 98* (1984) 146–55.
7 H.K. Kleinman, M.L. McGarvey, L.A. Liotta, P.G. Robey, K. Tryggvason and G.R. Martin, *Biochemistry 21* (1982) 6188–6193.
8 M.H. Barcellos-Hoff, J. Aggeler, T.G. Ram and M.J. Bissell, *Development 105* (1989) 223–235.
9 V.M. Weaver, O.W. Petersen, F. Wang, C.A. Larabell, P. Briand, C. Damsky and M.J. Bissell, *J. Cell Biol. 137* (1997) 231–245.
10 S.K. Muthuswamy, D. Li, S. Lelievre, M.J. Bissell and J.S. Brugge, *Nat. Cell Biol. 3* (2001) 785–792.
11 J. Debnath, S.K. Muthuswamy and J.S. Brugge, *Methods 30* (2003) 256–268.
12 J. Debnath, K.R. Mills, N.L. Collins, M.J. Reginato, S.K. Muthuswamy and J.S. Brugge, *Cell 111* (2002) 29–40.
13 P. Keely, A. Fong, M. Zutter and S. Santoro, *J. Cell Sci. 108* (1995) 595–607.
14 R.E. Hewitt, D.G. Powe, K. Morrell, E. Balley, I.H. Leach, I.O. Ellis and D.R. Turner, *Br. J. Cancer 75* (1997) 221–229.
15 M.J. Bissell, H.G. Hall and G. Parry, *J. Theoret. Biol. 99* (1982) 31–68.
16 H.G. Hall, D.A. Farson and M.J. Bissell, *Proc. Natl. Acad. Sci. USA 79* (1982) 4672–4676.
17 G. Parry, E.Y. Lee, D. Farson, M. Koval and M.J. Bissell, *Exp. Cell Res. 156* (1985) 487–499.
18 M.S. al Adnani, S. Taylor, A.A. al-Bader, A.G. al-Zuhair and J.O. McGee, *Histol. Histopathol. 2* (1987) 227–238.
19 F. Berdichevsky, D. Alford, B. D'Souza and J. Taylor-Papadimitriou, *J. Cell Sci. 107* (1994) 3557–3568.
20 F. Grinnell, *Trends Cell Biol. 10* (2000) 362–365.
21 G. Su, S.A. Blaine, D. Qiao and A. Friedl, *J. Biol. Chem. 282* (2007) 14906–14915.
22 E. Sahai, J. Wyckoff, U. Philippar, J. Segall, F. Gertler and J. Condeelis, *BMC Biotechnology 5* (2005) 14.
23 P.P. Provenzano, K.W. Eliceiri, J.M. Campbell, D.R. Inman, J.G. White and P.J. Keely, *BMC Med 4* (2006) 38.
24 K. Wolf, I. Mazo, H. Leung, K. Engelke, U.H. von Andrian, E.I. Deryugina, A.Y.

Strongin, E.-B. Brocker and P. Friedl, *J. Cell Biol.* 160 (2003) 267–277.

25 Y. Hegerfeldt, M. Tusch, E.B. Brocker and P. Friedl, *Cancer Res.* 62 (2002) 2125–2130.

26 F. Grinnell, L.B. Rocha, C. Iucu, S. Rhee and H. Jiang, *Exp. Cell Res.* 312 (2006) 86–94.

27 P.J. Campagnola, A.C. Millard, M. Terasaki, P.E. Hoppe, C.J. Malone and W.A. Mohler, *Biophys. J.* 82 (2002) 493–508.

28 E. Brown, T. McKee, E. diTomaso, A. Pluen, B. Seed, Y. Boucher and R.K. Jain, *Nat. Med.* 9 (2003) 796–800.

29 W.R. Zipfel, R.M. Williams, R. Christie, A.Y. Nikitin, B.T. Hyman and W.W. Webb, *Proc. Natl. Acad. Sci. USA* 100 (2003) 7075–7080.

30 F. Grinnell, C.H. Ho, E. Tamariz, D.J. Lee and G. Skuta, *Mol. Biol. Cell* 14 (2003) 384–395.

31 B.A. Roeder, K. Kokini, J.E. Sturgis, J.P. Robinson and S.L. Voytik-Harbin, *J. Biomech. Eng.* 124 (2002) 214–222.

32 M.A. Wozniak, R. Desai, P.A. Solski, C.J. Der and P.J. Keely, *J. Cell Biol.* 163 (2003) 583–595.

33 A. Stephanou, G. Meskaoui, B. Vailhe and P. Tracqui, *Microvasc. Res.* 73 (2007) 182–190.

34 H. Liu, S.F. Collins and L.J. Suggs, *Biomaterials* 27 (2006) 6004–6014.

35 D. Eyrich, A. Gopferich and T. Blunk, *Adv. Exp. Med. Biol.* 585 (2006) 379–392.

36 D. Eyrich, F. Brandl, B. Appel, H. Wiese, G. Maier, M. Wenzel, R. Staudenmaier, A. Goepferich and T. Blunk, *Biomaterials* 28 (2007) 55–65.

37 Y.E. Ju, P.A. Janmey, M.E. McCormick, E.S. Sawyer and L.A. Flanagan, *Biomaterials* 28 (2007) 2097–2108.

38 E. Alsberg, E. Feinstein, M.P. Joy, M. Prentiss and D.E. Ingber, *Tissue Eng.* 12 (2006) 3247–3256.

39 K.M. Yamada, R. Pankov and E. Cukierman, *Braz. J. Med. Biol. Res.* 36 (2003) 959–966.

40 J.A. McDonald, D.G. Kelley and T.J. Broekelmann, *J. Cell Biol.* 92 (1982) 485–492.

41 T. Velling, J. Risteli, K. Wennerberg, D.F. Mosher and S. Johansson, *J. Biol. Chem.* 277 (2002) 37377–37381.

42 J. Sottile and D.C. Hocking, *Mol. Biol. Cell* 13 (2002) 3546–3559.

43 E. Cukierman, R. Pankov, D.R. Stevens and K.M. Yamada, *Science* 294 (2001) 1708–1712.

44 J.E. Ahlfors and K.L. Billiar, *Biomaterials* 28 (2007) 2183–2191.

45 L.G. Griffith, *Ann. N. Y. Acad. Sci.* 961 (2002) 83–95.

46 L.G. Griffith and M.A. Swartz, *Nat. Rev. Mol. Cell Biol.* 7 (2006) 211–224.

47 R.J. Pelham, Jr. and Y. Wang, *Proc. Natl. Acad. Sci. USA* 94 (1997) 13661–13665.

48 Y.L. Wang and R.J. Pelham, Jr., *Methods Enzymol.* 298 (1998) 489–496.

49 W.L. Grayson, T. Ma and B. Bunnell, *Biotechnol. Prog.* 20 (2004) 905–912.

50 M.P. Lutolf, J.L. Lauer-Fields, H.G. Schmoekel, A.T. Metters, F.E. Weber, G.B. Fields and J.A. Hubbell, *Proc. Natl. Acad. Sci. USA* 100 (2003) 5413–5418.

51 G.M. Walker, H.C. Zeringue and D.J. Beebe, *Lab. Chip* 4 (2004) 91–97.

52 D.J. Beebe, G.A. Mensing and G.M. Walker, *Annu. Rev. Biomed. Eng.* 4 (2002) 261–286.

53 P. Lee, R. Lin, J. Moon and L.P. Lee, *Biomed. Microdevices* 8 (2006) 35–41.

54 C.L. Helm, A. Zisch and M.A. Swartz, *Biotechnol. Bioeng.* 96 (2007) 167–176.

55 C.P. Ng, B. Hinz and M.A. Swartz, *J. Cell Sci.* 118 (2005) 4731–4739.

56 V.M. Weaver, S. Lelievre, J.N. Lakins, M.A. Chrenek, J.C. Jones, F. Giancotti, Z. Werb and M.J. Bissell, *Cancer Cell* 2 (2002) 205–216.

57 F. Grinnell, *Trends Cell Biol.* 13 (2003) 264–269.

58 H. Koyama, E.W. Raines, K.E. Bornfeldt, J.M. Roberts and R. Ross, *Cell* 87 (1996) 1069–1078.

59 J.P. Stegemann and R.M. Nerem, *Exp. Cell Res.* 283 (2003) 146–155.

60 E. Cukierman, R. Pankov and K.M. Yamada, *Curr. Opin. Cell Biol.* 14 (2002) 633–639.

61 M.J. Paszek, N. Zahir, K.R. Johnson, J.N. Lakins, G.I. Rozenberg, A. Gefen, C.A. Reinhart-King, S.S. Margulies, M. Dembo, D. Boettiger, D.A. Hammer and V.M. Weaver, *Cancer Cell* 8 (2005) 241–254.

62 J. Fringer and F. Grinnell, *J. Biol. Chem.* 276 (2001) 31047–31052.

63 E. Tamariz and F. Grinnell, *Mol. Biol. Cell 13* (2002) 3915–3929.
64 A.J. Engler, S. Sen, H.L. Sweeney and D.E. Discher, *Cell 126* (2006) 677–689.
65 R. McBeath, D.M. Pirone, C.M. Nelson, K. Bhadriraju and C.S. Chen, *Dev. Cell 6* (2004) 483–495.
66 A.J. Engler, M.A. Griffin, S. Sen, C.G. Bonnemann, H.L. Sweeney and D.E. Discher, *J. Cell Biol. 166* (2004) 877–887.
67 D.E. Ingber and J. Folkman, *J. Cell Biol. 109* (1989) 317–330.
68 S. Huang, C.S. Chen and D.E. Ingber, *Mol. Biol. Cell 9* (1998) 3179–3193.
69 H.B. Wang, M. Dembo and Y.L. Wang, *Am. J. Physiol. Cell Physiol. 279* (2000) C1345–C1350.
70 J.E. Wagenseil, T. Wakatsuki, R.J. Okamoto, G.I. Zahalak and E.L. Elson, *J. Biomech. Eng. 125* (2003) 719–725.
71 N.F. Boyd, G.A. Lockwood, J.W. Byng, D.L. Tritchler and M.J. Yaffe, *Cancer Epidemiol. Biomarkers Prev. 7* (1998) 1133–1144.
72 D.E. Discher, P. Janmey and Y.L. Wang, *Science 310* (2005) 1139–1143.
73 T.S. Jou, E.E. Schneeberger and W.J. Nelson, *J. Cell Biol. 142* (1998) 101–115.
74 A.J. Ridley and A. Hall, *Cell 70* (1992) 389–399.
75 K. Kimura, M. Ito, M. Amano, K. Chihara, Y. Fukata, M. Nakafuku, B. Yamamori, J. Feng, T. Nakano, K. Okawa, A. Iwamatsu and K. Kaibuchi, *Science 273* (1996) 245–248.
76 Y. Kureishi, S. Kobayashi, M. Amano, K. Kimura, H. Kanaide, T. Nakano, K. Kaibuchi and M. Ito, *J. Biol. Chem. 272* (1997) 12257–12260.
77 N. Watanabe, P. Madaule, T. Reid, T. Ishizaki, G. Watanabe, A. Kakizuka, Y. Saito, K. Nakao, B.M. Jockusch and S. Narumiya, *EMBO J. 16* (1997) 3044–3056.
78 M. Amano, K. Chihara, K. Kimura, Y. Fukata, N. Nakamura, Y. Matsuura and K. Kaibuchi, *Science 275* (1997) 1308–1311.
79 G. Totsukawa, Y. Yamakita, S. Yamashiro, D.J. Hartshorne, Y. Sasaki and F. Matsumura, *J. Cell Biol. 150* (2000) 797–806.
80 D. Riveline, E. Zamir, N.Q. Balaban, U.S. Schwarz, T. Ishizaki, S. Narumiya, Z. Kam, B. Geiger and A.D. Bershadsky, *J. Cell Biol. 153* (2001) 1175–1186.
81 R. Kaunas, P. Nguyen, S. Usami and S. Chien, *Proc. Natl. Acad. Sci. USA 102* (2005) 15895–15900.
82 K.A. Beningo, K. Hamao, M. Dembo, Y.L. Wang and H. Hosoya, *Arch. Biochem. Biophys. 456* (2006) 224–231.
83 J. Folkman and A. Moscona, *Nature 273* (1978) 345–349.
84 C.S. Chen, M. Mrksich, S. Huang, G.M. Whitesides and D.E. Ingber, *Science 276* (1997) 1425–1428.
85 C.S. Chen, J.L. Alonso, E. Ostuni, G.M. Whitesides and D.E. Ingber, *Biochem. Biophys. Res. Commun. 307* (2003) 355–361.
86 D.M. Pirone, W.F. Liu, S.A. Ruiz, L. Gao, S. Raghavan, C.A. Lemmon, L.H. Romer and C.S. Chen, *J. Cell Biol. 174* (2006) 277–288.
87 K.A. Moore, T. Polte, S. Huang, B. Shi, E. Alsberg, M.E. Sunday and D.E. Ingber, *Dev. Dyn. 232* (2005) 268–281.
88 M. Chrzanowska-Wodnicka and K. Burridge, *J. Cell Biol. 133* (1996) 1403–1415.
89 Q. Zhang, M.K. Magnusson and D.F. Mosher, *Mol. Biol. Cell 8* (1997) 1415–1425.
90 D. Choquet, D.P. Felsenfeld and M.P. Sheetz, *Cell 88* (1997) 39–48.
91 C.G. Galbraith and M.P. Sheetz, *Curr. Opin. Cell Biol. 10* (1998) 566–571.
92 C.G. Galbraith, K.M. Yamada and M.P. Sheetz, *J. Cell Biol. 159* (2002) 695–705.
93 G.Y. Jiang, G. Giannone, D.R. Critchley, E. Fukumoto and M.P. Sheetz, *Nature 424* (2003) 334–337.
94 G. Giannone, G. Jiang, D.H. Sutton, D.R. Critchley and M.P. Sheetz, *J. Cell Biol. 163* (2003) 409–419.
95 B. Geiger and A. Bershadsky, *Cell 110* (2002) 139–142.
96 B.Z. Katz, E. Zamir, A. Bershadsky, Z. Kam, K.M. Yamada and B. Geiger, *Mol. Biol. Cell 11* (2000) 1047–1060.
97 N.Q. Balaban, U.S. Schwarz, D. Riveline, P. Goichberg, G. Tzur, I. Sabanay, D. Mahalu, S. Safran, A. Bershadsky, L.

Addadi and B. Geiger, *Nat. Cell Biol.* 3 (2001) 466–472.
98. C. Ballestrem, B. Hinz, B.A. Imhof and B. Wehrle-Haller, *J. Cell Biol.* 155 (2001) 1319–1332.
99. B. Wehrle-Haller and B. Imhof, *Trends Cell Biol.* 12 (2002) 382.
100. M. Wozniak, A. Fausto, C.P. Carron, D.M. Meyer and K.A. Hruska, *J. Bone Miner. Res.* 15 (2000) 1731–1745.
101. D.P. Felsenfeld, P.L. Schwartzberg, A. Venegas, R. Tse and M.P. Sheetz, *Nat. Cell Biol.* 1 (1999) 200–206.
102. M.C. Frame, *J. Cell Sci.* 117 (2004) 989–998.
103. S. Li, P. Butler, Y. Wang, Y. Hu, D.C. Han, S. Usami, J.L. Guan and S. Chien, *Proc. Natl. Acad. Sci. USA* 99 (2002) 3546–3551.
104. H.B. Wang, M. Dembo, S.K. Hanks and Y. Wang, *Proc. Natl. Acad. Sci. USA* 98 (2001) 11295–11300.
105. Q. Shi and D. Boettiger, *Mol. Biol. Cell* 14 (2003) 4306–43015.
106. A. Katsumi, J. Milanini, W.B. Kiosses, M.A. del Pozo, R. Kaunas, S. Chien, K.M. Hahn and M.A. Schwartz, *J. Cell Biol.* 158 (2002) 153–164.
107. D.T. Loo, S.B. Kanner and A. Aruffo, *J. Biol. Chem.* 273 (1998) 23304–23312.
108. M. Yamazaki, S. Furuike and T. Ito, *J. Muscle Res. Cell Motil.* 23 (2002) 525–534.
109. D.A. Calderwood, B. Yan, J.M. de Pereda, B.G. Alvarez, Y. Fujioka, R.C. Liddington and M.H. Ginsberg, *J. Biol. Chem.* 277 (2002) 21749–21758.
110. S. Tadokoro, S.J. Shattil, K. Eto, V. Tai, R.C. Liddington, J.M. de Pereda, M.H. Ginsberg and D.A. Calderwood, *Science* 302 (2003) 103–106.
111. K.L. Wegener, A.W. Partridge, J. Han, A.R. Pickford, R.C. Liddington, M.H. Ginsberg and I.D. Campbell, *Cell* 128 (2007) 171–182.
112. D.A. Calderwood, V. Tai, G. Di Paolo, P. De Camilli and M.H. Ginsberg, *J. Biol. Chem.* 279 (2004) 28889–28895.
113. D.A. Calderwood, A. Huttenlocher, W.B. Kiosses, D.M. Rose, D.G. Woodside, M.A. Schwartz and M.H. Ginsberg, *Nat. Cell Biol.* 3 (2001) 1060–1068.
114. S. Huang and D.E. Ingber, *Nat. Cell Biol.* 1 (1999) E131–E138.
115. M.F. Olson, *Trends Cell Biol.* 14 (2004) 111–114.

Part Two Cell–Cell Junctions

8
Armadillo Repeat Proteins at Epithelial Adherens Junctions
Laura J. Lewis-Tuffin and Panos Z. Anastasiadis

8.1
Introduction

The adherens junction (AJ, also known as the zonula adherens), along with tight junctions and desmosomes, is one of three types of membrane cytoskeletal-based cell–cell interacting regions, the maintenance of which is critical to normal polarized epithelial tissue organization (Figure 8.1). Adherens junctions have three main functions that are not mutually exclusive: (i) to provide structural adhesiveness in support of cell function; (ii) to coordinate the behavior of cell populations; and (iii) to regulate intracellular signaling in a context-dependent manner. The restriction of AJs to the apical region of polarized epithelial cells confers localized signaling which is important for maintaining epithelial polarity. In addition, the simultaneous contraction of such apically localized AJs can result in the coordinated movement of adjoining cells. AJs are not limited to polarized epithelia; rather, they mediate intercellular adhesion in a variety of cell and tissue types. For example, in cardiac tissue AJs are part of the intercalated disk structures which coordinate the synchronous contraction of cardiac myocytes.

Regardless of the cell type in which they reside, AJs have a similar basic protein composition. Adherens junctions consist of members of the cadherin family of adhesion molecules in complex with cytoplasmic proteins termed catenins (the name indicates that they associate with cadherins, not that they are related proteins) that mediate signaling between the cadherin complex and the actin cytoskeleton (Figure 8.2). Cadherins are transmembrane proteins that interact homotypically and in a Ca^{2+}-dependent manner with cadherins on adjacent cells to form cell–cell adhesions (for a review, see Ref. [1]). AJs typically utilize one of the classical type I cadherins, E-cadherin, N-cadherin, R-cadherin, or P-cadherin, all of which have five extracellular domains, four of which are Ca^{2+}-binding, cadherin repeat domains. Classical cadherins also have a single transmembrane domain and a highly conserved cytoplasmic domain. The cytoplasmic domain is further divided into two regions that interact with proteins of the armadillo repeat family – β-catenin, plakoglobin (also known as γ-catenin), and p120 catenin (p120). The cadherin carboxy terminal "catenin-binding domain" binds to either β-catenin or plakoglobin, while

Cell Junctions. Adhesion, Development, and Disease.
Edited by Susan E. LaFlamme and Andrew Kowalczyk
Copyright © 2008 WILEY-VCH Verlag GmbH & Co. KGaA, Weinheim
ISBN: 978-3-527-31882-7

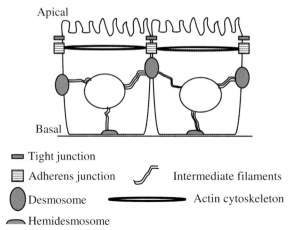

Figure 8.1 Epithelial adhesive junctions. Tight junctions, adherens junctions, and desmosomes are located apically to basally, respectively, and mediate cell–cell adhesion. Hemidesmosomes mediate cell adhesion to the basal laminar surface. All of these structures share a similar basic protein make-up consisting of one or more transmembrane proteins that mediate the interaction with the adjacent cell surface and a variety of associated cytoplasmic proteins that link the transmembrane proteins to the cytoskeleton (actin or intermediate filaments) and mediate signaling between the adhesion and the rest of the cell.

the cadherin juxtamembrane domain (JMD) interacts with p120. β-Catenin is further able to associate with the actin-binding protein α-catenin. Through their association/dissociation with the cadherin cytoplasmic domain, β-catenin, plakoglobin, and p120 mediate a variety of signaling events that functionally link the cadherin–catenin complex to the actin cytoskeleton and ultimately regulate the balance between cell–cell adhesion and cell differentiation on the one hand and cell growth and motility on the other.

The armadillo repeat family of proteins is defined by the existence in its members of a series of internal repeats that are conserved in length (42 amino acids) and spacing (typically end-to-end with no intervening sequences), though not necessarily highly conserved in sequence. Rather, it is the structure of the predominantly hydrophobic core of the arm motif that is conserved; repeats of this domain together form a positively charged groove that mediates protein–protein interactions [2]. This so-called "arm motif" was originally identified in the protein product of the *Drosophila* segment polarity gene known as armadillo [3]. It was subsequently determined that the vertebrate homologues of armadillo, β-catenin and plakoglobin, and the Src tyrosine kinase substrate p120 catenin, also contain multiple arm repeats. Together, these four proteins initially defined the armadillo repeat family of proteins (for reviews, see Refs. [4, 5]) (Figure 8.3). Although the function of these four proteins all coincidentally relate to cell–cell adhesion, it is now clear that members of the armadillo repeat family have widely varied roles in cellular physiology.

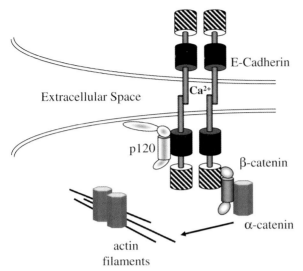

Figure 8.2 The cadherin/catenin complex at adherens junctions (AJs). Cadherins, such as the epithelial-cell expressed E-cadherin, are transmembrane proteins that interact with each other across the extracellular space to adhere adjacent cells to each other. The extracellular cadherin domain is responsible for cadherin-binding specificity. The highly conserved cytoplasmic domain has two regions: (i) a juxtamembrane (black) domain that binds p120 catenin; and (ii) a carboxy-terminal binding (striped) domain that binds β-catenin (or plakoglobin, not depicted). β-Catenin can also interact with the actin-binding protein α-catenin, thereby localizing α-catenin to the AJ where it can reorganize actin filaments in response to AJ formation/signaling. For simplicity, a variety of other proteins which regulate and/or are regulated by the cadherin/catenin complex, including kinases, phosphatases, Rho GTPases, and additional actin-binding proteins, are not depicted.

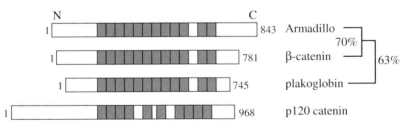

Figure 8.3 N → C terminus structural comparison of armadillo, β-catenin, plakoglobin, and p120-catenin proteins. Armadillo and β-catenin display the highest overall level of sequence homology at 70%; plakoglobin exhibits 63% sequence homology to both armadillo and β-catenin. These three proteins form one subfamily in the larger armadillo repeat protein family; the number and structural spacing of their armadillo repeat domains (gray boxes) is highly conserved, with the main areas of divergence being in their N- and C-terminal regions. p120 catenin is the founding member of a separate subfamily of armadillo repeat proteins; it differs significantly from the other three proteins in terms of its overall sequence and the number and structural spacing of its armadillo repeat domains. Sequence homology between p120 and the other three proteins is restricted to the armadillo repeat domain and is a low 22%, highlighting the importance of structural rather than sequence homology in this domain.

Armadillo was originally characterized by its role in the Wingless signaling pathway which controls changes in gene expression that specify cell fate during *Drosophila* development. β-Catenin was originally identified by its binding to vertebrate E-cadherin at AJs. With time, it has become clear that Armadillo/β-catenin proteins play dual roles in regulating cell–cell adhesion and gene expression; p120 catenin also has this dual function, though by different mechanisms. In this chapter, the dual roles played by these armadillo repeat family members will be examined, as they relate to AJ function in normal and diseased epithelial cells.

8.2
β-Catenin

β-Catenin is a 92 kDa protein that binds tightly to the carboxy-terminal "catenin-binding domain" (CBD) of cadherins. Through its binding to α-catenin, β-catenin is crucial for the reorganization of the actin cytoskeleton at AJs, and consequently for the stability of AJs. β-Catenin/cadherin binding is regulated by phosphorylation: for example, Src-mediated phosphorylation of β-catenin on tyrosine 654 disrupts the binding of β-catenin to E-cadherin [6]. Other tyrosine kinases, such as the epidermal growth factor (EGF) receptor and c-met, are similarly disruptive of β-catenin/cadherin interactions (for a review, see Ref. [7]). The β-catenin/α-catenin interaction is also regulated by phosphorylation: for example, casein kinase II phosphorylation of β-catenin disrupts β-catenin/α-catenin binding [8].

Traditionally, it was thought that β-catenin served as a direct link between cadherins and actin-bound α-catenin, such that the β-catenin/α-catenin interaction provided the physical link between AJs and the actin cytoskeleton. Indeed, disruption of β-catenin binding to the cadherin CBD results in cadherin uncoupling from the actin cytoskeleton and a loss of cadherin function. Recent evidence, however, suggests that this model is incorrect (for reviews, see Refs. [9, 10]). The binding of α-catenin to β-catenin and to F-actin is mutually exclusive [11, 12], making a direct link between the AJs and the actin cytoskeleton unlikely to be mediated by a β-catenin/α-catenin complex. Instead, the role of β-catenin, with respect to α-catenin, may be to recruit it to the AJ, where it can then dissociate from the cadherin/catenin complex and locally reorganize the actin cytoskeleton in response to AJ signaling. However, further investigations are required in order to understand the mechanisms by which dissociation of β-catenin from cadherins results in uncoupling of the actin cytoskeleton from AJs.

When β-catenin is released into the cytoplasm, it interacts with a protein complex consisting of a scaffold of axin and adenomatous polyposis coli (APC) proteins and the Ser/Thr kinase glycogen synthase 3β (GSK-3β) (Figure 8.4A) (see Ref. [13]). GSK-3β, with help from casein kinases I and II, phosphorylates β-catenin, thereby targeting it for ubiquitination and degradation in the 26S proteosome. This complex functions to maintain the cytoplasmic (and hence nuclear) levels of β-catenin low in the absence of Wnt signaling. Wnt is the vertebrate homologue of the *Drosophila* protein Wingless. In *Drosophila*, the invertebrate homologue of β-catenin, Armadillo, was originally identified as a segment polarity gene, regulated by Wingless signaling [14]. The inhibition of Armadillo by Zeste-white3 could be relieved by Wingless-mediated activation of Dishevelled, leading to changes in gene expression that directed cell fate choices during embryonic development. Vertebrate homologues of Dishevelled (Dsh), and Zeste-white3 (GSK-3β) were later identified, thus illuminating the mechanism by which β-catenin regulates gene expression (Figure 8.4B). Wnt signaling through its receptor, Frizzled, activates Dsh, which inhibits the Axin/APC/GSK-3β complex, thus preventing degradation of β-catenin. Intact cytoplasmic β-catenin can then translocate to the nucleus, where it acts as a coactivator to the architectural transcription factors of the T-cell factor/lymphoid enhancer factor (Tcf/Lef) family, thereby increasing transcription of target genes.

Historically, the two functions of β-catenin as a stabilizer of cadherins and as a Wnt-dependent regulator of transcription have been experimentally separated from one another, as if each cell contained two independent pools of β-catenin. Recently, however, it has become clear that the Wnt and cadherin signaling pathways can influence each other in multiple ways [15] (for reviews, see Refs. [16, 17]). For example, cadherins may function to sequester β-catenin from the cytoplasm, making it unavailable to mediate Wnt signals. Conversely, the Wnt-dependent regulation of the phosphorylation of β-catenin may regulate its ability to bind to

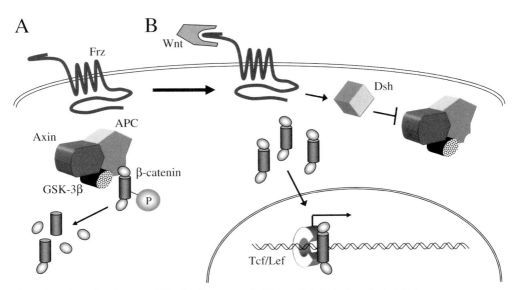

Figure 8.4 (A) In the absence of Wnt binding to the Frizzled (Frz) receptor, the APC/Axin/GSK-3β complex is active. β-Catenin that is released from the cadherin/catenin complex is immediately phosphorylated by GSK-3β and targeted for degradation. (B) When Wnt ligand binds to the Frizzled receptor, the protein Dishevelled (Dsh) is activated. Dsh inhibits the GSK-3β complex. As a result, cytoplasmic β-catenin levels increase and β-catenin can move into the nucleus, where it interacts with the Tcf/Lef family of transcription factors to activate transcription of target genes.

cadherins, and thus to stabilize AJs. This very simplified model becomes increasingly complex with the addition of the activity of various kinases, phosphatases, and transcription factors that can modulate AJ and Wnt signaling in response to other intracellular and extracellular signals. Clearly, the present understanding of the ways in which Wnt and cadherin/AJ signaling influence each other, and the role that β-catenin plays in such signaling, is at an early stage.

8.3
Plakoglobin

Plakoglobin is an 83 kDa protein that usually resides in desmosomes but can also be found at AJs. Plakoglobin is structurally and physically very similar to β-catenin, and links desmosomal or classical cadherins to either desmoplakin or α-catenin, respectively [18]. The α-catenin-binding domains of β-catenin and plakoglobin are highly similar, with 76% overall homology in this 129 amino acid domain and six of seven critical amino acids being identical [19]. Thus, it is highly likely that the interaction of plakoglobin with α-catenin at AJs causes the reorganization of the actin cytoskeleton similarly to the β-catenin/α-catenin interaction. At desmosomes, the interaction of plakoglobin with desmoplakin is thought to mediate the direct linkage of desmosomal cadherins to intermediate filaments.

The binding of plakoglobin and β-catenin to the CBD of classical cadherins is mutually exclusive. Although the circumstances that govern plakoglobin versus β-catenin binding at AJs are generally unclear, it seems that plakoglobin participation in AJs during development may be a prerequisite for the proper formation of desmosomes [20]. Additionally, the inappropriate substitution of plakoglobin for β-catenin at mature AJs may promote inappropriate β-catenin signaling that underlies some types of cancer [21, 22]. Thus, the importance of plakoglobin localization at AJs goes beyond simply its role in stabilizing cell–cell adhesion, and includes signaling events that can modify a cell's behavior.

Molecular signals that can regulate plakoglobin localization to desmosomes versus AJs have now begun to be elucidated [18]. For example, Src phosphorylation of plakoglobin Tyr643 decreases E-cadherin and α-catenin binding and increases desmoplakin binding. Conversely, Fer and Fyn phosphorylation of plakoglobin Tyr549 increases α-catenin binding and decreases desmoplakin binding. EGF receptor-dependent phosphorylation of plakoglobin has also been shown to impair desmoplakin binding. The latter is particularly interesting as epidermal growth factor receptor (EGFR) activity can also disrupt β-catenin–cadherin binding at AJs. However, it is still far from clear how these signals might integrate into a path that results in the replacement of β-catenin with plakoglobin at mature AJs.

Like β-catenin, plakoglobin is also targeted for degradation by interaction with the Axin/APC/GSK-3β complex. Additionally, the cytoplasmic accumulation of plakoglobin results in its translocation into the nucleus, where it interacts with members of the Tcf/Lef transcription factor family. However, that is where the similarities between the two proteins' signaling ends. Although both β-catenin and plakoglobin knockout mice exhibit embryonic lethality, the phenotypes that lead to death are very different. β-Catenin-null embryos are unable to form dorsal structures and die before gastrulation due to impaired Wnt pathway signaling [23, 24], whereas plakoglobin null-mice proceed to organogenesis but die at embryonic day (E)10.5–12.5 due to severe heart defects resulting from impaired desmosome formation [25, 26]. Similarly, transgenic mice overexpressing β-catenin or plakoglobin under control of the same, skin-specific keratin 14 promoter exhibit opposite hair-growth-rate phenotypes, indicating that β-catenin and plakoglobin signaling differs [27, 28].

It has proven to be rather difficult to tease apart the direct transcriptional actions of plakoglobin from the plakoglobin-dependent modulation of β-catenin-mediated transcription, partly because the signaling that governs the interaction of plakoglobin with mature AJs is not understood. One model postulates that substitution of plakoglobin for β-catenin at AJs and in the Axin/APC/GSK-3β degradation complex frees β-catenin to translocate to the nucleus and mediate Tcf/Lef dependent transcription [21]; in this straightforward scenario plakoglobin is an indirect activator of transcription.

The function of nuclear plakoglobin is much less straightforward, however. While β-catenin interactions with Tcf/Lef family members generally result in robust DNA binding and transcriptional activation, studies suggest that plakoglobin/Tcf/Lef complexes form inefficiently, bind DNA weakly, and are less effective at activating transcription [29]. In this scenario nuclear plakoglobin accumulation has no effect on, or may even enhance (by preventing its degradation), β-catenin-mediated transcription. However, other studies suggest an alternative scenario in which plakoglobin binds Tcf (bound or not to β-catenin), but because the complex is unable to bind to DNA the β-catenin-mediated transcription is prevented [30, 31]. Finally, plakoglobin is able directly to induce Tcf/Lef-mediated transcription in β-catenin-deficient cells as well as in a transformed rat kidney epithelial cell line [32, 33]. Thus, in some circumstances plakoglobin may be able to regulate transcription directly.

Taken together, it is clear that: (i) crosstalk between β-catenin and plakoglobin signaling exists and affects gene expression; and (ii) plakoglobin itself is able to directly regulate gene expression. The cell context, including the state of differentiation and/or transformation as well as extracellular signaling events, most likely determines the ability of plakoglobin to regulate gene expression, both indirectly through β-catenin and directly. Clearly, many more investigations must be conducted in order to understand the circumstances in which plakoglobin associates with AJs and/or localizes to the nucleus, as well as how plakoglobin affects gene transcription.

8.4
p120-Catenin

The p120 catenin (p120) protein was originally identified in a screen for substrates of the Src tyrosine kinase as a 120 kDa protein, the phosphorylation of which on tyrosine residues correlated with cellular transformation [34, 35]. Cloning and analysis of the p120 gene revealed four alternately spliced exons and four transcriptional start sites that together can potentially produce 64 different isoforms of p120 [36–38]. In addition, p120 has several serine, threonine and tyrosine residues that can be phosphorylated and potentially regulate its function. The potential of this variation in protein sequence and phosphorylation status makes p120 an interesting – but also a very complex – member of the AJs. p120 is the prototypic and better-studied member of a subfamily of armadillo repeat proteins; the other members are also found at cell–cell junctions and can participate in signaling: armadillo repeat gene deleted in VeloCardioFacial syndrome (ARVCF) and δ-catenin at AJs; plakophilins at desmosomes, and p0071 at both [5, 38, 39]. Although p120 is superficially similar to β-catenin in the sense of being a member of the armadillo repeat family of proteins that interacts with cadherins at AJs, is regulated by phosphorylation, and has transcriptional regulatory functions, in reality the functions of p120 are very different from those of β-catenin, and indeed are only just beginning to be appreciated [35]. Perhaps the most important difference is that p120 has a well-established role in regulating members of the Rho family of small GTPases, which are key mediators of cytoskeletal dynamics and cadherin-mediated cell–cell adhesion [40].

It is well established that p120 binding to the JMD of the cadherin cytoplasmic tail is directly responsible for stabilizing cadherin expression at the cell surface, as well as for inducing cadherin clustering that promotes the formation of AJs [38, 41] (Figure 8.5). Unlike β-catenin's interaction with the CBD, the interaction

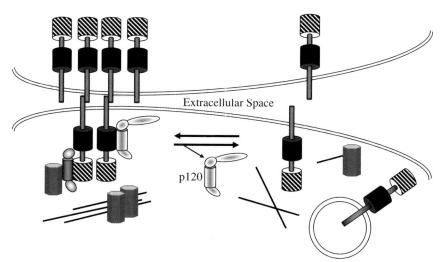

Figure 8.5 Regulation of cadherin function by p120. p120 binding to the cadherin juxtamembrane domain (JMD) results in stabilized cadherin expression at the cell surface and facilitates cadherin clustering that, along with β-catenin binding, recruitment of α-catenin, and local actin cytoskeletal rearrangement, results in the formation of mature adherens junctions (AJs). When p120 dissociates from the cadherin JMD, cadherins are destabilized and undergo rapid endocytosis, resulting in the dissolution of AJs.

between p120 and the JMD is relatively low affinity. Despite this, cadherin levels are tightly regulated by competition for the limited pool of p120. Even if β-catenin is bound to the cadherin CBD, the dissociation of p120 from the cadherin complex results in the endocytic internalization of cadherins, from which they can be targeted for lysosomal degradation. Thus, p120 is a master regulator (a rheostat) that controls cadherin stabilization and assembly into AJs.

The signaling that regulates the interaction of p120 with cadherins likely includes dynamic phosphorylation of p120 and cadherin [42]. p120 can recruit Src-family kinases, such as Fyn and Yes, and a variety of protein phosphatases, including PTPμ and SHP-1, to cadherin complexes. Indeed, only membrane-associated p120 is phosphorylated by Src. p120 is also the target of serine/threonine kinases and phosphatases, although these remain to be identified. p120 may also serve as a scaffold for these proteins, enabling them to regulate the function of other cadherin/catenin complex proteins, although the details of such interactions are at present poorly understood.

p120 does not associate with APC which suggests that, unlike β-catenin, it is not rapidly targeted for degradation [43]. Thus, when p120 is not associated with the cadherin cytoplasmic domain, it is still relatively stable and able to mediate additional signaling. One type of signaling in which p120 has a major role is regulating the activity of the Rho GTPases RhoA, Rac1, and Cdc42. These proteins are in turn involved in the stabilization of cadherins and rearrangement of the actin cytoskeleton to regulate either cell adhesion or motility. The precise mechanisms by which p120 regulates Rho GTPases are still unclear, partly because they seem to differ depending on cell context [40]. In some cells, the regulation of RhoA seems to occur in the cytoplasm, whereas in others it can occur at the membrane. When not bound to cadherins, p120 inhibits RhoA activity directly by acting as a guanine-nucleotide dissociation inhibitor (GDI). An alternative mechanism by which cadherin-associated p120 regulates RhoA involves signaling downstream of Rac1, which promotes the p120-mediated recruitment of p190RhoGAP and subsequent inhibition of RhoA activity. Rac1 and Cdc42 activation most likely occurs at the membrane by cadherin-associated p120 through p120-mediated regulation of the activity of Rho guanine nucleotide exchange factors (GEFs).

The downstream effects of Rho GTPase activation/inhibition also depend on cell context, with the outcome of p120's regulation of Rho GTPases differing depending on the type of cadherin that a cell expresses [40, 44] (Figure 8.6). At nascent AJs in E-cadherin-expressing cells, cadherin-associated p120 locally activates Rac1 and Cdc42 to promote the reorganization of the actin cytoskeleton, the maturation of AJs, and increased cell–cell adhesion. In contrast, in cells expressing mesenchymal cadherins, cadherin-associated p120 promotes Rac1 and Cdc42 activation, reorganization of the actin cytoskeleton, formation of lamellipodia and filopodia, and increased cell motility. It is not clear why p120 promotes such different cell behavior when bound to different cadherins. However, the association of p120 with mesenchymal cadherins is low affinity compared to E-cadherin, and thus mesenchymal cells exhibit a large pool of cytoplasmic p120. Cytoplasmic p120 reduces RhoA activity, possibly allowing the dissolution of stress fibers and

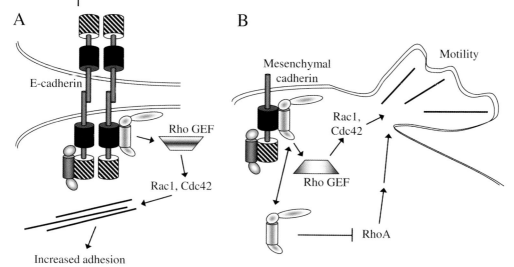

Figure 8.6 p120 regulates sessile versus motile behavior. (A) When p120 is bound to the E-cadherin juxtamembrane domain (JMD), it activates RhoGEFs that locally activate Rac1 and Cdc42, resulting in changes to the actin cytoskeleton that promote stabilization of the cadherin/catenin complex and increased cell–cell adhesion. (B) When p120 is associated with mesenchymal cadherins, increased Rac1 and Cdc42 activation via Rho GEFs stimulates rearrangement of the actin cytoskeleton and results in increased formation of lamellipodia and filopodia. In addition, the increased cytoplasmic localization of p120 in mesenchymal cells inhibits RhoA activity, which also results in actin cytoskeleton rearrangement and promotes cell motility. It is unclear if the same Rho GEFs mediate p120's action in the two scenarios.

reorganization of the actin cytoskeleton at the leading cell edges. Despite such limited understanding it is clear that, in addition to directly regulating cadherin stabilization, p120 is also a regulator of the balance between a cell's sessile versus motile behavior through its regulation of Rho GTPases.

Another signaling function of p120 is the regulation of gene transcription, particularly the convergent regulation of canonical Wnt/β-catenin target genes such as cyclin D1 and Wnt-11. Regulation of transcription by p120 occurs through its interaction with the transcription factor Kaiso [45, 46] (Figure 8.7). Kaiso recognizes two DNA sites: a sequence-specific DNA consensus site (5′CTGCNA3′) and methylated CpG dinucleotides. Many Wnt/β-catenin target genes contain the sequence-specific Kaiso binding site in their promoters, and Kaiso acts as a transcriptional repressor toward these genes, antagonizing β-catenin-mediated transcriptional activation. p120 interacts with Kaiso's zinc finger domain to prevent Kaiso-DNA binding; thus p120 antagonizes Kaiso's transcriptional repression, facilitating Wnt/β-catenin signaling.

The signals that govern the p120–Kaiso interaction and their combined regulation of transcription are largely unclear. However, the results of a recent study in *Xenopus* indicated that p120 may be stabilized by Frodo, a downstream target of Wnt/Dsh signaling [47]. This stabilization resulted in the inhibition of Kaiso-

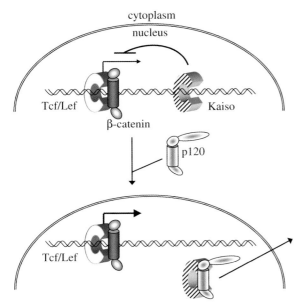

Figure 8.7 Transcriptional regulation by p120. Top: The transcription factor Kaiso can bind to specific DNA sequences located in the promoters of many genes that are targets of Wnt/β-catenin signaling. Kaiso recruits corepressors and consequently inhibits transcription of these genes. Bottom: When p120 enters the nucleus, it binds to Kaiso's zinc finger domain, thus preventing Kaiso-DNA binding. The p120–Kaiso complex can then be exported to the cytoplasm. This relieves repression of the promoters of Wnt/β-catenin target genes, allowing transcription.

mediated repression of Wnt/β-catenin target genes. Thus, p120's transcriptional activity may be regulated by Wnt signaling, another point of signaling convergence between p120 and β-catenin. The signals that govern the cytoplasmic versus nuclear localization of Kaiso are also not well characterized, although it is suggested that the cell's microenvironment plays a role in this respect. As with p120's other roles in mediating cadherin stabilization and Rho GTPase signaling, cell context undoubtedly affects the ability of p120 to regulate transcription, though the details regarding how this is achieved remain to be elucidated.

8.5
The Role of Armadillo Repeat Proteins and AJs in Cancer

The correct establishment and maintenance of AJs has a critical role during normal development and tissue morphogenesis. On the flip side, disruption of AJs in a mature organism is frequently associated with disease, specifically a loss of contact inhibition of proliferation and the development of invasive/metastatic cancer. It is well established in a variety of epithelial cancers (e.g., invasive lobular breast cancer and diffuse-type gastric adenocarcinoma) that the loss of E-cadherin expression is correlated with increased infiltrative growth and tumor cell motility [48, 49]. Importantly, it is not simply changes in adhesion, but also the changes in cadherin/catenin signaling that promote tumor cell proliferation and motility.

With respect to cadherins, altering the cellular localization of E-cadherin and/or loss of E-cadherin expression is one way to change this signaling. However this is frequently combined with the upregulation of so-called mesenchymal cadherins, the expression of which is normally restricted from epithelial tissues. In addition, the cadherin/catenin complex typically includes growth factor receptors (e.g., EGFR, or the hepatocyte growth factor (HGF) receptor c-met), the normal or aberrant signaling of which impacts upon the activity of β-catenin, plakoglobin, and p120, as well as many other kinases, phosphatases, and AJ proteins not mentioned here. All of these changes together can result in significant alterations in armadillo repeat family member function that result in tumor development and progression.

Tumorigenic β-catenin signaling primarily occurs in the nucleus via its interaction with Tcf/Lef transcription factors and subsequent inappropriate regulation of Wnt pathway genes such as c-myc and cyclin D1 [13]. Thus, anything that promotes inappropriate β-catenin nuclear localization is potentially tumorigenic. Mutations in the proteins of the Wnt signaling pathway upstream of β-catenin that result in inhibition of the Axin/APC/GSK-3β degradation machinery are well documented promoters of tumorigenesis [13]. Since β-catenin is usually sequestered at AJs by its interaction with cadherins, an additional tumorigenic mechanism is the disruption of AJs, such as by the downregulation of E-cadherin expression [48]. Likewise, the replacement of β-catenin with plakoglobin at AJs can lead to increased nuclear β-catenin and β-catenin-mediated transcription [30, 50].

Tumorigenic p120 signaling may result from any of its three main functions. Downregulation of E-cadherin expression and increased β-catenin signaling as a result of p120 JMD dissociation is an obvious mechanism by which tumorigenesis could be promoted. This β-catenin-dependent transcriptional regulation may be enhanced by p120 interactions with Kaiso (as discussed above), resulting in signaling that regulates cell proliferation. Equally important, however, is p120's modulation of Rho GTPase activity, which can directly affect tumor cell growth and motility. Recent evidence suggests that the interplay between cadherin expression and the p120-dependent regulation of Rho GTPase activity is more complicated than is suggested by the model of a simple loss of E-cadherin leading to cytoplasmic p120 localization and inhibition of RhoA. Experiments in breast and kidney cancer cell lines have revealed that, in addition to a loss of E-cadherin, an accompanying increase in mesenchymal cadherin expression, stabilized by p120-binding, and mesenchymal cadherin-associated Rac1 signaling is required for the invasiveness of E-cadherin-deficient tumor cells [44]. This mechanism for regulating invasiveness is almost certainly influenced by cell context-dependent signaling through receptor tyrosine kinases as well as integrin-mediated crosstalk between AJs and focal adhesions.

8.6
Conclusions

Although there is now a solid foundation of knowledge regarding the role of armadillo repeat family proteins in establishing and maintaining AJs, the details of these interactions – and particularly the specifics of regulatory signaling – remain far from clear. In addition, the roles that AJs play in epithelial cell function – from providing a physical link between cells, to establishing and maintaining epithelial cell polarity, to serving as a disbursement platform for a variety of extracellular and intracellular signals – are only just beginning to be understood. Complicating matters here are the differences in armadillo repeat family function that exist as a result of differences in cell context. The challenge for the future is to more completely understand how β-catenin, plakoglobin and p120 signaling differs between developmental and mature epithelial states, and how such signaling is modified by and contributes to disease processes such as tumorigenesis.

References

1. Yagi, T. and Takeichi, M. Cadherin superfamily genes: functions, genomic organization, and neurologic diversity. *Genes Dev.* 2000, *14*(10), 1169–1180.
2. Huber, A. H., Nelson, W. J., and Weis, W. I. Three-dimensional structure of the armadillo repeat region of beta-catenin. *Cell* 1997, *90*(5), 871–882.
3. Riggleman, B., Wieschaus, E., and Schedl, P. Molecular analysis of the armadillo locus: uniformly distributed transcripts and a protein with novel internal repeats are associated with a Drosophila segment polarity gene. *Genes Dev.* 1989, *3*(1), 96–113.
4. Peifer, M., Berg, S., and Reynolds, A. B. A repeating amino acid motif shared by proteins with diverse cellular roles. *Cell* 1994, *76*(5), 789–791.
5. Hatzfeld, M. The armadillo family of structural proteins. *Int. Rev. Cytol.* 1999, *186*, 179–224.
6. Roura, S., Miravet, S., Piedra, J., Garcia de Herreros, A., and Dunach, M. Regulation of E-cadherin/Catenin association by tyrosine phosphorylation. *J. Biol. Chem.* 1999, *274*(51), 36734–36740.
7. Lilien, J., Balsamo, J., Arregui, C., and Xu, G. Turn-off, drop-out: functional state switching of cadherins. *Dev. Dyn.* 2002, *224*(1), 18–29.
8. Bek, S. and Kemler, R. Protein kinase CKII regulates the interaction of beta-catenin with alpha-catenin and its protein stability. *J. Cell Sci.* 2002, *115*(Pt 24), 4743–4753.
9. Scott, J. A. and Yap, A. S. Cinderella no longer: alpha-catenin steps out of cadherin's shadow. *J. Cell Sci.* 2006, *119*(Pt 22), 4599–4605.
10. Weis, W. I. and Nelson, W. J. Re-solving the cadherin-catenin-actin conundrum. *J. Biol. Chem.* 2006, *281*(47), 35593–35597.
11. Drees, F., Pokutta, S., Yamada, S., Nelson, W. J., and Weis, W. I. Alpha-catenin is a molecular switch that binds E-cadherin-beta-catenin and regulates actin-filament assembly. *Cell* 2005, *123*(5), 903–915.
12. Yamada, S., Pokutta, S., Drees, F., Weis, W. I., and Nelson, W. J. Deconstructing the cadherin-catenin-actin complex. *Cell* 2005, *123*(5), 889–901.
13. Polakis, P. Wnt signaling and cancer. *Genes Dev.* 2000, *14*(15), 1837–1851.
14. Peifer, M. Cell adhesion and signal transduction: the Armadillo connection. *Trends Cell Biol.* 1995, *5*(6), 224–229.
15. Gottardi, C. J. and Gumbiner, B. M. Distinct molecular forms of beta-catenin are targeted to adhesive or transcriptional complexes. *J. Cell Biol.* 2004, *167*(2), 339–349.

16. Nelson, W. J. and Nusse, R. Convergence of Wnt, beta-catenin, and cadherin pathways. *Science* 2004, *303*, 1483–1487.
17. Wodarz, A., Stewart, D. B., Nelson, W. J., and Nusse, R. Wingless signaling modulates cadherin-mediated cell adhesion in Drosophila imaginal disc cells. *J. Cell Sci.* 2006, *119*(Pt 12), 2425–2434.
18. Yin, T. and Green, K. J. Regulation of desmosome assembly and adhesion. *Semin. Cell Dev. Biol.* 2004, *15*(6), 665–677.
19. Aberle, H., Schwartz, H., Hoschuetzky, H., and Kemler, R. Single amino acid substitutions in proteins of the armadillo gene family abolish their binding to alpha-catenin. *J. Biol. Chem.* 1996, *271*(3), 1520–1526.
20. Huber, O. Structure and function of desmosomal proteins and their role in development and disease. *Cell. Mol. Life Sci.* 2003, *60*(9), 1872–1890.
21. Ben-Ze'ev, A. The dual role of cytoskeletal anchor proteins in cell adhesion and signal transduction. *Ann. N. Y. Acad. Sci.* 1999, *886*, 37–47.
22. Zhurinsky, J., Shtutman, M., and Ben-Ze'ev, A. Plakoglobin and beta-catenin: protein interactions, regulation and biological roles. *J. Cell Sci.* 2000, *113*(Pt 18), 3127–3139.
23. Haegel, H., Larue, L., Ohsugi, M., Fedorov, L., Herrenknecht, K., and Kemler, R. Lack of beta-catenin affects mouse development at gastrulation. *Development* 1995, *121*(11), 3529–3537.
24. Huelsken, J., Vogel, R., Brinkmann, V., Erdmann, B., Birchmeier, C., and Birchmeier, W. Requirement for beta-catenin in anterior-posterior axis formation in mice. *J. Cell Biol.* 2000, *148*(3), 567–578.
25. Bierkamp, C., McLaughlin, K. J., Schwarz, H., Huber, O., and Kemler, R. Embryonic heart and skin defects in mice lacking plakoglobin. *Dev. Biol.* 1996, *180*(2), 780–785.
26. Ruiz, P., Brinkmann, V., Ledermann, B., Behrend, M., Grund, C., Thalhammer, C., Vogel, F., Birchmeier, C., Gunthert, U., Franke, W. W., and Birchmeier, W. Targeted mutation of plakoglobin in mice reveals essential functions of desmosomes in the embryonic heart. *J. Cell Biol.* 1996, *135*(1), 215–225.
27. Charpentier, E., Lavker, R. M., Acquista, E., and Cowin, P. Plakoglobin suppresses epithelial proliferation and hair growth in vivo. *J. Cell Biol.* 2000, *149*(2), 503–520.
28. Gat, U., DasGupta, R., Degenstein, L., and Fuchs, E. De novo hair follicle morphogenesis and hair tumors in mice expressing a truncated beta-catenin in skin. *Cell* 1998, *95*(5), 605–614.
29. Zhurinsky, J., Shtutman, M., and Ben-Ze'ev, A. Differential mechanisms of LEF/TCF family-dependent transcriptional activation by beta-catenin and plakoglobin. *Mol. Cell. Biol.* 2000, *20*(12), 4238–4252.
30. Miravet, S., Piedra, J., Castano, J., Raurell, I., Franci, C., Dunach, M., and Garcia de Herreros, A. Tyrosine phosphorylation of plakoglobin causes contrary effects on its association with desmosomes and adherens junction components and modulates beta-catenin-mediated transcription. *Mol. Cell. Biol.* 2003, *23*(20), 7391–7402.
31. Williams, B. O., Barish, G. D., Klymkowsky, M. W., and Varmus, H. E. A comparative evaluation of beta-catenin and plakoglobin signaling activity. *Oncogene* 2000, *19*(50), 5720–5728.
32. Kolligs, F. T., Kolligs, B., Hajra, K. M., Hu, G., Tani, M., Cho, K. R., and Fearon, E. R. gamma-catenin is regulated by the APC tumor suppressor and its oncogenic activity is distinct from that of beta-catenin. *Genes Dev.* 2000, *14*(11), 1319–1331.
33. Maeda, O., Usami, N., Kondo, M., Takahashi, M., Goto, H., Shimokata, K., Kusugami, K., and Sekido, Y. Plakoglobin (gamma-catenin) has TCF/LEF family-dependent transcriptional activity in beta-catenin-deficient cell line. *Oncogene* 2004, *23*(4), 964–972.
34. Reynolds, A. B., Roesel, D. J., Kanner, S. B., and Parsons, J. T. Transformation-specific tyrosine phosphorylation of a novel cellular protein in chicken cells expressing oncogenic variants of the avian cellular src gene. *Mol. Cell. Biol.* 1989, *9*(2), 629–638.
35. Reynolds, A. B. p120-catenin: Past and present. *Biochim. Biophys. Acta* 2007, *1773*(1), 2–7.

36 Aho, S., Rothenberger, K., and Uitto, J. Human p120ctn: tissue-specific expression of isoforms and molecular interactions with BP180/type XVII collagen. *J. Cell Biochem.* 1999, *73*(3), 390–399.

37 Keirsebilck, A., Bonne, S., Staes, K., van Hengel, J., Nollet, F., Reynolds, A., and van Roy, F. Molecular cloning of the human p120ctn catenin gene (CTNND1): expression of multiple alternatively spliced isoforms. *Genomics* 1998, *50*(2), 129–146.

38 Anastasiadis, P. Z. and Reynolds, A. B. The p120 catenin family: complex roles in adhesion, signaling and cancer. *J. Cell Sci.* 2000, *113*(Pt 8), 1319–1334.

39 Hatzfeld, M. Plakophilins: Multifunctional proteins or just regulators of desmosomal adhesion? *Biochim. Biophys. Acta* 2007, *1773*(1), 69–77.

40 Anastasiadis, P. Z. p120-ctn: A nexus for contextual signaling via Rho GTPases. *Biochim. Biophys. Acta* 2007, *1773*(1), 34–46.

41 Xiao, K., Oas, R. G., Chiasson, C. M., and Kowalczyk, A. P. Role of p120-catenin in cadherin trafficking. *Biochim. Biophys. Acta* 2007, *1773*(1), 8–16.

42 Alema, S. and Salvatore, A. M. p120 catenin and phosphorylation: Mechanisms and traits of an unresolved issue. *Biochim. Biophys. Acta* 2007, *1773*(1), 47–58.

43 Daniel, J. M. and Reynolds, A. B. The catenin p120(ctn) interacts with Kaiso, a novel BTB/POZ domain zinc finger transcription factor. *Mol. Cell. Biol.* 1999, *19*(5), 3614–3623.

44 Yanagisawa, M. and Anastasiadis, P. Z. p120 catenin is essential for mesenchymal cadherin-mediated regulation of cell motility and invasiveness. *J. Cell Biol.* 2006, *174*(7), 1087–1096.

45 Daniel, J. M. Dancing in and out of the nucleus: p120(ctn) and the transcription factor Kaiso. *Biochim. Biophys. Acta* 2007, *1773*(1), 59–68.

46 van Roy, F. M. and McCrea, P. D. A role for Kaiso-p120ctn complexes in cancer? *Nat. Rev. Cancer* 2005, *5*(12), 956–964.

47 Park, J. I., Ji, H., Jun, S., Gu, D., Hikasa, H., Li, L., Sokol, S. Y., and McCrea, P. D. Frodo links Dishevelled to the p120-catenin/Kaiso pathway: distinct catenin subfamilies promote Wnt signals. *Dev. Cell* 2006, *11*(5), 683–695.

48 Nollet, F., Berx, G., and van Roy, F. The role of the E-cadherin/catenin adhesion complex in the development and progression of cancer. *Mol. Cell. Biol. Res. Commun.* 1999, *2*(2), 77–85.

49 Wijnhoven, B. P., Dinjens, W. N., and Pignatelli, M. E-cadherin-catenin cell-cell adhesion complex and human cancer. *Br. J. Surg.* 2000, *87*(8), 992–1005.

50 Li, L., Chapman, K., Hu, X., Wong, A., and Pasdar, M. Modulation of the oncogenic potential of beta-catenin by the subcellular distribution of plakoglobin. *Mol. Carcinogen.* 2007, *46*(10), 824–838.

9
Signaling To and Through The Endothelial Adherens Junction
Deana M. Ferreri and Peter A. Vincent

9.1
Introduction

Endothelial cells form a monolayer that lines the entire circulation, where they perform a number of specialized functions that differ from one organ to another. This monolayer of endothelial cells is maintained by a large number of transmembrane cell adhesion proteins found at the cell–cell junction. The transmembrane proteins bind to intracellular proteins to form dynamic structures that are responsible for regulating the movement of fluid, macromolecules, and white blood cells across the endothelial barrier, as well as regulating communication between endothelial cells within the monolayer. For example, the transmembrane proteins claudin-1 and -5 bind with ZO-1, ZO-2 and 7H-6 to form tight junctions that regulate permeability (for a review, see Ref. [1]), whereas connexins bind in hexamer structures to form gap junctions that allow intercellular communication between endothelial cells [2]. Adherens junctions (AJs) are defined as a cell junction in which the cytoplasmic face of the membrane is linked to the actin cytoskeleton. This includes the interface of the plasma membrane between two cells (cell–cell) or the interface between the plasma membrane of the cell and the extracellular matrix (ECM) (cell–matrix). In this chapter, attention is focused on the endothelial cell–cell AJ and the mechanisms that regulate the assembly and disassembly of this adhesion complex.

The AJ is a complex of proteins that attaches one cell to another and serves as a point for actin attachment. The complex consists of a cadherin, a single pass transmembrane protein, which interacts indirectly with the actin cytoskeleton through intracellular binding partners called catenins. The AJ complex is found in a number of cell types including epithelial cells, where they have been extensively studied. Although there are similarities between the AJ found in endothelial and epithelial cells, there are also specific differences between the cell–cell junctions of these two cell types. For example, the prominent cadherin in endothelial cells is VE-cadherin, whereas E-cadherin is the prominent cadherin in epithelial cells. In addition, Braga et al. [3] found that VE-cadherin required Rho GTPase for the development of an AJ when expressed in Chinese hamster ovary (CHO) cells

Cell Junctions. Adhesion, Development, and Disease.
Edited by Susan E. LaFlamme and Andrew Kowalczyk
Copyright © 2008 WILEY-VCH Verlag GmbH & Co. KGaA, Weinheim
ISBN: 978-3-527-31882-7

but that this cadherin was not as sensitive to Rho GTPase in endothelial cells. Thus, cellular context is important for the function of cadherins within the AJ. For this reason, although the information in this chapter is focused on studies conducted in endothelial cells, many of the references cited relate to the large body of literature centered around other cell types.

A number of functions performed by the endothelium require the cell–cell junction to disassemble and then subsequently reassemble to re-establish a restrictive monolayer. For example, white blood cells must pass from the vascular space into tissues through the cell–cell junction as part of the inflammatory response. Investigations have shown that the AJ proteins are lost from the cell–cell junction as the leukocyte passes through the endothelial monolayer [4, 5]. Once the leukocyte has traversed through the monolayer, the endothelium re-establishes the cell–cell junction. This rapid disassembly also occurs in response to inflammatory mediators such as histamine and thrombin or angiogenic stimuli as the endothelial cells detach to migrate and proliferate. The endothelial monolayer must then re-establish an intact junction as the vessel either matures to become part of the circulation in the later stages of angiogenesis or repairs following inflammation. A number of factors, including angiopoietin-1 [6] and sphingosine-1-phosphate [7], have been implicated in monolayer restoration. This chapter will focus on studies that have enhanced the present understanding of the organization and functions of endothelial AJ components with respect to the dynamic nature of endothelial cell–cell adhesion.

9.2
Cadherins

Cadherins are a superfamily containing at least 50 members [8], all of which possess cadherin domains in their extracellular region that mediate cell–cell adhesion. The cadherin domain is characterized by a seven-stand β-sheet with the amino and carboxy termini located at opposite ends of this domain. Nollet et al. [9] divided the cadherins into five subfamilies based on the domain structure, genomic organization and phylogenetic analysis of the cadherin molecule. The classical cadherins are responsible for forming the AJs in endothelial cells. The classical cadherins have a number of identical structural features, such as five cadherin repeats in their extracellular domain, and possess a single spanning transmembrane region; in addition, all share a highly conserved cytoplasmic tail [8, 9]. These cadherins are also synthesized as precursor molecules containing signal peptide and pro-peptide regions that are both removed by proteolytic cleavage. A number of classical cadherins have been found in endothelial cells, two of which – R- and E-cadherin – have not been widely studied. Indeed, only two reports have been made of the expression of E-cadherin in endothelial cells, specifically rat lung microvascular endothelial cells [10, 11]. R-cadherin has been shown to be present in the blood vessels of the eye, where it plays an important role in guidance of the developing vascular bed in the retina [12, 13].

VE-cadherin and N-cadherin are the two major classical cadherins found in most, if not all, endothelial cells, and both of these molecules are critical for proper vascular development. Two transgenic mouse models of VE-cadherin deficiency have been developed, both embryonic lethal due to vascular defects [14, 15]. Both groups reported normal differentiation of angioblasts into endothelial cells, irregular vessel size, and endothelial cell death. Interestingly, further studies using one transgenic model revealed that, although vasculogenesis occurred, nascent vessels collapsed or disassembled in the absence of VE-cadherin [16]. Thus, it was concluded that VE-cadherin's role in vascular development is the maintenance of nascent vessels. Not surprisingly, VE-cadherin is most well known for its critical role in maintaining a restrictive endothelial monolayer [17, 18]; additionally, VE-cadherin is thought to mediate contact inhibition of endothelial cell growth via an interaction with vascular endothelial growth factor receptor-2 (VEGFR2) [19].

There is strong support for the importance of N-cadherin in the recruitment of mural cells to nascent vessels during angiogenesis, which is key to vessel maturation [20, 21]. An endothelial-specific N-cadherin knockout, similar to the VE-cadherin knockouts, was embryonic lethal due to vascular defects, including abnormally organized vessels and pericardial effusion [22]. Endothelial cell death was also observed, consistent with N-cadherin's proposed role in survival [23]. It is interesting to note that N-cadherin is thought to promote survival signaling via an interaction with fibroblast growth factor receptor-1 (FGFR1); thus, both VE- and N-cadherin take part in signaling though growth factor receptors in the endothelium. Also of interest is the fact that endothelial N-cadherin-null embryos were lethal before the investment of mural cells in the developing vasculature, implying that N-cadherin is involved in earlier events of vasculogenesis in addition to its role in endothelial–mural cell interactions leading to vessel maturation. These additional roles for N-cadherin have yet to be identified, and as yet the specific role of N-cadherin-based AJs in endothelial cell–cell adhesion remains unclear. Although both N- and VE-cadherin play a critical role in endothelial cells, attention will be focused here on AJs containing VE-cadherin, as this cadherin is specific for endothelial cells and has been implicated in the regulation of cell–cell adhesion, as well as other processes that require the assembly and disassembly of endothelial intercellular contacts.

9.2.1
VE-Cadherin Extracellular Domain

The binding and formation of a mature cell–cell junction requires that the extracellular domain of a cadherin on one cell binds to the extracellular domain of a cadherin on an adjacent cell, a process referred to as *trans* dimer formation. Once *trans* dimers are formed, the cadherins cluster to form lateral complexes or clusters that become associated with the actin cytoskeleton, thereby enhancing the strength of the cell–cell adhesion (Figure 9.1). *Trans* dimer formation is mediated by the extracellular domain, and clustering and actin attachment occur via domains in the cytoplasmic region of the cadherin.

The specifics of how the extracellular domains of cadherins interact to form *trans* dimers are the subject of controversy [24, 25]. A recent series of studies have suggested that VE-cadherin forms a hexamer consisting of three cadherins on one cell and three cadherins on an adjacent cell [26–28]. In this model, the EC1 domains are responsible for *trans* dimer formation, and the inner domains 2–5 are responsible for *cis* dimer formation. The hexamer model, developed using VE-cadherin expressed from bacteria, has been questioned by the experiments of Ahrens et al. [29], the results of whose studies have suggested that VE-cadherin uses a strand exchange model of the EC1 domains. It should be noted here that the determination of which portions of the extracellular domain are critical to adhesion will be important when designing therapeutic treatments that target cadherins. Indeed, monoclonal antibodies (mAbs) to different regions of the extracellular domain of VE-cadherin have been shown to have different functional consequences [17, 30, 31]. For example, antibodies to the EC1 region of VE-cadherin inhibits vessel formation in angiogenesis assays and disrupts endothelial barrier function, leading to increased protein permeability [30]. In contrast, antibodies to the EC4 domain do not alter the endothelial barrier properties but do prevent vessel formation in angiogenesis assays [30]. These data suggest that the latter antibody may be of therapeutic value in treating tumor growth, as it would prevent angiogenesis without the side effect of the development of pulmonary edema.

9.2.2
Interaction of Cadherin Cytoplasmic Domain and Catenins

The cytoplasmic tail of cadherin contains two important domains, the juxtamembrane domain (JMD) and the C-terminal catenin-binding domain (CBD), both of which bind to catenins to form a cadherin–catenin complex. The catenins, with the exception of α-catenin, are a family of related proteins that share a central domain known as the armadillo-repeat domain (ARM domain) [32]. This domain contains 10 to 13 armadillo (ARM) repeats, so named because of their sequence homology to the armadillo gene product first identified in *Drosophila*. The ARM repeats have been shown to create a tertiary structure consisting of a positively charged groove that binds with a negatively charged region of cadherins, although electrostatic complementarity is not a major contributor to the stability of the cadherin–catenin complex [33, 34].

9.2.2.1 VE-Cadherin Juxtamembrane Region and p120
Members of the p120 family, which includes p120, p0071, δ-catenin, plakophilins and ARVCF (armadillo repeat gene deleted VeloCardioFacial syndrome), bind to the JMD of cadherins. These proteins have greater than 45% identity in the ARM domains, and bind to cadherins via the interaction of the ARM domains with a highly conserved region in the JMD [35]. To date, only p120 (Figure 9.1) and p0071 are known to bind the JMD of VE-cadherin [36]. Using polymerase chain reaction (PCR) primers to different isoforms of p120 [37], it has been found that the 1A

and 3A isoforms correspond to the two bands that appear on Western blots of endothelial cell lysates (D. M. Ferreri and P. A. Vincent, unpublished data).

In endothelial cells, p120 has been shown to play an important role in maintaining VE-cadherin at the plasma membrane. Early studies using recombinant fragments of cadherins showed that the expression of these mutant cadherins decreased endogenous cadherin levels. For example, Iyer et al. [38] found that expression of a fragment consisting of only the JMD resulted in a decrease in the level of VE-cadherin, which could be restored by coexpressing p120 with the JMD. At this same time, Xiao et al. [39] found that mutating the p120 binding site on dominant-negative VE-cadherin mutants abrogated the ability of the mutant to cause downregulation of the endogenous cadherin. These investigators further demonstrated that the coexpression of p120 with the mutant cadherin prevented downregulation of endogenous cadherin. The dependence of VE-cadherin on p120 for retention at the plasma membrane was further confirmed using siRNA to knockdown p120 [38, 39]. The results of these two studies showed that cadherin mutants cause a downregulation of endogenous cadherins by competing for p120. It has now been found that overexpressed N-cadherin can compete with VE-cadherin for p120, resulting in a decrease in VE-cadherin. Conversely, overexpression of VE-cadherin decreases N-cadherin, and thus the levels of VE- and N-cadherin are reciprocally regulated via competition for p120 binding (D. M. Ferreri and P. A. Vincent, unpublished data).

The mechanism by which p120 controls the level of VE-cadherin is through prevention of VE-cadherin endocytosis. Endocytosed VE-cadherin is targeted to the lysosome, as was demonstrated under normal conditions by the prevention of VE-cadherin degradation with the addition of the lysosomal inhibitor chloroquine [40]. Decreasing the level of p120 with either siRNA or dominant-negative VE-cadherin enhanced VE-cadherin endocytosis, as evidenced by an increase in the number of vesicles containing VE-cadherin [39]. Cadherin endocytosis is mediated by several different endocytic pathways, including both clathrin-dependent and -independent mechanisms. Xiao et al. [41] showed that K^+ depletion and cytosol acidification – two mechanisms known to inhibit clathrin-dependent endocytosis – prevented VE-cadherin endocytosis, thus suggesting that this is a clathrin-dependent process. The interaction between clathrin and the JMD region of VE-cadherin is mediated through binding of the clathrin adaptor protein complex AP-2 (A. Kowalczyk, personal communication). In this model, p120 is thought to interfere with the interaction between AP-2 and the VE-cadherin JMD. The regulation of cadherin endocytosis by p120 occurs in a number of different cell types, and the pathways in each cell type appear to have unique aspects [42].

In addition to regulating endocytosis, p120 also serves as a scaffold for a number of proteins that regulate the assembly and disassembly of the AJ. Protein kinases, phosphatases, and small GTPases have all been shown to interact with p120 and to regulate either the phosphorylation of other AJ proteins or the actin cytoskeleton [35, 43]. Many of these studies have been performed in epithelial or fibroblast cells. For example, Wildenberg et al. [44] recently demonstrated in NIH3T3 fibroblasts that p190RhoGAP is localized to the AJ following receptor-induced increases in

Rac activity, and that this localization is mediated through p190RhoGAP interaction with p120. However, a number of searches failed to identify any studies that demonstrated a role for p120 as a scaffold specifically in endothelial cells. It has also been shown that the overexpression of p120 will decrease thrombin-induced increases in myosin light chain phosphorylation, a process mediated through the Rho-Rho Kinase pathway [38]. This suggests that p120 inhibits Rho in endothelial cells, similar to what has been found in other cell types. It has also been found that the exogenous expression of p120 increases the activity of Rac1 in endothelial cells (D. M. Ferreri and P. A. Vincent, unpublished data). Thus, although many of the properties of p120 observed with p120 overexpression in other cell types may be applicable to endothelial cells, further research is needed to determine the mechanism by which these occur, and whether specific endothelial cell functions are regulated by these p120-mediated effects.

9.2.2.2 VE-Cadherin Catenin Binding Domain

The carboxy terminus of the cadherin cytoplasmic tail contains the catenin-binding domain (CBD). This region of the cadherin mediates attachment to the actin cytoskeleton (see Figure 9.1). Attachment to the actin cytoskeleton reduces the lateral mobility of the cadherin, thereby allowing for cadherin clustering and a subsequent increase in the strength of adhesion. This was demonstrated in endothelial cells using laser tweezer capture of beads coated with the extracellular domain of VE-cadherin [45, 46]. When endothelial cells were treated with agents that disrupted actin polymerization (e.g., cytochalasin D), the lateral mobility of the beads was increased and the adhesion strength decreased. In contrast, stabilization of the actin cytoskeleton, by treatment with jasplakinolide once the beads were attached, decreased the lateral mobility and increased the adhesion strength. The immobilization of cadherins (i.e., decreased lateral mobility) has been shown to increase the number of bonds per unit surface area, resulting in an increased adhesion strength [47]. Interestingly, the lifetime of *trans* interaction events and the disassociation rate constant (K_{off}) determined using the extracellular portion of VE-cadherin were not altered by disruption of the actin cytoskeleton, which suggested that the affinity of individual VE-cadherin homophilic interactions is not dependent on cytoskeletal attachment [47]. This would also suggest that inside-out modulation of affinity – as seen in integrins – is not altered by attachment to the actin cytoskeleton, but rather that lateral clustering of VE-cadherin dimers mediated by the actin cytoskeleton is responsible for the increasing adhesion strength.

The CBD of VE-cadherin binds to either β-catenin or γ-catenin (plakoglobin) (see Figure 9.1). Plakoglobin and β-catenin share a similar structure consisting of ten ARM repeats found in the central portion of the molecule (ARM domain) that are flanked on either side by amino and carboxy domains [48]. Similar to p120, the binding of β-catenin and plakoglobin to cadherins is mediated through the ARM domains, and these two proteins bind to the CBD in a mutually exclusive fashion. In addition, the localization of β-catenin and plakoglobin at the cell periphery are dependent on the time of confluence. This was demonstrated in an

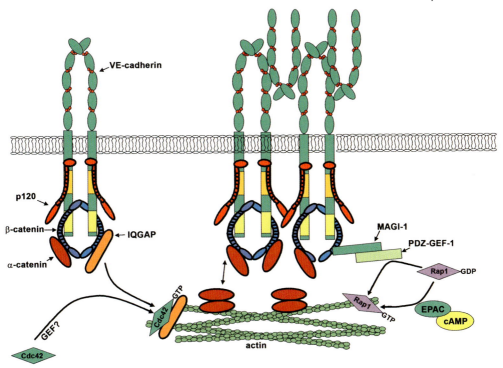

Figure 9.1 Model of the endothelial adherens junction (AJ). The AJ complex in endothelial cells consists of VE-cadherin, a single-span transmembrane protein, which binds to p120 at the juxtamembrane domain (JMD) and either β-catenin or plakoglobin at the catenin-binding domain (CBD). Indirect association of α-catenin with VE-cadherin occurs via β-catenin or plakoglobin. The means by which the VE-cadherin–catenin complex is linked to the actin cytoskeleton remain controversial. Originally, α-catenin was thought to be a direct link between β-catenin and actin; however, more recent studies have revealed that α-catenin cannot simultaneously bind β-catenin and actin. Two other possible mechanisms of how actin associates with the VE-cadherin–catenin complex may involve binding of β-catenin to either α-catenin, IQGAP, or MAGI-1. These β-catenin-associated proteins may then interact with small GTPases to control actin polymerization (see text for details).

early study by Lampugnani et al. [49], in which plakoglobin was noted to localize at the junctions of endothelial monolayers once the cells had been confluent for 48 hours, whereas β-catenin was found upon initial junction formation, even in subconfluent cells. The presence of plakoglobin in the mature cell junction has been associated with functional changes within the endothelial monolayer. Indeed, the overexpression of plakoglobin was found to enhance the barrier function [50] of human microvascular endothelial cell-1 (HMEC-1) monolayers, whereas a decrease in plakoglobin induced by antisense oligonucleotides was found to decrease resistance of the junction to disruption by shear stress [51].

The functional consequences of plakoglobin localization in the mature junction may stem from plakoglobin-mediated attachment of VE-cadherin to intermediate filaments. Desmosomal junctions are found in epithelial cells, and are composed of desmosomal cadherins (desmoglein and desmocollin) that are attached to intermediate filaments by a protein complex that includes desmoplakin and plakoglobin. Endothelial cells do not form desmosomes as they do not express desmosomal cadherins; however, Valiron et al. [52] noted that desmoplakin can be found at the endothelial cell–cell junctions. Kowalczyk et al. [53] extended these findings by showing that plakoglobin served to link VE-cadherin to desmoplakin, which in turn bound to intermediate filaments. Thus, plakoglobin may enhance barrier function and create a stronger cell junction by complexing VE-cadherin to the intermediate filament cytoskeleton. This junction of VE-cadherin–plakoglobin–desmoplakin has been termed *complexus adherens*, and its importance in vascular development has been demonstrated in a desmoplakin knockout mouse model, which showed a vascular defect [54], and by siRNA knockdown of desmoplakin, which inhibited microvessel formation in culture [55]. In the adult vasculature, the *complexus adherens* is found primarily in the endothelial cells of the lymphatic system [56].

β-Catenin is also required for the proper development of the vasculature, as conditional knockout of β-catenin in endothelial cells causes embryonic death between embryonic day (E) 11.5 and 13.5 [57]. Although endothelial cells and vessels begin to develop, vascular patterning and blood vessel strength are defective. Although the role of β-catenin has not been extensively studied in endothelial cells, β-catenin has been shown in other cell types to participate in transcriptional regulation via the TCF/LEF signaling pathway [58], and to act as a scaffolding protein. The majority of β-catenin in a cell is found at the AJ, and any free β-catenin is usually rapidly degraded through a complex series of reactions that lead to the ubiquitin–proteasome pathway. However, activation of the Wnt signaling pathway stabilizes cytoplasmic β-catenin, which then translocates into the nucleus, complexes with TCF/LEF transcription factors, and activates the TCF/LEF target genes that encode proteins involved in cell proliferation, apoptosis, and remodeling of the ECM. Studies have now suggested that series of events also takes place in endothelial cells. In addition, proteins known to participate in the canonical Wnt/β-catenin pathway are expressed in endothelial cells [59]. Venkititeswaran et al. [50] showed that the expression of stable mutants of β-catenin stimulated proliferation of HDMEC-1, whereas the inhibition of β-catenin signaling using a dominant-negative mutant of TCF-4 decreased proliferation. This finding was recently confirmed in human umbilical vein endothelial cells (HUVEC), and was extended by the observation that stable expression of Wnt resulted in the stabilization of cytosolic β-catenin and increased proliferation [60]. Stable expression of Wnt also increased cell survival and expression of the proangiogenic factor interleukin-8 [60]. On the topic of transcriptional regulation by β-catenin, it is also of note that p120 has been documented to interact with and inhibit the effects of the transcriptional repressor Kaiso, although this interaction as not yet been confirmed to occur in endothelial cells [61].

β-Catenin also acts as a scaffolding protein in endothelial cells. As mentioned above, the VEGF receptor 2 has been shown to bind to the VE-cadherin–catenin complex, and this interaction modulates VEGF signaling through VEGFR2 [19]. β-Catenin is confirmed to be required for the binding of VEGFR2 to the VE-cadherin–catenin complex upon stimulation of endothelial cells with VEGF [19], although the binding sites for this interaction have not been identified. Traditionally, β-catenin has been thought to play a major role in the attachment of cadherins to the actin cytoskeleton, serving as a bridge between cadherin and α-catenin. However, recent findings have shown that attachment of the cadherin–catenin complex to the actin cytoskeleton is more complicated than the previous models had suggested.

Two proteins known to bind to cadherin-bound β-catenin are α-catenin and IQGAP. Unlike β- and γ-catenin, α-catenin is not a member of the ARM family but is an actin-binding protein similar to vinculin [62]. IQGAP is an IQ domain-containing protein that has a RasGAP homology region but does not have GAP activity. In fact, IQGAP has the opposite effect of GTPase-activating proteins (GAPs), which decrease the activity of small GTPases by enhancing the hydrolysis of GTP to GDP; binding of the small RhoGTPases Cdc42 or Rac1 (the participation of which in endothelial AJ dynamics will be discussed later) to IQGAP inhibits their intrinsic GTPase activity, thus stabilizing them in the active GTP-bound form [63, 64]. IQGAP, similar to α-catenin, can bind to actin, and also binds a large number of kinases, phosphatases, receptors and other cytoskeletal-associated proteins, thereby implicating its involvement in a great number of signaling pathways in the cell. The means by which IQGAP and α-catenin regulate attachment of the VE-cadherin–catenin complex to the actin cytoskeleton is only now beginning to be understood.

α-Catenin binds to β-catenin and was previously thought to be the link between the actin cytoskeleton and β-catenin. Three recent studies, one of which was conducted in endothelial cells, have challenged this notion. Yamada et al. [65] and Drees et al. [66] found surprising results: binding of α-catenin to cadherin-bound β-catenin and actin was mutually exclusive. Thus, α-catenin cannot simultaneously bind to β-catenin and actin, and therefore cannot serve as a link between the actin cytoskeleton and the cadherin complex. Yamaoka-Tojo et al. [67] found that IQGAP knockdown in endothelial cells led to an increased association of α-catenin with the VE-cadherin–β-catenin complex and, interestingly, a concomitant decrease in junctional localization of VE-cadherin. These data make two important implications about the endothelial adherens junction: (i) that IQGAP is required for localization of the VE-cadherin complex at the cell–cell junction and connection to the actin cytoskeleton; and (ii) that α-catenin is not sufficient to maintain the interaction between the actin cytoskeleton and the VE-cadherin–catenin complex.

9.3
Phosphorylation and Junction Assembly/Disassembly

The phosphorylation of AJ proteins has been implicated in the processes of disassembly and assembly of cell–cell junctions (see Table 9.1). There are two postulated mechanisms by which phosphorylation can contribute to these processes:

- That the phosphorylation of AJ proteins disrupts tethering of the cadherin to the actin cytoskeleton. For example, β-catenin is known to possess two phosphorylation sites, tyrosines 142 and 654, that regulate the binding of β-catenin to α-catenin and cadherin, respectively [68]. Phosphorylation of the two tyrosines results in detachment of β-catenin from each of these two proteins, thereby disrupting cadherin linkage to the actin cytoskeleton [69]. These phosphorylation events have been shown to mediate tumor cell–cell junctions in response to Ras [70].

- That phosphorylation changes the scaffolding properties of AJ components, allowing kinases and phosphatases to be localized to the cell–cell junction and resulting in downstream changes in signaling pathways that alter cell function. For example, scaffolding of Fer and Fyn kinase by p120 has been implicated in regulating the phosphorylation state of β-catenin and, as described above, the linking of β-catenin to α-catenin and E-cadherin [70, 71].

Investigations that support each of these postulated mechanisms have implicated the phosphorylation of AJ proteins in the regulation of endothelial cell function.

9.3.1
Tyrosine Phosphorylation of Endothelial AJ Proteins

Lampungani et al. [72] provided the first evidence that phosphorylation of AJ proteins may participate in changes in the endothelial cell–cell junction. This study correlated the intensity of phosphotyrosine staining at endothelial cell–cell junctions with level of confluence, suggesting a role for phosphorylation in the assembly and/or maturity of the AJ. High levels of phosphotyrosine signal were seen at the cell–cell junctions of recently confluent, or "loose," cells as compared to long confluent cells, indicating that tyrosine phosphorylation decreases with junction maturity. Specifically, tyrosine phosphorylation of VE-cadherin, α-catenin, β-catenin, and plakoglobin was greater in recently confluent as compared to long confluent cells [72]. Since this investigation, numerous studies (see Table 9.1) have associated changes in the tyrosine phosphorylation of AJ proteins with barrier dysregulation brought about by treatment with inflammatory mediators. Inflammatory mediators and growth factors, such as tumor necrosis factor-α (TNF-α), histamine, thrombin, and VEGF, have all been shown to have deleterious effects on barrier function and to induce some degree of AJ protein phosphorylation. It should be noted that some studies employed direct measures of monolayer integrity, such as protein permeability or electrical resistance, whereas others noted AJ disassembly based on decreased localization of AJ proteins at the cell–cell junction. Importantly, these studies have correlated the phosphorylation of AJ proteins with inflammation-induced or growth factor-induced breakdown of the AJ. Interestingly, integrin engagement, as modeled by attachment of fibronectin-coated beads to endothelial monolayers, also induces phosphorylation and disassembly of the AJ, implying a negative regulation of cell–cell adhesion by cell–matrix adhesion in the endothelium [94].

The above-described studies have only shown an association between tyrosine phosphorylation of AJ proteins and decreases in barrier function/monolayer integrity. However, direct cause-and-effect relationships between AJ protein phosphorylation and changes in endothelial function will require studies that use point mutations to identify which phosphorylation sites are critical to specific endothelial cell functions. Two such studies have implicated different tyrosine residues of VE-cadherin in mediating Src kinase-induced effects on endothelial cell function [90, 91]. Src – or more generally, the Src family of kinases – have been implicated in the loss of endothelial barrier function following treatment with mediators known to disrupt monolayer integrity, such as VEGF [85, 98]. A recent study from the Cheresh laboratory [91] has identified two specific tyrosines in the cytoplasmic region of VE-cadherin, Y658 and Y731, that are phosphorylated by Src. In this study, wild-type or one of several phosphomimetic mutants of VE-cadherin was transfected into CHO cells, after which assays were performed to determine catenin association and monolayer permeability. The mutation at Y658 prevented the co-immunoprecipitation of p120 with VE-cadherin, whereas the mutation at Y731 prevented the co-immunoprecipitation of β-catenin. Both phosphomimetic mutants produced an increase in monolayer permeability, correlating VE-cadherin phosphorylation, AJ disassembly, and decreased barrier function. The cotransfection of CHO cells with constitutively active Src and wild-type VE-cadherin resulted in phosphorylation of these two sites, providing further support for their importance. A caveat regarding these data is that the studies were conducted in CHO cells rather than endothelial cells and, as shown by Braga et al. [3], VE-cadherin may have different responses in CHO cells as compared to endothelial cells. In contrast to this study, Wallez et al. [90] provided evidence that only Y685 in VE-cadherin is phosphorylated by Src. Using purified active Src kinase and the cytoplasmic tail of VE-cadherin, these investigators performed *in-vitro* phosphorylation assays which found that Src directly phosphorylated VE-cadherin at Y685. They then went on to show that Y685 is the only tyrosine residue phosphorylated on VE-cadherin when endothelial cells are transfected with active Src or treated with VEGF. Interestingly, Y685 is the site that mediates binding of VE-cadherin to C-terminal Src kinase (Csk), a protein that performs an inhibitory phosphorylation of Src kinase, and association of Csk with VE-cadherin is dependent on phosphorylation at Y685 [99]. The discrepancy with respect to sites phosphorylated on VE-cadherin by active Src suggests that further studies are warranted to clarify the role of tyrosine phosphorylation in VE-cadherin function.

Consistent with a model in which tyrosine phosphorylation of AJ proteins promotes AJ disassembly, tyrosine phosphatase activity has been implicated in the formation and maintenance of barrier function. Several phosphatases have been found which localize to the endothelial cell–cell junction, including PTPμ, SHP-2,

Table 9.1 Phosphorylation of adherens junction proteins.

Mediator	Reference	Residue[b]	VE-cadherin	β-Catenin	Plakoglobin	p120	Function	Signaling pathway implicated/Notes
Cell confluence	72	Y	X	X	–	X	–	–
Angiogenic versus quiescent	73	Y	X	–	–	–	–	Increased VE-cadherin-P in angiogenic vascular beds (ovary, uterus)
Histamine	74	Y	X	X	X	–	Permeability	–
	75	Y	X	X	X	–	Permeability	Cell confluence-dependent
	76	S/T	–	–	–	Dephos	–	–
PMN	77	Y	X	X	–	–	Permeability	–
TNF-α	78	Y	X	–	–	–	–	Src
	79	Y	X	–	–	–	–	NADPH, PAK1, JNK
	80	Y	X	X	X	X	Permeability	Src family
Antibodies to VE-cadherin	81	Y	–	X	–	–	–	Pyk2, Rac1, ROS
PAF	82	Y	X	–	–	–	–	–

9.3 Phosphorylation and Junction Assembly/Disassembly

Stimulus	Ref.	Residue[b]	VE-cadherin	β-catenin	Plakoglobin	p120		Effect	Kinase/Notes
VEGF	83	Y	X	X	–	–	X	–	–
	84	Y	X	–	–	–	–	–	Src
	85	Y	X	X	–	–	–	Permeability	Src *in vivo*
	86	Y	–	X	–	–	–	Permeability	PAF *in vivo*
	87	Y	X	–	X	–	–	–	SHC
	88	S/T	–	–	–	Dephos	–	–	–
	89[a]	S	X	–	–	–	–	Permeability	PAK1; endocytosis
	90[a]	Y	X	–	–	–	–	Migration	Src
	91[a]	Y	X	–	–	–	–	Permeability	Src
	73	Y	X	–	–	–	–	–	Src
Thrombin	92	S/T	Dephos	–	–	–	X	Permeability	PKC
	93	Y	X	X	X	X	–	–	SHP-2
Integrin engagement	94	Y	–	X	–	X	–	–	Src
Tumor cell adhesion	95	Y	X	–	–	–	–	–	–
Fluid shear stress	96	Y	–	X	–	–	–	–	SHP-2
Mannitol (BBB)	97	Y	–	X	–	–	–	–	Src

a Studies that identified specific residues important to the regulation of junctional integrity. See text for details.
b Y = tyrosine; S = serine; T = threonine. X indicates which molecule (VE-cadherin, β-catenin, plakoglobin, or p120) was phosphorylated. Dephos = dephosphorylation.

VE-PTP, and DEP-1. The original studies that implicated phosphatases relied on the use of global inhibitors of tyrosine phosphatase activity, such as diperoxovanadate (DPV) [100–102]. More recently, connections have been made between specific phosphatases and AJ protein phosphorylation state. The receptor protein tyrosine phosphatase PTPµ localizes to endothelial cell–cell junctions and coprecipitates with VE-cadherin. The overexpression of wild-type PTPµ decreased both VE-cadherin tyrosine phosphorylation and the permeability of monolayers to albumin, whereas a catalytically inactive mutant increased permeability [103]. SHP-2 associates specifically with β-catenin, and is most likely responsible for maintaining β-catenin in its dephosphorylated state [93, 96]. VE-PTP is a receptor protein tyrosine phosphatase that is expressed exclusively in the endothelium [104], associates extracellularly with VE-cadherin, promotes a restrictive monolayer, and dephosphorylates VE-cadherin [105]. A transgenic mouse model expressing a truncated, non-membrane-targeted mutant form of VE-PTP has been developed [106]. These embryos are embryonic lethal due to cardiac and vascular defects, and show striking resemblance to the VE-cadherin-deficient mouse model [15, 16], as the initial vasculogenesis occurs normally but nascent vessels are not subsequently maintained. Therefore, it is likely that regulation of the VE-cadherin phosphorylation state by VE-PTP is a factor in maintenance of the developing vasculature. Density-enhanced phosphatase 1 (DEP-1) is a protein tyrosine phosphatase that is upregulated with increasing levels of cell confluence, and is implicated in vasculogenesis and angiogenesis [107, 108]. The expression of DEP-1 in the endothelium is positively regulated by VE-cadherin expression, and plays a part in VE-cadherin-induced contact inhibition of embryonic cell growth via dephosphorylation of VEGFR2 [19].

9.3.2
Serine/Threonine Phosphorylation of Endothelial AJ Proteins

Although not as abundant as the tyrosine phosphorylation data, limited information is available on the serine/threonine phosphorylation of AJ proteins in endothelial cells in response to inflammatory mediators. Similar to many studies on tyrosine phosphorylation, the bulk of the literature to date has shown a correlation between changes in serine/threonine phosphorylation of AJ proteins and changes in AJ protein localization with associated changes in barrier function. For example, VEGF and histamine – both of which decrease barrier function – have been reported to induce dephosphorylation of p120 on serine/threonine residues [76, 88]. Treatment with thrombin – a mediator having well-documented barrier disruptive effects including the formation of intercellular gaps and central actin stress fibers – results in phosphorylation of p120, but in dephosphorylation of VE-cadherin and β-catenin on serine/threonine residues via a protein kinase C (PKC)-dependent pathway [92]. Today, research groups are beginning to use recombinant proteins that target specific serine residues to determine if these phosphorylation events indeed contribute to the observed changes in function. Gavard and Gutkind [89] have conducted such a study and identified Ser665 of VE-cadherin as an important residue that regulates VE-cadherin endocytosis following VEGF treatment. VEGF-induced VE-cadherin endocytosis was dependent on Src-mediated phosphorylation of the Rac GEF Vav2 and Rac activation, and VE-cadherin Ser665 was found to be phosphorylated by PAK (p21-activated kinase), a downstream effector of Rac. A non-phosphorylatable VE-cadherin mutant displayed an attenuated permeability response to VEGF. Notably, this has been the only study [89] to identify a direct link between VE-cadherin phosphorylation and barrier function in endothelial cells.

9.4
Small GTPases and Junction Assembly

Small GTPases act as molecular switches that stimulate downstream signals which control cellular processes [109]. Several subfamilies of GTPases have been identified, including Rho, Ras, Rab, Arf, and Ran. Although there are several sets of these proteins, each is controlled similarly, existing in either a GDP-bound inactive state or a GTP-bound active state. A guanine nucleotide exchange factor (GEF) dissociates the GDP from the G-protein to allow GTP binding, producing the active form of the G-protein. The inactivation of the small G-protein is controlled by GTPase-activating proteins (GAPs) that enhance the hydrolysis of GTP to GDP, thus producing the inactive form of the small G-protein [109]. GTPases perform a wide array of functions, some of which involve regulation of, or by, the AJ.

9.4.1
Rho GTPases

The small GTPases Rho, Rac, and Cdc42, are signaling molecules that affect cell morphology, adhesion and motility, primarily via the regulation of actin dynamics [110]. In short, Rho activation produces central actin stress fibers, which are associated with cell contraction via actin–myosin interaction; Rac activation produces lamellipodia, which are characterized by increased peripheral actin assembly and are often associated with cell motility; Cdc42 activation produces filopodia, cytoskeletal processes projecting out from the cell border. Given that the AJ is intimately associated with the actin cytoskeleton, it is not surprising that the Rho-family GTPases are involved in assembly/disassembly of the AJ and can be directly activated by cadherin engagement.

The association between the AJ and Rho GTPases in endothelial cells was first implied by studies that documented the effects of Rho, Rac, and Cdc42 activation on endothelial barrier function. A specific link between the AJ and the activity of small GTPases has been made in studies showing that barrier dysfunction induced by recombinant forms of Rac and Rho or inhibition of small GTPases correspond with decreased junctional localization of VE-cadherin [111, 112]. A number of studies have also shown that signaling via Rho GTPases participates in the barrier-disruptive effects of several mediators, including thrombin [113–116], TNF-α [112, 117], histamine [111, 116], and transforming growth factor-β (TGF-β) [118]. Wojciak-Stothard and Ridley have recently provided a detailed review regarding the regulation of endothelial barrier function by Rho, Rac, and Cdc42 [119], in which many of the aforementioned studies are discussed, and a table summarizes the involvement of Rho GTPases mediator-induced alterations in endothelial permeability.

Several studies are in agreement that a baseline threshold of Rac activation is required for control levels of monolayer integrity to be maintained [45, 46, 111]; this is a delicate balance as both dominant-negative and constitutively active Rac increase endothelial monolayer permeability and are associated with a loss of VE-cadherin at the cell–cell junction [111, 120]. A connection between Rac1 activity and VE-cadherin was made by showing that steady-state Rac1 activity was increased in VE-cadherin-null endothelial cells when VE-cadherin was re-expressed [121]. The increase in Rac1 activity was associated with an increase in the localization of Tiam, a Rac GEF, with the VE-cadherin–catenin complex, suggesting that recruitment of this GEF may be the mechanism of activation of Rac by VE-cadherin. In a series of reports, Waschke et al. [45, 46] directly tested VE-cadherin-dependent adhesion using laser tweezers to displace beads coated with the extracellular domain of VE-cadherin from endothelial monolayers. The inhibition of Rac1 by *Clostridium sordelli* lethal toxin decreased the percentage of beads that resisted displacement by the laser tweezers, thereby implicating Rac in maintaining basal VE-cadherin-mediated adhesion [45, 46]. Interestingly, inhibition of the Rho effector Rho kinase did not decrease VE-cadherin-dependent adhesion, which suggested that – unlike Rac – a threshold Rho GTPase activity is not required for basal levels of VE-cadherin-mediated adhesion [45, 46]. This is consistent with other studies in which expression of dominant-negative Rho did not alter baseline permeability [111, 120].

The addition [120] or overexpression [111, 118] of active Rac or active Rho resulted in increased monolayer permeability and an associated loss of AJ proteins from the cell–cell junction. In contrast to these findings, Waschke et al. [122] observed a stabilization of the microvascular barrier when Rac1 and Cdc42 were simultaneously activated using *Escherichia coli* cytotoxic necrotizing factor (CNF-1). These investigators noted an increase in cortical actin band size following increased Rac and Cdc42 activity, but did not observe any alteration in VE-cadherin-mediated adhesion. However, this may have been due to the method used to assess VE-cadherin-dependent adhesion. As described above, beads coated with the extracellular domain of VE-cadherin were considered tightly bound when they resisted lateral displacement by laser tweezers at a single laser setting. This assay would detect losses in adhesion strength when VE-cadherin attachment could no longer prevent displacement by the laser tweezer. However, an increase in strength would not be detected, as both control cells and cells with increased VE-cadherin-mediated adhesion would inhibit displacement of the VE-cadherin-coated beads by the laser tweezer.

The simultaneous activation of Rac and Cdc42 was also investigated by Birukov et al. [123], who showed that cotransfection of constitutively active V12Rac and

L61Cdc42 increased cortical actin, in agreement with the results of Waschke et al. The finding that active Rac decreases junctional integrity, but that simultaneous activation of Rac and Cdc42 increases junctional integrity, suggests two different hypotheses. The first hypothesis is that junctional integrity is dependent on both the level of Rac activity as well as localized control. For example, CNF-1 treatment may have produced a more subtle and more localized increase in Rac activation that was barrier-protective as compared to the overexpression of constitutively active Rac. This hypothesis is supported by the findings that active Rac1 is required for basal barrier integrity [111] and VE-cadherin-mediated adhesion [45, 46], as discussed above. Recent data from the Malik laboratory may support the second hypothesis: that Cdc42 activity was responsible for the CNF-1-induced barrier stabilization – that is, that Cdc42 plays a more prominent role than Rac1 in the regulation of junctional integrity. Studies from this laboratory have also shown that the overexpression of either wild-type VE-cadherin or a mutant consisting of only the cytoplasmic domain of VE-cadherin results in the activation of Cdc42 [124, 125]. This finding led the authors to conclude that the disruption of cell–cell junctions may result in the release of VE-cadherin from the junction where it initiates activation of Cdc42, which in turn promotes the reformation of the junction (see Figure 9.1) [126]. This is consistent with the finding that Cdc42 activation plays a role in the reannealing of the cell–cell junctions in endothelial cells following thrombin treatment [127]. Interestingly, active Cdc42 does not appear to be required for basal barrier function, as dominant-negative Cdc42 did not increase permeability in an isolated lung system [124] or cultured endothelial cells [111].

Expression of the cytoplasmic domain of VE-cadherin also increased the interaction between α-catenin with β-catenin [124]. The increase in the association of α- and β-catenin was due to the increased activity of Cdc42 activity, as dominant-negative Cdc42 inhibited this response and overexpression of constitutively active Cdc42 in endothelial cells enhanced the binding of α-catenin with β-catenin [124]. An increased α-catenin binding to β-catenin with Cdc42 activation in this model may involve IQGAP. Fukata et al. [128] showed that binding of active Cdc42, as well as active Rac, to IQGAP will inhibit the binding of IQGAP to β-catenin, and when this occurs there is an increase in α-catenin–β-catenin binding. This finding is consistent with competition between IQGAP and α-catenin for the same binding site on β-catenin. The finding of Yamaoka-Tojo et al. [67], however, would suggest that the interaction of IQGAP and Cdc42 plays a greater role than simply enhancing the binding of β-catenin to α-catenin. As noted previously, decreasing the levels of IQGAP by siRNA resulted in an increase in the binding of α-catenin to β-catenin and a decrease in the localization of VE-cadherin at the cell–cell junction [67]. Thus, simply enhancing the binding of α-catenin is not sufficient to enhance barrier function, which suggests that the interaction of IQGAP and Cdc42 regulates other processes other than simply freeing up β-catenin for α-catenin. These data demonstrate that the interaction of the cadherin–catenin complex with the actin cytoskeleton is more complicated than previously thought, and that further research in this area is warranted.

9.4.2
Rap1 GTPase

Another small GTPase that has more recently been credited with regulating endothelial permeability – and in particular VE-cadherin-mediated adhesion – is Rap1, a member of the Ras family of small GTPases [129]. The importance of Rap1 was first noted in studies investigating the role of cAMP in the maintenance of endothelial barrier function. Increases in cAMP have long been known to increase the barrier function of endothelial monolayers, and many investigations have tried to establish that protein kinase A would be part of the signaling pathway leading to enhanced barrier function. In late 2004, two laboratories reported studies in which a modified cAMP [8-(4-chlorophenylthio)-2'-O-methyladenosine-3',5'-cyclic monophosphate; O-me-cAMP] was used. This is a form of cAMP that activates a Rap1 GEF known as Epac, but does not activate PKA [130, 131]. The addition of O-me-cAMP was shown to activate Rap1, to stimulate actin remodeling (specifically increased cortical actin), and to promote the junctional localization of AJ proteins [130, 131]. Using substrates coated with the extracellular domain of VE-cadherin, activation of Epac through the addition of O-me-cAMP was shown to increase VE-cadherin-dependent adhesion [131]. Using this same assay, constitutively active Rap1 increased VE-cadherin-dependent adhesion, whereas decreasing Rap1 activity via expression of a RapGAP decreased VE-cadherin-mediated adhesion.

Further support that Rap1 regulates barrier function through VE cadherin was provided by the observation that the activation of Epac-Rap1 did not enhance barrier properties in VE-cadherin-null endothelial cells, and that the restoration of VE-cadherin expression also restored the Rap1-mediated barrier enhancement [132]. Subsequent studies have confirmed and extended these findings by showing that Rap1 activation improves basal barrier function [130–133], which may be important in the barrier-protective effect of the lipid mediators prostaglandin E_2 (PGE2) and prostacyclin (PGI2) that activate the Epac-Rap1 pathway [133]. In addition, Rap1 activation attenuates barrier dysfunction mediated by thrombin [130, 131, 133]. This response may involve crosstalk between Rap1 and the Rho family GTPases in the endothelium, as Rap1 activation attenuated Rho activity [130, 133] in the context of thrombin treatment and promoted activation of Rac by PGE2 [133].

A number of studies have demonstrated that the engagement of cadherins can trigger the activation of small GTPases [134], though these data are scarce in endothelial cells. In an elegant study, Sakurai et al. [135] demonstrated that VE-cadherin homophilic binding can initiate outside-in signals that activate Rap1. By using fluorescence resonance energy transfer (FRET) to monitor Rap1 activity, these investigators found that formation of endothelial cell–cell junctions increased Rap1 activity. Knockdown of VE-cadherin by siRNA decreased the amount of FRET, thus demonstrating that VE-cadherin contributed to Rap1 activation. In addition, the treatment of endothelial cells with chimeric proteins consisting of the extracellular domain of VE-cadherin and the Fc region of IgG caused a transient increase in Rap1 activity which was not observed with addition of the Fc region alone. The activation of Rap1 was dependent upon the localization of

MAGI-1 to the AJ via binding to β-catenin. The results of this study demonstrated that VE-cadherin – similar to E-, N-, and C-cadherin – can induce outside-in signals that alter the actin cytoskeleton, and may also serve to alter cell function.

9.5
Conclusions

The above discussion highlights the numerous signal transduction pathways that can act through, be initiated by, or act directly on, the VE-cadherin-mediated AJ in endothelial cells. Although it is clear that the association between the actin cytoskeleton and the AJ is important, the nature of the link between these is not understood. As shown in Figure 9.1, numerous adaptor molecules (IQGAP, MAGI-1, α-catenin) interact with the actin cytoskeleton either directly or indirectly through the small GTPases Rap1, Rac1 and Cdc42 known to regulate actin cytoskeletal dynamics. These adaptor molecules also interact with β-catenin. Yet, determining how these molecules function together to regulate the AJ and ultimately endothelial cell–cell adhesion will require further investigation. In addition, determining how the phosphorylation of specific serine, threonine, and tyrosine residues on VE-cadherin, p120, and β-catenin alters the scaffolding properties of these molecules – not only with each other but also with signaling molecules such as kinases and phosphatases – will increase the present understanding of the dynamic properties of endothelial cell–cell adhesion.

One final area not discussed here is how the VE-cadherin–catenin complex interacts with other adhesion proteins. For example, nectins are transmembrane proteins that mediate cell–cell adhesion and also form a complex that attaches to the actin cytoskeleton by complexing with afadin; they have also been implicated in contributing to AJ formation and function [136, 137]. Rap1 is activated by engagement of nectin. Although endothelial cells express nectin, as yet no studies have reported the function of this protein in regulating endothelial cell–cell adhesion. Hence, further investigation of the nectin–afadin complex is warranted as it may play a part in the regulation of the AJ and thus participate in the regulation of endothelial cell functions that are mediated by, or require changes in, cell–cell adhesion.

References

1 Bazzoni G. Endothelial tight junctions: permeable barriers of the vessel wall 2. *Thromb. Haemost.* **95**: 36–42, 2006.
2 Herve JC, Bourmeyster N, Sarrouilhe D and Duffy HS. Gap junctional complexes: from partners to functions. *Prog. Biophys. Mol. Biol.* **94**: 29–65, 2007.
3 Braga VM, Del MA, Machesky L and Dejana E. Regulation of cadherin function by Rho and Rac: modulation by junction maturation and cellular context. *Mol. Biol. Cell* **10**: 9–22, 1999.
4 Shaw SK, Bamba PS, Perkins BN and Luscinskas FW. Real-time imaging of vascular endothelial-cadherin during

leukocyte transmigration across endothelium. *J. Immunol.* 167: 2323–2330, 2001.

5 Allport JR, Muller WA and Luscinskas FW. Monocytes induce reversible focal changes in vascular endothelial cadherin complex during transendothelial migration under flow. *J. Cell Biol.* 148: 203–216, 2000.

6 Baffert F, Le T, Thurston G and McDonald DM. Angiopoietin-1 decreases plasma leakage by reducing number and size of endothelial gaps in venules. *Am. J. Physiol. Heart Circ. Physiol.* 290: H107–H118, 2006.

7 Xu M, Waters CL, Hu C, Wysolmerski RB, Vincent PA and Minnear FL. Sphingosine 1-phosphate rapidly increases endothelial barrier function independently of VE-cadherin but requires cell spreading and Rho kinase. *Am. J. Physiol. Cell Physiol.* 293(4): C1309–C1318, 2007.

8 Angst BD, Marcozzi C and Magee AI. The cadherin superfamily: diversity in form and function. *J. Cell Sci.* 114: 629–641, 2001.

9 Nollet F, Kools P and van RF. Phylogenetic analysis of the cadherin superfamily allows identification of six major subfamilies besides several solitary members. *J. Mol. Biol.* 299: 551–572, 2000.

10 Adkison JB, Miller GT, Weber DS, Miyahara T, Ballard ST, Frost JR and Parker JC. Differential responses of pulmonary endothelial phenotypes to cyclical stretch. *Microvasc. Res.* 71: 175–184, 2006.

11 Quadri SK, Bhattacharjee M, Parthasarathi K, Tanita T and Bhattacharya J. Endothelial barrier strengthening by activation of focal adhesion kinase. *J. Biol. Chem.* 278: 13342–13349, 2003.

12 Dorrell MI, Otani A, Aguilar E, Moreno SK and Friedlander M. Adult bone marrow-derived stem cells use R-cadherin to target sites of neovascularization in the developing retina. *Blood* 103: 3420–3427, 2004.

13 Dorrell MI, Aguilar E and Friedlander M. Retinal vascular development is mediated by endothelial filopodia, a preexisting astrocytic template and specific R-cadherin adhesion. *Invest. Ophthalmol. Vis. Sci.* 43: 3500–3510, 2002.

14 Gory-Faure S, Prandini MH, Pointu H, Roullot V, Pignot-Paintrand I, Vernet M and Huber P. Role of vascular endothelial-cadherin in vascular morphogenesis. *Development* 126: 2093–2102, 1999.

15 Carmeliet P, Lampugnani MG, Moons L, Breviario F, Compernolle V, Bono F, Balconi G, Spagnuolo R, Oostuyse B, Dewerchin M, Zanetti A, Angellilo A, Mattot V, Nuyens D, Lutgens E, Clotman F, De Ruiter MC, Gittenberger-de GA, Poelmann R, Lupu F, Herbert JM, Collen D and Dejana E. Targeted deficiency or cytosolic truncation of the VE-cadherin gene in mice impairs VEGF-mediated endothelial survival and angiogenesis. *Cell* 98: 147–157, 1999.

16 Crosby CV, Fleming PA, Argraves WS, Corada M, Zanetta L, Dejana E and Drake CJ. VE-cadherin is not required for the formation of nascent blood vessels but acts to prevent their disassembly. *Blood* 105: 2771–2776, 2005.

17 Corada M, Liao F, Lindgren M, Lampugnani MG, Breviario F, Frank R, Muller WA, Hicklin DJ, Bohlen P and Dejana E. Monoclonal antibodies directed to different regions of vascular endothelial cadherin extracellular domain affect adhesion and clustering of the protein and modulate endothelial permeability. *Blood* 97: 1679–1684, 2001.

18 Corada M, Mariotti M, Thurston G, Smith K, Kunkel R, Brockhaus M, Lampugnani MG, Martin-Padura I, Stoppacciaro A, Ruco L, McDonald DM, Ward PA and Dejana E. Vascular endothelial-cadherin is an important determinant of microvascular integrity in vivo. *Proc. Natl. Acad. Sci. USA* 96: 9815–9820, 1999.

19 Grazia LM, Zanetti A, Corada M, Takahashi T, Balconi G, Breviario F, Orsenigo F, Cattelino A, Kemler R, Daniel TO and Dejana E. Contact inhibition of VEGF-induced proliferation requires vascular endothelial cadherin, beta-catenin, and the phosphatase DEP-1/CD148. *J. Cell Biol.* 161: 793–804, 2003.

20 Paik JH, Skoura A, Chae SS, Cowan AE, Han DK, Proia RL and Hla T. Sphingosine 1-phosphate receptor regulation of N-cadherin mediates vascular stabilization. *Genes Dev.* 18: 2392–2403, 2004.

21 Tillet E, Vittet D, Feraud O, Moore R, Kemler R and Huber P. N-cadherin deficiency impairs pericyte recruitment, and not endothelial differentiation or sprouting, in embryonic stem cell-derived angiogenesis. *Exp. Cell Res.* 310: 392–400, 2005.

22 Luo Y and Radice GL. N-cadherin acts upstream of VE-cadherin in controlling vascular morphogenesis. *J. Cell Biol.* 169: 29–34, 2005.

23 Erez N, Zamir E, Gour BJ, Blaschuk OW and Geiger B. Induction of apoptosis in cultured endothelial cells by a cadherin antagonist peptide: involvement of fibroblast growth factor receptor-mediated signalling. *Exp. Cell Res.* 294: 366–378, 2004.

24 Troyanovsky S. Cadherin dimers in cell-cell adhesion. *Eur. J. Cell Biol.* 84: 225–233, 2005.

25 Leckband D and Prakasam A. Mechanism and dynamics of cadherin adhesion. *Annu. Rev. Biomed. Eng.* 8: 259–287, 2006.

26 Al-Kurdi R, Gulino-Debrac D, Martel L, Legrand JF, Renault A, Hewat E and Venien-Bryan C. A soluble VE-cadherin fragment forms 2D arrays of dimers upon binding to a lipid monolayer. *J. Mol. Biol.* 337: 881–892, 2004.

27 Hewat EA, Durmort C, Jacquamet L, Concord E and Gulino-Debrac D. Architecture of the VE-cadherin hexamer. *J. Mol. Biol.* 365: 744–751, 2007.

28 Lambert O, Taveau JC, Him JL, Al KR, Gulino-Debrac D and Brisson A. The basic framework of VE-cadherin junctions revealed by cryo-EM. *J. Mol. Biol.* 346: 1193–1196, 2005.

29 Ahrens T, Lambert M, Pertz O, Sasaki T, Schulthess T, Mege RM, Timpl R and Engel J. Homoassociation of VE-cadherin follows a mechanism common to 'classical' cadherins. *J. Mol. Biol.* 325: 733–742, 2003.

30 Corada M, Zanetta L, Orsenigo F, Breviario F, Lampugnani MG, Bernasconi S, Liao F, Hicklin DJ, Bohlen P and Dejana E. A monoclonal antibody to vascular endothelial-cadherin inhibits tumor angiogenesis without side effects on endothelial permeability. *Blood* 100: 905–911, 2002.

31 Liao F, Doody JF, Overholser J, Finnerty B, Bassi R, Wu Y, Dejana E, Kussie P, Bohlen P and Hicklin DJ. Selective targeting of angiogenic tumor vasculature by vascular endothelial-cadherin antibody inhibits tumor growth without affecting vascular permeability. *Cancer Res.* 62: 2567–2575, 2002.

32 Hatzfeld M. The armadillo family of structural proteins. *Int. Rev. Cytol.* 186: 179–224, 1999.

33 Huber AH, Nelson WJ and Weis WI. Three-dimensional structure of the armadillo repeat region of β-catenin. *Cell* 90: 871–882, 1997.

34 Huber AH, Stewart DB, Laurents DV, Nelson WJ and Weis WI. The cadherin cytoplasmic domain is unstructured in the absence of β-catenin. A possible mechanism for regulating cadherin turnover. *J. Biol. Chem.* 276: 12301–12309, 2001.

35 Anastasiadis PZ and Reynolds AB. The p120 catenin family: complex roles in adhesion, signaling and cancer. *J. Cell Sci.* 113 (Pt 8): 1319–1334, 2000.

36 Calkins CC, Hoepner BL, Law CM, Novak MR, Setzer SV, Hatzfeld M and Kowalczyk AP. The Armadillo family protein p0071 is a VE-cadherin- and desmoplakin-binding protein. *J. Biol. Chem.* 278: 1774–1783, 2003.

37 Keirsebilck A, Bonne S, Staes K, van HJ, Nollet F, Reynolds A and van RF. Molecular cloning of the human p120ctn catenin gene (CTNND1): expression of multiple alternatively spliced isoforms. *Genomics* 50: 129–146, 1998.

38 Iyer S, Ferreri DM, DeCocco NC, Minnear FL and Vincent PA. VE-cadherin-p120 interaction is required for maintenance of endothelial barrier function. *Am. J. Physiol. Lung Cell. Mol. Physiol.* 286: L1143–L1153, 2004.

39 Xiao K, Allison DF, Buckley KM, Kottke MD, Vincent PA, Faundez V and

Kowalczyk AP. Cellular levels of p120 catenin function as a set point for cadherin expression levels in microvascular endothelial cells. *J. Cell. Biol.* 163: 535–545, 2003.

40. Xiao K, Allison DF, Kottke MD, Summers S, Sorescu GP, Faundez V and Kowalczyk AP. Mechanisms of VE-cadherin processing and degradation in microvascular endothelial cells. *J. Biol. Chem.* 278: 19199–19208, 2003.

41. Xiao K, Garner J, Buckley KM, Vincent PA, Chiasson CM, Dejana E, Faundez V and Kowalczyk AP. p120-Catenin regulates clathrin-dependent endocytosis of VE-cadherin. *Mol. Biol. Cell* 16: 5141–5151, 2005.

42. Xiao K, Oas RG, Chiasson CM and Kowalczyk AP. Role of p120-catenin in cadherin trafficking. *Biochim. Biophys. Acta* 1773: 8–16, 2007.

43. Alema S and Salvatore AM. p120 catenin and phosphorylation: Mechanisms and traits of an unresolved issue. *Biochim. Biophys. Acta* 1773: 47–58, 2007.

44. Wildenberg GA, Dohn MR, Carnahan RH, Davis MA, Lobdell NA, Settleman J and Reynolds AB. p120-catenin and p190RhoGAP regulate cell-cell adhesion by coordinating antagonism between Rac and Rho. *Cell* 127: 1027–1039, 2006.

45. Waschke J, Drenckhahn D, Adamson RH and Curry FE. Role of adhesion and contraction in Rac 1-regulated endothelial barrier function in vivo and in vitro. *Am. J. Physiol. Heart Circ. Physiol.* 287: H704–H711, 2004.

46. Waschke J, Baumgartner W, Adamson RH, Zeng M, Aktories K, Barth H, Wilde C, Curry FE and Drenckhahn D. Requirement of Rac activity for maintenance of capillary endothelial barrier properties. *Am. J. Physiol. Heart Circ. Physiol.* 286: H394–H401, 2004.

47. Baumgartner W, Schutz GJ, Wiegand J, Golenhofen N and Drenckhahn D. Cadherin function probed by laser tweezer and single molecule fluorescence in vascular endothelial cells. *J. Cell Sci.* 116: 1001–1011, 2003.

48. Zhurinsky J, Shtutman M and Ben-Ze'ev A. Plakoglobin and β-catenin: protein interactions, regulation and biological roles. *J. Cell Sci.* 113 (Pt 18): 3127–3139, 2000.

49. Lampugnani MG, Corada M, Caveda L, Breviario F, Ayalon O, Geiger B and Dejana E. The molecular organization of endothelial cell to cell junctions: differential association of plakoglobin, β-catenin, and alpha-catenin with vascular endothelial cadherin (VE-cadherin). *J. Cell Biol.* 129: 203–217, 1995.

50. Venkiteswaran K, Xiao K, Summers S, Calkins CC, Vincent PA, Pumiglia K and Kowalczyk AP. Regulation of endothelial barrier function and growth by VE-cadherin, plakoglobin, and beta-catenin. *Am. J. Physiol. Cell Physiol.* 283: C811–C821, 2002.

51. Schnittler HJ, Puschel B and Drenckhahn D. Role of cadherins and plakoglobin in interendothelial adhesion under resting conditions and shear stress. *Am. J. Physiol. Heart Circ. Physiol.* 273: H2396–H2405, 1997.

52. Valiron O, Chevrier V, Usson Y, Breviario F, Job D and Dejana E. Desmoplakin expression and organization at human umbilical vein endothelial cell-to-cell junctions. *J. Cell Sci.* 109 (Pt 8): 2141–2149, 1996.

53. Kowalczyk AP, Navarro P, Dejana E, Bornslaeger EA, Green KJ, Kopp DS and Borgwardt JE. VE-cadherin and desmoplakin are assembled into dermal microvascular endothelial intercellular junctions: a pivotal role for plakoglobin in the recruitment of desmoplakin to intercellular junctions. *J. Cell Sci.* 111 (Pt 20): 3045–3057, 1998.

54. Gallicano GI, Bauer C and Fuchs E. Rescuing desmoplakin function in extra-embryonic ectoderm reveals the importance of this protein in embryonic heart, neuroepithelium, skin and vasculature. *Development* 128: 929–941, 2001.

55. Zhou X, Stuart A, Dettin LE, Rodriguez G, Hoel B and Gallicano GI. Desmoplakin is required for microvascular tube formation in culture. *J. Cell Sci.* 117: 3129–3140, 2004.

56. Hammerling B, Grund C, Boda-Heggemann J, Moll R and Franke WW. The complexus adhaerens of mammalian

lymphatic endothelia revisited: a junction even more complex than hitherto thought. *Cell Tissue Res. 324*: 55–67, 2006.
57. Cattelino A, Liebner S, Gallini R, Zanetti A, Balconi G, Corsi A, Bianco P, Wolburg H, Moore R, Oreda B, Kemler R and Dejana E. The conditional inactivation of the β-catenin gene in endothelial cells causes a defective vascular pattern and increased vascular fragility. *J. Cell Biol. 162*: 1111–1122, 2003.
58. Moon RT, Kohn AD, De Ferrari GV and Kaykas A. WNT and β-catenin signalling: diseases and therapies. *Nat. Rev. Genet. 5*: 691–701, 2004.
59. Goodwin AM, Sullivan KM and D'Amore PA. Cultured endothelial cells display endogenous activation of the canonical Wnt signaling pathway and express multiple ligands, receptors, and secreted modulators of Wnt signaling. *Dev. Dyn. 235*: 3110–3120, 2006.
60. Masckauchan TN, Shawber CJ, Funahashi Y, Li CM and Kitajewski J. Wnt/β-catenin signaling induces proliferation, survival and interleukin-8 in human endothelial cells. *Angiogenesis 8*: 43–51, 2005.
61. Daniel JM. Dancing in and out of the nucleus: p120(ctn) and the transcription factor Kaiso. *Biochim. Biophys. Acta 1773*: 59–68, 2007.
62. Scott JA and Yap AS. Cinderella no longer: α-catenin steps out of cadherin's shadow. *J. Cell Sci. 119*: 4599–4605, 2006.
63. Briggs MW and Sacks DB. IQGAP proteins are integral components of cytoskeletal regulation. *EMBO Rep. 4*: 571–574, 2003.
64. Brown MD and Sacks DB. IQGAP1 in cellular signaling: bridging the GAP. *Trends Cell Biol. 16*: 242–249, 2006.
65. Yamada S, Pokutta S, Drees F, Weis WI and Nelson WJ. Deconstructing the cadherin-catenin-actin complex. *Cell 123*: 889–901, 2005.
66. Drees F, Pokutta S, Yamada S, Nelson WJ and Weis WI. Alpha-catenin is a molecular switch that binds E-cadherin-β-catenin and regulates actin-filament assembly. *Cell 123*: 903–915, 2005.
67. Yamaoka-Tojo M, Tojo T, Kim HW, Hilenski L, Patrushev NA, Zhang L, Fukai T and Ushio-Fukai M. IQGAP1 mediates VE-cadherin-based cell-cell contacts and VEGF signaling at adherence junctions linked to angiogenesis. *Arterioscler. Thromb. Vasc. Biol. 26*: 1991–1997, 2006.
68. Piedra J, Martinez D, Castano J, Miravet S, Dunach M and de Herreros AG. Regulation of β-catenin structure and activity by tyrosine phosphorylation. *J. Biol. Chem. 276*: 20436–20443, 2001.
69. Castano J, Raurell I, Piedra JA, Miravet S, Dunach M and Garcia de HA. β-catenin N- and C-terminal tails modulate the coordinated binding of adherens junction proteins to β-catenin. *J. Biol. Chem. 277*: 31541–31550, 2002.
70. Piedra J, Miravet S, Castano J, Palmer HG, Heisterkamp N, Garcia de HA and Dunach M. p120 Catenin-associated Fer and Fyn tyrosine kinases regulate β-catenin Tyr-142 phosphorylation and β-catenin-alpha-catenin interaction. *Mol. Cell. Biol. 23*: 2287–2297, 2003.
71. Xu G, Craig AW, Greer P, Miller M, Anastasiadis PZ, Lilien J and Balsamo J. Continuous association of cadherin with β-catenin requires the non-receptor tyrosine-kinase Fer. *J. Cell Sci. 117*: 3207–3219, 2004.
72. Lampugnani MG, Corada M, Andriopoulou P, Esser S, Risau W and Dejana E. Cell confluence regulates tyrosine phosphorylation of adherens junction components in endothelial cells. *J. Cell Sci. 110 (Pt 17)*: 2065–2077, 1997.
73. Lambeng N, Wallez Y, Rampon C, Cand F, Christe G, Gulino-Debrac D, Vilgrain I and Huber P. Vascular endothelial-cadherin tyrosine phosphorylation in angiogenic and quiescent adult tissues. *Circ. Res. 96*: 384–391, 2005.
74. Shasby DM, Ries DR, Shasby SS and Winter MC. Histamine stimulates phosphorylation of adherens junction proteins and alters their link to vimentin. *Am. J. Physiol. Lung Cell. Mol. Physiol. 282*: L1330–L1338, 2002.
75. Andriopoulou P, Navarro P, Zanetti A, Lampugnani MG and Dejana E. Histamine induces tyrosine phosphorylation of endothelial cell-to-cell

adherens junctions. *Arterioscler. Thromb. Vasc. Biol.* 19: 2286–2297, 1999.

76 Ratcliffe MJ, Smales C and Staddon JM. Dephosphorylation of the catenins p120 and p100 in endothelial cells in response to inflammatory stimuli. *Biochem. J.* 338 (Pt 2): 471–478, 1999.

77 Tinsley JH, Wu MH, Ma W, Taulman AC and Yuan SY. Activated neutrophils induce hyperpermeability and phosphorylation of adherens junction proteins in coronary venular endothelial cells. *J. Biol. Chem.* 274: 24930–24934, 1999.

78 Nwariaku FE, Liu Z, Zhu X, Turnage RH, Sarosi GA and Terada LS. Tyrosine phosphorylation of vascular endothelial cadherin and the regulation of microvascular permeability. *Surgery* 132: 180–185, 2002.

79 Nwariaku FE, Liu Z, Zhu X, Nahari D, Ingle C, Wu RF, Gu Y, Sarosi G and Terada LS. NADPH oxidase mediates vascular endothelial cadherin phosphorylation and endothelial dysfunction. *Blood* 104: 3214–3220, 2004.

80 Angelini DJ, Hyun SW, Grigoryev DN, Garg P, Gong P, Singh IS, Passaniti A, Hasday JD and Goldblum SE. TNF-α increases tyrosine phosphorylation of vascular endothelial cadherin and opens the paracellular pathway through fyn activation in human lung endothelia. *Am. J. Physiol. Lung Cell. Mol. Physiol.* 291: L1232–L1245, 2006.

81 van Buul JD, Anthony EC, Fernandez-Borja M, Burridge K and Hordijk PL. Proline-rich tyrosine kinase 2 (Pyk2) mediates vascular endothelial-cadherin-based cell-cell adhesion by regulating β-catenin tyrosine phosphorylation. *J. Biol. Chem.* 280: 21129–21136, 2005.

82 Hudry-Clergeon H, Stengel D, Ninio E and Vilgrain I. Platelet-activating factor increases VE-cadherin tyrosine phosphorylation in mouse endothelial cells and its association with the PtdIns3′-kinase. *FASEB J.* 19: 512–520, 2005.

83 Esser S, Lampugnani MG, Corada M, Dejana E and Risau W. Vascular endothelial growth factor induces VE-cadherin tyrosine phosphorylation in endothelial cells. *J. Cell Sci.* 111 (Pt 13): 1853–1865, 1998.

84 Ali N, Yoshizumi M, Yano S, Sone S, Ohnishi H, Ishizawa K, Kanematsu Y, Tsuchiya K and Tamaki T. The novel Src kinase inhibitor M475271 inhibits VEGF-induced vascular endothelial-cadherin and beta-catenin phosphorylation but increases their association. *J. Pharmacol. Sci.* 102: 112–120, 2006.

85 Weis S, Shintani S, Weber A, Kirchmair R, Wood M, Cravens A, McSharry H, Iwakura A, Yoon YS, Himes N, Burstein D, Doukas J, Soll R, Losordo D and Cheresh D. Src blockade stabilizes a Flk/cadherin complex, reducing edema and tissue injury following myocardial infarction. *J. Clin. Invest.* 113: 885–894, 2004.

86 Brkovic A and Sirois MG. Vascular permeability induced by VEGF family members in vivo: role of endogenous PAF and NO synthesis. *J. Cell Biochem.* 100: 727–737, 2007.

87 Zanetti A, Lampugnani MG, Balconi G, Breviario F, Corada M, Lanfrancone L and Dejana E. Vascular endothelial growth factor induces SHC association with vascular endothelial cadherin: a potential feedback mechanism to control vascular endothelial growth factor receptor-2 signaling. *Arterioscler. Thromb. Vasc. Biol.* 22: 617–622, 2002.

88 Wong EY, Morgan L, Smales C, Lang P, Gubby SE and Staddon JM. Vascular endothelial growth factor stimulates dephosphorylation of the catenins p120 and p100 in endothelial cells. *Biochem. J.* 346 (Pt 1): 209–216, 2000.

89 Gavard J and Gutkind JS. VEGF controls endothelial-cell permeability by promoting the β-arrestin-dependent endocytosis of VE-cadherin. *Nat. Cell Biol.* 8: 1223–1234, 2006.

90 Wallez Y, Cand F, Cruzalegui F, Wernstedt C, Souchelnytskyi S, Vilgrain I and Huber P. Src kinase phosphorylates vascular endothelial-cadherin in response to vascular endothelial growth factor: identification of tyrosine 685 as the unique target site. *Oncogene* 26: 1067–1077, 2007.

91 Potter MD, Barbero S and Cheresh DA. Tyrosine phosphorylation of VE-cadherin

prevents binding of p120- and β-catenin and maintains the cellular mesenchymal state. *J. Biol. Chem.* 280: 31906–31912, 2005.

92 Konstantoulaki M, Kouklis P and Malik AB. Protein kinase C modifications of VE-cadherin, p120, and β-catenin contribute to endothelial barrier dysregulation induced by thrombin. *Am. J. Physiol. Lung Cell. Mol. Physiol.* 285: L434–L442, 2003.

93 Ukropec JA, Hollinger MK, Salva SM and Woolkalis MJ. SHP2 association with VE-cadherin complexes in human endothelial cells is regulated by thrombin. *J. Biol. Chem.* 275: 5983–5986, 2000.

94 Wang Y, Jin G, Miao H, Li JY, Usami S and Chien S. Integrins regulate VE-cadherin and catenins: dependence of this regulation on Src, but not on Ras. *Proc. Natl. Acad. Sci. USA* 103: 1774–1779, 2006.

95 Cai J, Jiang WG and Mansel RE. Phosphorylation and disorganization of vascular-endothelial cadherin in interaction between breast cancer and vascular endothelial cells. *Int J. Mol. Med.* 4: 191–195, 1999.

96 Ukropec JA, Hollinger MK and Woolkalis MJ. Regulation of VE-cadherin linkage to the cytoskeleton in endothelial cells exposed to fluid shear stress. *Exp. Cell Res.* 273: 240–247, 2002.

97 Farkas A, Szatmari E, Orbok A, Wilhelm I, Wejksza K, Nagyoszi P, Hutamekalin P, Bauer H, Bauer HC, Traweger A and Krizbai IA. Hyperosmotic mannitol induces Src kinase-dependent phosphorylation of β-catenin in cerebral endothelial cells. *J. Neurosci. Res.* 80: 855–861, 2005.

98 Paul R, Zhang ZG, Eliceiri BP, Jiang Q, Boccia AD, Zhang RL, Chopp M and Cheresh DA. Src deficiency or blockade of Src activity in mice provides cerebral protection following stroke. *Nat. Med.* 7: 222–227, 2001.

99 Baumeister U, Funke R, Ebnet K, Vorschmitt H, Koch S and Vestweber D. Association of Csk to VE-cadherin and inhibition of cell proliferation. *EMBO J.* 24: 1686–1695, 2005.

100 Garcia JG, Schaphorst KL, Verin AD, Vepa S, Patterson CE and Natarajan V. Diperoxovanadate alters endothelial cell focal contacts and barrier function: role of tyrosine phosphorylation. *J. Appl. Physiol.* 89: 2333–2343, 2000.

101 Shi S, Garcia JG, Roy S, Parinandi NL and Natarajan V. Involvement of c-Src in diperoxovanadate-induced endothelial cell barrier dysfunction. *Am. J. Physiol. Lung Cell. Mol. Physiol.* 279: L441–L451, 2000.

102 Young BA, Sui X, Kiser TD, Hyun SW, Wang P, Sakarya S, Angelini DJ, Schaphorst KL, Hasday JD, Cross AS, Romer LH, Passaniti A and Goldblum SE. Protein tyrosine phosphatase activity regulates endothelial cell-cell interactions, the paracellular pathway, and capillary tube stability. *Am. J. Physiol. Lung Cell. Mol. Physiol.* 285: L63–L75, 2003.

103 Sui XF, Kiser TD, Hyun SW, Angelini DJ, Del Vecchio RL, Young BA, Hasday JD, Romer LH, Passaniti A, Tonks NK and Goldblum SE. Receptor protein tyrosine phosphatase micro regulates the paracellular pathway in human lung microvascular endothelia. *Am. J. Pathol.* 166: 1247–1258, 2005.

104 Baumer S, Keller L, Holtmann A, Funke R, August B, Gamp A, Wolburg H, Wolburg-Buchholz K, Deutsch U and Vestweber D. Vascular endothelial cell-specific phosphotyrosine phosphatase (VE-PTP) activity is required for blood vessel development. *Blood* 107: 4754–4762, 2006.

105 Nawroth R, Poell G, Ranft A, Kloep S, Samulowitz U, Fachinger G, Golding M, Shima DT, Deutsch U and Vestweber D. VE-PTP and VE-cadherin ectodomains interact to facilitate regulation of phosphorylation and cell contacts. *EMBO J.* 21: 4885–4895, 2002.

106 Dominguez MG, Hughes VC, Pan L, Simmons M, Daly C, Anderson K, Noguera-Troise I, Murphy AJ, Valenzuela DM, Davis S, Thurston G, Yancopoulos GD and Gale NW. Vascular endothelial tyrosine phosphatase (VE-PTP)-null mice undergo vasculogenesis but die embryonically because of defects in angiogenesis. *Proc. Natl. Acad. Sci. USA* 104: 3243–3248, 2007.

107 Takahashi T, Takahashi K, Mernaugh RL, Tsuboi N, Liu H and Daniel TO. A monoclonal antibody against CD148, a receptor-like tyrosine phosphatase, inhibits endothelial-cell growth and angiogenesis. *Blood* 108: 1234–1242, 2006.

108 Takahashi T, Takahashi K, St JP, Fleming PA, Tomemori T, Watanabe T, Abrahamson DR, Drake CJ, Shirasawa T and Daniel TO. A mutant receptor tyrosine phosphatase, CD148, causes defects in vascular development. *Mol. Cell. Biol.* 23: 1817–1831, 2003.

109 Takai Y, Sasaki T and Matozaki T. Small GTP-binding proteins. *Physiol. Rev.* 81: 153–208, 2001.

110 Jaffe AB and Hall A. Rho GTPases: biochemistry and biology. *Annu. Rev. Cell Dev. Biol.* 21: 247–269, 2005.

111 Wojciak-Stothard B, Potempa S, Eichholtz T and Ridley AJ. Rho and Rac but not Cdc42 regulate endothelial cell permeability. *J. Cell Sci.* 114: 1343–1355, 2001.

112 Wojciak-Stothard B, Entwistle A, Garg R and Ridley AJ. Regulation of TNF-α-induced reorganization of the actin cytoskeleton and cell-cell junctions by Rho, Rac, and Cdc42 in human endothelial cells. *J. Cell Physiol.* 176: 150–165, 1998.

113 Birukova AA, Smurova K, Birukov KG, Kaibuchi K, Garcia JG and Verin AD. Role of Rho GTPases in thrombin-induced lung vascular endothelial cells barrier dysfunction. *Microvasc. Res.* 67: 64–77, 2004.

114 Essler M, Amano M, Kruse HJ, Kaibuchi K, Weber PC and Aepfelbacher M. Thrombin inactivates myosin light chain phosphatase via Rho and its target Rho kinase in human endothelial cells. *J. Biol. Chem.* 273: 21867–21874, 1998.

115 van Nieuw Amerongen GP, Van DS, Vermeer MA, Collard JG and van H, V. Activation of RhoA by thrombin in endothelial hyperpermeability: role of Rho kinase and protein tyrosine kinases. *Circ. Res.* 87: 335–340, 2000.

116 van Nieuw Amerongen GP, Draijer R, Vermeer MA and van H, V. Transient and prolonged increase in endothelial permeability induced by histamine and thrombin: role of protein kinases, calcium, and RhoA. *Circ. Res.* 83: 1115–1123, 1998.

117 McKenzie JA and Ridley AJ. Roles of Rho/ROCK and MLCK in TNF-α-induced changes in endothelial morphology and permeability. *J. Cell Physiol.* 213: 221–228, 2007.

118 Clements RT, Minnear FL, Singer HA, Keller RS and Vincent PA. RhoA and Rho-kinase dependent and independent signals mediate TGF-β-induced pulmonary endothelial cytoskeletal reorganization and permeability. *Am. J. Physiol. Lung Cell. Mol. Physiol.* 288: L294–L306, 2005.

119 Wojciak-Stothard B and Ridley AJ. Rho GTPases and the regulation of endothelial permeability. *Vasc. Pharmacol.* 39: 187–199, 2002.

120 van WS, Van Buul JD, Quik S, Mul FP, Anthony EC, ten Klooster JP, Collard JG and Hordijk PL. Reactive oxygen species mediate Rac-induced loss of cell-cell adhesion in primary human endothelial cells. *J. Cell Sci.* 115: 1837–1846, 2002.

121 Lampugnani MG, Zanetti A, Breviario F, Balconi G, Orsenigo F, Corada M, Spagnuolo R, Betson M, Braga V and Dejana E. VE-cadherin regulates endothelial actin activating Rac and increasing membrane association of Tiam. *Mol. Biol. Cell* 13: 1175–1189, 2002.

122 Waschke J, Burger S, Curry FR, Drenckhahn D and Adamson RH. Activation of Rac-1 and Cdc42 stabilizes the microvascular endothelial barrier. *Histochem. Cell. Biol.* 125: 397–406, 2006.

123 Birukov KG, Bochkov VN, Birukova AA, Kawkitinarong K, Rios A, Leitner A, Verin AD, Bokoch GM, Leitinger N and Garcia JG. Epoxycyclopentenone-containing oxidized phospholipids restore endothelial barrier function via Cdc42 and Rac 13. *Circ. Res.* 95: 892–901, 2004.

124 Broman MT, Kouklis P, Gao X, Ramchandran R, Neamu RF, Minshall RD and Malik AB. Cdc42 regulates adherens junction stability and endothelial permeability by inducing alpha-catenin interaction with the vascular endothelial cadherin complex. *Circ. Res.* 98: 73–80, 2006.

125. Kouklis P, Konstantoulaki M and Malik AB. VE-cadherin-induced Cdc42 signaling regulates formation of membrane protrusions in endothelial cells. *J. Biol. Chem.* 278: 16230–16236, 2003.
126. Broman MT, Mehta D and Malik AB. Cdc42 regulates the restoration of endothelial adherens junctions and permeability. *Trends Cardiovasc. Med.* 17: 151–156, 2007.
127. Kouklis P, Konstantoulaki M, Vogel S, Broman M and Malik AB. Cdc42 regulates the restoration of endothelial barrier function. *Circ. Res.* 94: 159–166, 2004.
128. Fukata M, Kuroda S, Nakagawa M, Kawajiri A, Itoh N, Shoji I, Matsuura Y, Yonehara S, Fujisawa H, Kikuchi A and Kaibuchi K. Cdc42 and Rac1 regulate the interaction of IQGAP1 with β-catenin. *J. Biol. Chem.* 274: 26044–26050, 1999.
129. Fukuhra S, Sakurai A, Yamagishi A, Sako K and Mochizuki N. Vascular endothelial cadherin-mediated cell-cell adhesion regulated by a small GTPase, Rap1. *J. Biochem. Mol. Biol.* 39: 132–139, 2006.
130. Cullere X, Shaw SK, Andersson L, Hirahashi J, Luscinskas FW and Mayadas TN. Regulation of vascular endothelial barrier function by Epac, a cAMP-activated exchange factor for Rap GTPase. *Blood* 105: 1950–1955, 2005.
131. Fukuhara S, Sakurai A, Sano H, Yamagishi A, Somekawa S, Takakura N, Saito Y, Kangawa K and Mochizuki N. Cyclic AMP potentiates vascular endothelial cadherin-mediated cell-cell contact to enhance endothelial barrier function through an Epac-Rap1 signaling pathway. *Mol. Cell. Biol.* 25: 136–146, 2005.
132. Kooistra MR, Corada M, Dejana E and Bos JL. Epac1 regulates integrity of endothelial cell junctions through VE-cadherin. *FEBS Lett.* 579: 4966–4972, 2005.
133. Birukova AA, Zagranichnaya T, Fu P, Alekseeva E, Chen W, Jacobson JR and Birukov KG. Prostaglandins PGE(2) and PGI(2) promote endothelial barrier enhancement via PKA- and Epac1/Rap1-dependent Rac activation. *Exp. Cell Res.* 313: 2504–2520, 2007.
134. Yap AS and Kovacs EM. Direct cadherin-activated cell signaling: a view from the plasma membrane. *J. Cell Biol.* 160: 11–16, 2003.
135. Sakurai A, Fukuhara S, Yamagishi A, Sako K, Kamioka Y, Masuda M, Nakaoka Y and Mochizuki N. MAGI-1 is required for Rap1 activation upon cell-cell contact and for enhancement of vascular endothelial cadherin-mediated cell adhesion. *Mol. Biol. Cell* 17: 966–976, 2006.
136. Ogita H and Takai Y. Nectins and nectin-like molecules: roles in cell adhesion, polarization, movement, and proliferation. *IUBMB Life* 58: 334–343, 2006.
137. Takai Y, Irie K, Shimizu K, Sakisaka T and Ikeda W. Nectins and nectin-like molecules: roles in cell adhesion, migration, and polarization. *Cancer Sci.* 94: 655–667, 2003.

10
Gap Junctions: Connexin Functions and Roles in Human Disease
Michael Koval

10.1
Introduction

Gap junction channels interconnect cells by forming a direct link to enable the diffusion of small aqueous molecules and ions from one cell to its nearest neighbor. This enables both the flow of specific intercellular signals and metabolic cooperation between communicating cells in a tissue [1–4]. Intercellular communication was initially correlated with sites identified by electron microscopy, where two adjacent cells in close contact were separated by a uniform 2- to 3-nm "gap", which led to the term "gap junction". Once techniques were developed to isolate junctions from rat liver and heart, this enabled the identification and subsequent cloning of the channel-forming proteins of gap junctions, known as connexins [5, 6].

Currently, 20 different human connexins have been identified, and cells frequently express multiple isoforms. To date, the connexin protein nomenclature most typically used is to denote a specific connexin by "Cx" plus a number corresponding to the predicted molecular mass based on the amino acid sequence. Unfortunately, this method has some pitfalls, particularly when comparing connexin orthologues between species. For instance, mammalian Cx46 is the orthologue to chick Cx56, and mammalian Cx50 is the orthologue to chick Cx44. Adding to the confusion, the current connexin gene nomenclature uses a different scheme, based in part, on amino acid homology which subdivides connexins into alpha and beta subgroups [7]. For instance, the gene name for Cx43 is GJA1, while the gene name for Cx32 is GJB1. One flaw with this approach is that six connexins do not fit well into either the alpha or beta homology groups (Figure 10.1). For instance, it is an open question whether it is appropriate to assign a gene name such as GJA9 to Cx36. Also, whether the alpha and beta subgroups have physiological significance is not known at present. Although the need for a unified connexin gene nomenclature has been recognized, at the present time a consensus has not been achieved. Any changes to connexin gene nomenclature will be posted on the Human Genome Organization (HUGO) website (www.gene.ucl.ac.uk/nomenclature/).

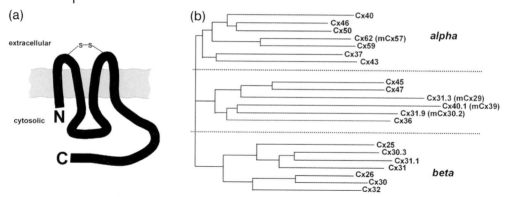

Figure 10.1 The connexin family of gap junction proteins. (a) A line diagram showing connexin orientation in the membrane. The extracellular loop domains are linked by one to three covalent disulfide bonds. (b) A dendrogram arranged using protein homology. Shown here are human connexin protein names; when mouse names differ, they are shown in parentheses. Connexins separate into two major subfamilies, alpha and beta, with several other members showing intermediate homology between these two subfamilies.

Another mammalian gene family, known as pannexins, has recently been discovered that is homologous with the invertebrate gap junction protein family, the innexins [8–10]. Despite the intriguing possibility that there might be two distinct classes of gap junction proteins in mammals, evidence is emerging that pannexins do not efficiently form intercellular channels [11]. Instead, it is far more likely that pannexins have a prominent role in forming large conductance channels [12]. Of particular note, it was found that pannexin-1 channels are the large conductance pore activated by P_2X_7 purinergic receptors [13]. The notion of pannexin channels complements the increasing body of evidence that free connexin hemichannels dispersed throughout the plasma membrane can act as *bona fide* plasma membrane channels, enabling the exchange of aqueous molecules between the cytoplasm to the extracellular environment [14, 15]. However, determining the relative contribution of pannexins as opposed to connexins for high-conductance plasma membrane channels may prove challenging, since pannexin channels have properties comparable to those of open connexin hemichannels [16]. For instance, Cx43-deficient astrocytes have hemichannel activity that is pharmacologically indistinguishable from connexin hemichannels, and that is not present in astrocytes isolated from P_2X_7-deficient mice [17]. Also, it is conceivable that gap junctional communication may be required to regulate pannexins, as opposed to directly forming active connexin hemichannels.

10.2
Connexin Structure and Assembly

Connexins are transmembrane proteins that span the membrane bilayer four times with both the N and C termini oriented towards the cytoplasm and two extracellular loop domains interlinked by one or more disulfide bridges [18–20] (see Figure 10.1). Unlike many plasma membrane transmembrane proteins, connexins are not glycosylated. Within the connexin gene family, the cytoplasmic loop and C-terminal tail domains show the greatest level of sequence divergence, with other regions of the protein showing considerably higher levels of conservation. Connexins show cell- and tissue-specific expression patterns and different tissues frequently express multiple connexin isoforms.

Connexins have a relatively short half-life of 1–5 h, which suggests that gap junction turnover is a constant process [21, 22]. Newly synthesized connexins first oligomerize into hexameric hemichannels, which are subsequently transported to the plasma membrane. Channels at the cell surface can then dock with hemichannels on neighboring cells to form complete intercellular channels. Although there are free hemichannels present at the cell surface, the intercellular channels are typically arranged in semicrystalline arrays, known as gap junction plaques, at sites of cell–cell contact where most intercellular communication occurs (Figure 10.2). Gap junctions are turned over by a unique endocytic mechanism shared by tight junctions, where one cell engulfs a portion of the other cell to internalize sections of plaques [23–25]. Once internalized, connexins are degraded by a combination of proteosomal and lysosomal hydrolysis [26, 27].

Defining the intracellular compartments where connexin oligomerization occurs has proven challenging [28]. In contrast to most multimeric transmembrane complexes, Cx43 oligomerizes into hexamers following exit from the endoplasmic reticulum (ER) in a secretory membrane compartment, which is most likely the trans Golgi network (TGN) [29, 30]. This initially surprising finding is consistent with the notion of a quality control system in the Golgi apparatus [31, 32] that complements ER quality control pathways [33]. Specific components of the quality control apparatus that regulates connexin oligomerization have not yet been identified. Moreover, increasing evidence suggests that different classes of connexins oligomerize in distinct intracellular compartments [28]. For instance, although Cx43 and Cx46 oligomerize in the Golgi apparatus, Cx32 oligomerizes in either the ER or ER-Golgi intermediate compartment (ERGIC) [34, 35]. Since a relatively small subset of connexins has been studied in any detail, it is difficult to generalize these findings. Nonetheless, differences in the intracellular site of connexin oligomerization may be relevant to human diseases involving the mistargeting of mutant connexins, as mutations that change the site of oligomerization may have deleterious effects or promote connexin misfolding.

10.3
Interactions Between Different Connexins

Cells which express multiple connexins have the capability to form heteromeric channels consisting of one or more different connexins. This, in turn, provides a mechanism to fine-tune channel permeability and function [36]. Connexins do not ubiquitously intermix; instead, they fall into innate compatibility groups based on their protein structure. When two connexins expressed in the same cell

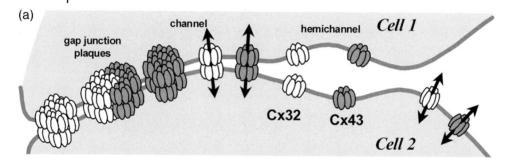

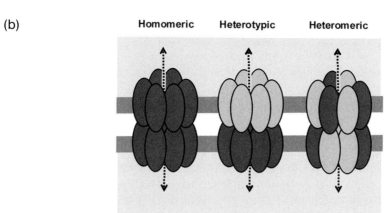

Figure 10.2 The arrangement of connexins at the plasma membrane. (a) Diagram showing different states of connexins at the plasma membrane, ranging from hemichannels to gap junction channels. In this case, the light ovals represent Cx32, and dark ovals represent Cx43, as examples of two innately incompatible connexins. Note that gap junction plaques can contain regions composed of different types of homomeric channels. (b) Classes of connexin channels formed by cells expressing multiple types of connexins. Here, the two different shades of ovals indicate connexins with the ability to interact (e.g., Cx43 and Cx46). Cells can form channels composed of either a single type connexin (homomeric), two different types of connexins in two different cells (heterotypic), or in the same cell (heteromeric). Another possible combination includes heterotypic–heteromeric channels formed by mixed heteromers with different composition. The arrows indicate the path of the aqueous pore of the channel.

heterooligomerize into a mixed gap junction channel, this is referred to as heteromeric compatibility (Figure 10.2). Examples of heteromeric gap junctions formed by endogenously expressed connexins include Cx26 + Cx32 in liver [4, 37], Cx46 + Cx50 in lens [38], and Cx43 + Cx46 in lung [39]. Heteromeric connexin compatibility has also been tested using transfected cell models; however, these are subject to the concern that overexpression could potentially force the aberrant formation of heterooligomers. The heteromeric compatibility groups correspond loosely to the alpha- and beta-connexin subfamilies [4, 36]. Connexins that do not fit into either the alpha or beta subgroups have been found to form heteromers with either alpha connexins (e.g., Cx45 + Cx43) [40] or beta connexins (e.g., mCx29 + Cx32) [41]. However, no connexin has been identified that forms normal heteromeric channels with both alpha and beta subfamily connexins. In fact, mutations that switch connexin heteromeric compatibility can lead to human disease (see Section 10.6).

Cells also regulate the formation of mixed gap junctions by compatible connexins [28]. For instance, two of the compatible connexins expressed by alveolar epithelial cells are Cx43 and Cx46. Type I alveolar epithelial cells form mixed gap junctions composed of Cx43 and Cx46, whereas type II cells prevent Cx43 and Cx46 from intermixing [39]. Other examples of cells which regulate connexin assembly include endothelial cells which restrict the formation of gap junctions containing Cx37, despite expressing two other compatible connexins, Cx40 and Cx43 [42]. Although the mechanisms which regulate connexin heterooligomerization are not known at present, one ramification of differential connexin expression and regulated assembly is that cells within a given tissue can create distinct cell–cell interfaces to preferentially allow or restrict intercellular communication.

Head-to-head heterotypic docking is also restricted, but it does not fall cleanly along the alpha/beta class distinction [4]. For instance, Cx32 and Cx43 are heterotypically incompatible, whereas Cx46 forms heterotypic channels with both [43, 44]. Also, whether homomeric Cx40 and homomeric Cx43 form functional channels is controversial, with evidence both for [45, 46] and against [43, 44, 47, 48]. Since heterotypic Cx40-Cx43 compatibility is regulated in part by intracellular Cx40 residues [48], changes in cell phenotype and/or expression of other connexins may influence the head-to-head binding of Cx40 to Cx43, suggesting a potential explanation for these conflicting results. Whether heterotypic Cx40-Cx43 coupling occurs is particularly relevant to the formation of functional myoendothelial gap junctions between endothelium and vascular smooth muscle, since Cx40 is a major connexin expressed by endothelial cells and Cx43 is the main connexin associated with vascular smooth muscle cells [49, 50].

10.4
Connexin Binding Proteins and Phosphorylation

The extracellular portion of connexins exclusively interacts with other connexins. However, the cytoplasmic C terminus interacts with several different protein cofactors, which interlink them with the cytoskeleton and/or regulate connexin function [51]. As there is considerable variability in the C terminus, different connexins interact with different proteins. One of the first connexin-interacting proteins identified is the zona occludens-1 (ZO-1) scaffold protein, which links junction proteins to the actin cytoskeleton and binds to Cx43, Cx45 and Cx46, but not to Cx32 [52–57]. Connexin–ZO-1 interactions have been shown to enhance connexin targeting to the plasma membrane, and may regulate plaque size and/or geometry. ZO-2 also binds to Cx43, and there are suggestions that ZO-2 may compete with ZO-1 binding as a mode to downregulate gap junction formation during mitosis

[52]. Cx43 has also been found to bind directly to tubulin via the C terminus, an interaction which can help target intracellular vesicles containing Cx43 directly to adherens junctions at the plasma membrane [58, 59].

Most connexins are phosphorylated [60–63]. Cx43 is the best-characterized connexin, with 12 known phosphoserine sites and two phosphotyrosine sites. Connexin phosphorylation has the capacity to regulate interactions with scaffold proteins. For instance, Cx43 phosphorylation by c-Src inhibits binding to ZO-1 which, in turn, decreases Cx43 channel localization at the plasma membrane [64]. Conversely, other kinases upregulate Cx43-mediated communication, including casein kinase I and protein kinase A, both of which act to enhance Cx43 assembly into gap junctions [62].

Differential phosphorylation also specifically regulates gap junctional communication by altering channel function (gating). For instance, vascular endothelial growth factor (VEGF) phosphorylates Cx43 via a MAP kinase cascade to inhibit intercellular coupling [65]. Also, phorbol esters stimulate protein kinase C (PK-C), which inhibits both Cx43 assembly and channel function [66, 67]. The observation that kinase cascades linked to cell growth control tend to attenuate Cx43 gap junctional communication fits well with studies demonstrating that cancer cells usually show low levels of connexin expression and that loss of intercellular communication may promote tumor cell growth [68].

Some clues to the complexity of connexin phosphorylation are provided by studies of cardiac ischemia [66, 69]. In response to ischemia, PK-C is upregulated; however, casein kinase I activity decreased, resulting in simultaneous increases and decreases in Cx43 phosphorylation which act in concert to downregulate gap junctional communication. Thus, the regulation of connexins by phosphorylation is not a simple on/off switch. Instead, connexin function is regulated by the overall balance of phosphorylation.

10.5
Channel Permeability and Signaling

A primary function of gap junction channels is to interconnect cells electrically and/or biochemically. In addition, intercellular channels play a central role in the transmission of transient increases in cytosolic calcium (calcium "waves") throughout a tissue. For instance, calcium signaling through gap junctions regulates vascular tone, particularly in small vessels where gap junctions interconnect endothelial cells to smooth muscle cells [50]. In airway epithelium, the intercellular transmission of inositol triphosphate creates oscillatory waves that synchronize ciliary beat frequencies in order to coordinate the outward flow of mucus [70]. In the terminal airway, gap junctions enable type I alveolar epithelial cells to act as mechanosensors, to help regulate surfactant secretion by type II cells [71], although paracrine stimulation of purinergic receptors also plays a role in this process [72].

With respect to small signaling molecules and metabolites, intercellular channels have relative selectivity, as channels composed of different connexins show different rates of diffusion for a given molecule. For instance, diffusion of ATP is approximately 10-fold faster through Cx43 channels as compared to Cx32, whereas glutathione flux through Cx32 and Cx43 channels is roughly comparable [73]. By and large, connexin hemichannel permeability and regulation appears to be comparable to the characteristics of complete intercellular channels of the same composition, with subtle differences [74, 75]. The mechanisms that underlie channel specificity are not well understood at present, but one possibility is that interactions of molecules with surfaces of the aqueous pores may play a role.

One consequence of gap junction formation by two or more connexins is that the cells can form channels with unique permeability characteristics. For instance, Cx26 has a fairly restrictive permeability and forms heteromeric gap junction channels with Cx32. Cx32 is fully permeable to cAMP and cGMP, while channels containing both Cx26 and Cx32 are preferentially permeable to cGMP as compared to cAMP [76]. Other examples of mixed channels which have unique permeability include Cx37 + Cx43, Cx40 + Cx43 and Cx43 + Cx45 [40, 77–79].

10.6
Connexins in Human Disease

Human diseases associated with connexin dysfunction show a considerable range in pathology [80]. In addition to identifying diseases due to the absence of connexin expression, considerable progress has been made in identifying connexin mutations associated with human disease [1, 81] (Table 10.1). However, there have been instances where allelic variability may be incorrectly linked to a disease state. Two examples of this include studies refuting roles for Cx43 mutations associated with cardiac heterotaxy [82, 83], and Cx37 mutations associated with lung cancer [84]. This underscores the value of using *in-vitro* and transgenic mouse models to verify the effect of a connexin mutation on its function as an adjunct to epidemiological approaches.

Although mutations which inhibit connexin expression can have a clear effect on intercellular communication, mutations can also alter connexin function and processing in several different ways (Figure 10.3). For instance, connexin mutations can effect channel permeability, either by producing channels with limited permeability or by producing channels that are misregulated. Aside from mutations that affect intercellular coupling, several disease-associated mutations have also been identified that either inhibit [85] or enhance [86–88] hemichannel activity. It has been suggested that aberrantly open hemichannels might promote apoptosis as a pathological mechanism [88], although further investigation is required to determine whether this is the case.

Connexin mutations can also reduce the efficiency of assembly and targeting to the plasma membrane. The most typical hallmark of this class of mutation is prominent localization in the perinuclear region of the cell, reflecting retention in an aspect of the Golgi apparatus [108, 121–123]. Although connexin misfolding is a likely cause for intracellular retention, in at least one case mislocalization was

Table 10.1 Connexins associated with human disease.

Connexin	Reference(s)
Cx26 (GJB2)[a]	
Hearing impairment	
Deafness, autosomal dominant non-syndromic sensorineural	89, 90
Deafness, neurosensory, autosomal recessive	89–92
Skin disease	
Erythrokeratodermia variabilis	93
Hearing impairment + skin disease	
Keratitis–ichthyosis–deafness syndrome, autosomal dominant	94
Keratoderma, palmoplantar, with deafness	95, 96
Vohwinkel syndrome	97, 98
Bart–Pumphrey syndrome	99, 100
Cx30 (GJB6)	
Deafness, autosomal dominant non-syndromic sensorineural	101
Ectodermal dysplasia, hidrotic	102
Cx30.3 (GJB4)	
Erythrokeratodermia variabilis	103, 104
Cx31 (GJB3)	
Deafness, autosomal dominant non-syndromic sensorineural	105
Erythrokeratodermia variabilis	93, 106
Cx32 (GJB1)	
Charcot–Marie–Tooth disease, X-linked	107–109
Cx37 (GJA4)	
Atherosclerosis, myocardial infarction	110, 111
Cx40 (GJA5)	
Atrial fibrillation, idiopathic	112
Cx43 (GJA1)	
Oculodentodigital dysplasia	113, 114
Cx46 (GJA3)	
Cataract, zonular pulverulent	115, 116
Cx47 (GJA12)	
Pelizaeus–Merzbacher-like disease, autosomal recessive	117, 118
Cx50 (GJA8)	
Cataract, zonular pulverulent	119, 120

a Human connexin gene names listed in parentheses reflect current nomenclature and may change (see text).

due to a frameshift mutation in Cx46 that alters the C terminus to create an inappropriate diphenylalanine-based ERGIC/Golgi retention signal [124]. Aggresome-like inclusions have also been observed for mutant Cx50 [125].

One consequence of connexin mutations that can have an effect on channel function and/or targeting is that they may have autosomal dominant effects on normal connexins. Several of the Cx43 mutations associated with oculodentodigi-

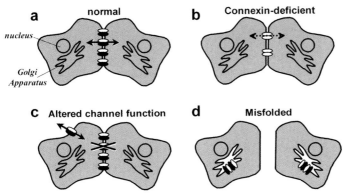

Figure 10.3 Classes of connexin defects that lead to misfunction. (a) A cartoon indicating normal placement of connexins into functional gap junction channels. (b) Connexin-deficient cells have reduced or completely inhibited intercellular communication, depending on the other connexins expressed by the cells. (c) Connexin mutations can lead to aberrant function, including changes in the ability to mediate gap junctional communication or hemichannel activity. Mutations can either increase, decrease, or change channel function. (d) Mutations can also interfere with trafficking which causes connexins to be retained in intracellular compartments (here shown as the Golgi apparatus). This, in turn, decreases channel formation and intercellular communication.

tal dysplasia (ODDD) have the ability to inhibit the function of the normal Cx43 gene product by interacting to form non-functional channels transported to the plasma membrane [121, 126]. Many of these ODDD mutations create Cx43 variants that are not properly trafficked when expressed alone in a transfected cell model, which suggests that they are prone to misfolding. Dominant-negative mutations are not restricted to Cx43. For instance, non-functional mutants of Cx40 associated with atrial fibrillation heterooligomerize with both normal Cx40 and heteromeric Cx40 + Cx43 to effectively inhibit channel activity [112]. Another non-functional mutant, Cx26-M34T associated with hearing impairment, heterooligomerizes with wild-type Cx26 to form channels that are functional but are misregulated as they have a reversed response to changes in membrane voltage potential [127].

The potential for dominant-negative interactions also provides a potential basis for the remarkable observation that different mutations in Cx26, Cx30 or Cx31 genes can cause either hearing impairment, skin disease, or both [96, 128]. While Cx26 is normally incompatible to heterooligomerize with Cx43 [129], several Cx26 mutants aberrantly heterooligomerize with Cx43 and form inactive channels [130]. Notably, Cx26 mutations that interfere with Cx43 are associated with the skin disease palmoplantar keratoderma, whereas Cx26 mutations that do not affect Cx43 are associated with non-syndromic hearing impairment [128]. Whether a comparable mechanism underlies differences in the pathology of mutant Cx30 and Cx31 remains to be determined, although this seems likely.

Connexin-deficient transgenic mouse models frequently mimic the corresponding human disease. Aside from their utility in studying connexin-based diseases, these models have also been used to evaluate the physiological requirement for a specific type of connexin. Connexin-replacement models are relatively easy to develop as almost all connexin genes lack introns [131]. For instance, mice in which Cx43 was replaced with either Cx32 or Cx40 are viable, but they suffer from other defects, including cardiac malformations and secretion defects not present in Cx43-deficient mice [132]. Also, mice where Cx50 is replaced with Cx46 no longer show cataracts; however, the lens growth is not normal and resembles the decreased lens growth seen in Cx50-deficient mice [133]. Connexin overexpression can also be deleterious, as changes in the levels of Cx43 expression interfere with cardiac development [134], and overexpression of Cx50 causes cataract [135]. However, an example where one connexin can compensate for another is derived from the studies of Ahmad et al. [136], who found that Cx26 can functionally replace Cx30 in hair cell function and restore hearing in mice. This latter result is consistent with the fact that Cx26 and Cx30 share considerable amino acid homology and are over 77% identical, although the C terminus of Cx30 is 35 amino acids longer than Cx26.

Since several connexins, including Cx26 and Cx43, are required for neonatal development and post-natal survival [137, 138], targeted deletions using the cre-lox system have frequently been employed to produce tissue-specific knockout models [139–141]. However, given the dominant-negative effect of several connexin-based diseases, an alternative approach is to develop mice expressing a copy of a mutant connexin. One example of this is a G60S mutation in Cx43 which was discovered as a mouse exhibiting the characteristics of ODDD in a *N*-ethyl-*N*-nitrosourea mutagenesis screen [142]. Transgenic mice expressing the R75W Cx26 mutation have also been generated which show hearing impairment comparable to mice with a targeted deletion of Cx26 [139, 143, 144]. The R75W mice were also used for a proof of principle study, where small interfering RNA (siRNA) was used to partially knock down the mutant transgene and restore cochlear function, suggesting a possible therapeutic approach [144]. Undoubtedly, transgenic mice expressing siRNAs also will be employed for tissue-specific connexin knockdown in the near future.

10.7
Role of Connexins in Vessel Inflammation and Atherosclerosis

In addition to diseases caused by mutant connexins, vascular disease may be exacerbated by gap junctional communication mediated by normal connexins in the vessel wall and by circulating leukocytes. Consistent with this, gap junction inhibitors are by and large anti-inflammatory, suggesting that connexins can be pro-inflammatory [145–147]. A direct role for endothelial cell–cell communication in regulating inflammation is suggested by the studies of endothelial-specific Cx43-deficient mice which were unable to transmit calcium waves between lung microvessel endothelial cells [145]. By contrast, in normal mice, both thrombin and excitation of caged calcium compounds caused increases in endothelial calcium that were transmitted through Cx43 which, in turn, induced an increase in plasma membrane P-selectin expression and leukocyte recruitment.

Consistent with a role in atherosclerotic plaque formation, endothelial Cx43 expression is upregulated in response to disturbed flow during plaque formation [148, 149], and in response to balloon distention injury [150, 151]. Kwak et al. further explored this by modifying an LDL receptor-deficient mouse model that is normally prone to atherosclerosis [152]. These mice were bred with $Cx43^{+/-}$-deficient mice to obtain mice which express half the normal level of Cx43 and are LDL-receptor deficient. $Cx43^{+/-}$: LDL receptor-deficient mice showed significantly less endothelial Cx43 expression and fewer atherosclerotic plaques than mice expressing normal levels of Cx43 [152]. In addition, $Cx43^{+/-}$:LDL receptor-deficient mice were also less susceptible to balloon distension injury, further indicating that endothelial Cx43 was required for the inflammatory response [153].

In addition to forming intercellular communication pathways between endothelial cells, connexins also enable the vascular endothelium to form gap junctions with circulating cells, such as neutrophils expressing Cx43 [146, 154] and macrophages expressing Cx37 [152]. Thus, gap junctional communication could also enhance the attachment of peripheral blood cells to the vessel wall. In particular, coupling between foam cells and the vascular wall is likely to contribute to atherosclerotic plaque formation by inhibiting foam cell migration [152, 155].

The results of recent studies using knockout mice have suggested an additional gap junction-independent role for Cx37 in modulating inflammation and atherosclerosis [85, 156]. Instead of mediating cell–cell interactions, Cx37 forms hemichannels on peripheral blood cells that enable the secretion of cytosolic ATP [14]. The extracellular ATP subsequently reduces leukocyte adhesion to the vessel wall. In contrast, monocytes and macrophages that do not express Cx37 readily flatten and attach to endothelial cells. Cx37 expressed by endothelial cells does not perform a comparable function [85]. Interestingly, macrophages expressing a human Cx37 polymorphism associated with enhanced atherosclerosis (C1019-T) [157] were found to be less efficient in both ATP secretion and regulation of cell adhesion, which further underscores the importance of leukocyte Cx37 in regulating atherosclerosis.

10.8 Conclusions

During the past decade, considerable progress has been made in identifying new physiological roles for gap junction channels and defining the mechanisms that underlie the misfunction of mutant connexins. However, the present understanding of how cells regulate connexin function and assembly remains incomplete. Roles for the differentiated state of cells and the microenvironment in regulating intercellular communication require further study. In addition, relatively few studies have been conducted to determine the extent of the coordinate regulation of hemichannels, gap junctions and other classes of cell–cell junctions. An understanding of how connexins interact with other types of protein to alter intercellular communication has the potential to help identify new functions for connexins as modifiers of cell and organ function.

References

1 D. W. Laird. Life cycle of connexins in health and disease, *Biochem. J.* 2006, 394, 527–543.
2 G. S. Goldberg, V. Valiunas, and P. R. Brink. Selective permeability of gap junction channels, *Biochim. Biophys. Acta* 2004, 1662, 96–101.
3 J. C. Saez, V. M. Berthoud, M. C. Branes, A. D. Martinez, and E. C. Beyer. Plasma membrane channels formed by connexins: their regulation and functions, *Physiol. Rev.* 2003, 83, 1359–1400.
4 A. L. Harris. Emerging issues of connexin channels: biophysics fills the gap, *Q. Rev. Biophys.* 2001, 34, 325–472.
5 E. L. Hertzberg and N. B. Gilula. Isolation and characterization of gap junctions from rat liver, *J. Biol. Chem.* 1979, 254, 2138–2147.
6 E. C. Beyer, D. L. Paul, and D. A. Goodenough. Connexin43: a protein from rat heart homologous to a gap junction protein from liver, *J. Cell Biol.* 1987, 105 (Pt 1), 2621–2629.
7 G. Sohl and K. Willecke. Gap junctions and the connexin protein family, *Cardiovasc. Res.* 2004, 62, 228–232.
8 Y. V. Panchin. Evolution of gap junction proteins – the pannexin alternative, *J. Exp. Biol.* 2005, 208, 1415–1419.
9 G. Dahl and S. Locovei. Pannexin: to gap or not to gap, is that a question? *IUBMB Life* 2006, 58, 409–419.
10 M. T. Barbe, H. Monyer, and R. Bruzzone. Cell–cell communication beyond connexins: the pannexin channels, *Physiology (Bethesda)* 2006, 21, 103–114.
11 Y. Huang, J. B. Grinspan, C. K. Abrams, and S. S. Scherer. Pannexin1 is expressed by neurons and glia but does not form functional gap junctions, *Glia* 2007, 55, 46–56.
12 F. Vanden Abeele, G. Bidaux, D. Gordienko, B. Beck, Y. V. Panchin, A. V. Baranova, D. V. Ivanov, R. Skryma, and N. Prevarskaya. Functional implications of calcium permeability of the channel formed by pannexin 1, *J. Cell Biol.* 2006, 174, 535–546.
13 P. Pelegrin and A. Surprenant. Pannexin-1 mediates large pore formation and interleukin-1beta release by the ATP-gated P2X7 receptor, *EMBO J.* 2006, 25, 5071–5082.
14 D. A. Goodenough and D. L. Paul. Beyond the gap: functions of unpaired connexon channels, *Nat. Rev. Mol. Cell. Biol.* 2003, 4, 285–294.
15 J. X. Jiang and S. Gu. Gap junction- and hemichannel-independent actions of connexins, *Biochim. Biophys. Acta* 2005, 1711, 208–214.
16 D. C. Spray, Z. C. Ye, and B. R. Ransom. Functional connexin "hemichannels": a critical appraisal, *Glia* 2006, 54, 758–773.
17 S. O. Suadicani, C. F. Brosnan, and E. Scemes. P2X7 receptors mediate ATP release and amplification of astrocytic intercellular Ca2+ signaling, *J. Neurosci.* 2006, 26, 1378–1385.
18 C. I. Foote, L. Zhou, X. Zhu, and B. J. Nicholson. The pattern of disulfide linkages in the extracellular loop regions of connexin 32 suggests a model for the docking interface of gap junctions, *J. Cell Biol.* 1998, 140, 1187–1197.
19 V. M. Unger, N. M. Kumar, N. B. Gilula, and M. Yeager. Three-dimensional structure of a recombinant gap junction membrane channel, *Science* 1999, 283, 1176–1180.
20 S. J. Fleishman, V. M. Unger, M. Yeager, and N. Ben-Tal. A C(alpha) model for the

transmembrane alpha helices of gap junction intercellular channels, *Mol. Cell* 2004, *15*, 879–888.
21 D. W. Laird, K. L. Puranam, and J. P. Revel. Turnover and phosphorylation dynamics of connexin43 gap junction protein in cultured cardiac myocytes, *Biochem. J.* 1991, *273*, 67–72.
22 R. F. Fallon and D. A. Goodenough. Five-hour half-life of mouse liver gap-junction protein, *J. Cell Biol.* 1981, *90*, 521–526.
23 K. Jordan, R. Chodock, A. R. Hand, and D. W. Laird. The origin of annular junctions: a mechanism of gap junction internalization, *J. Cell Sci.* 2001, *114*, 763–773.
24 U. Lauf, B. N. Giepmans, P. Lopez, S. Braconnot, S. C. Chen, and M. M. Falk. Dynamic trafficking and delivery of connexons to the plasma membrane and accretion to gap junctions in living cells, *Proc. Natl. Acad. Sci. USA* 2002, *99*, 10446–10451.
25 M. Matsuda, A. Kubo, M. Furuse, and S. Tsukita. A peculiar internalization of claudins, tight junction-specific adhesion molecules, during the intercellular movement of epithelial cells, *J. Cell Sci.* 2004, *117*, 1247–1257.
26 J. G. Laing, P. N. Tadros, E. M. Westphale, and E. C. Beyer. Degradation of connexin43 gap junctions involves both the proteasome and the lysosome, *Exp. Cell Res.* 1997, *236*, 482–492.
27 J. G. Laing, P. N. Tadros, K. Green, J. E. Saffitz, and E. C. Beyer. Proteolysis of connexin43-containing gap junctions in normal and heat-stressed cardiac myocytes, *Cardiovasc. Res.* 1998, *38*, 711–718.
28 M. Koval. Pathways and control of connexin oligomerization, *Trends Cell Biol.* 2006, *16*, 159–166.
29 L. S. Musil and D. A. Goodenough. Multisubunit assembly of an integral plasma membrane channel protein, gap junction connexin43, occurs after exit from the ER, *Cell* 1993, *74*, 1065–1077.
30 M. Koval, J. E. Harley, E. Hick, and T. H. Steinberg. Connexin46 is retained as monomers in a trans-Golgi compartment of osteoblastic cells, *J. Cell Biol.* 1997, *137*, 847–857.
31 P. Arvan, X. Zhao, J. Ramos-Castaneda, and A. Chang. Secretory pathway quality control operating in Golgi, plasmalemmal, and endosomal systems, *Traffic* 2002, *3*, 771–780.
32 E. S. Trombetta and A. J. Parodi. Quality control and protein folding in the secretory pathway, *Annu. Rev. Cell Dev. Biol.* 2003, *19*, 649–676.
33 L. Ellgaard and A. Helenius. Quality control in the endoplasmic reticulum, *Nat. Rev. Mol. Cell. Biol.* 2003, *4*, 181–191.
34 J. Maza, J. Das Sarma, and M. Koval. Defining a minimal motif required to prevent connexin oligomerization in the endoplasmic reticulum, *J. Biol. Chem.* 2005, *280*, 21115–21121.
35 J. Maza, M. Mateescu, J. D. Sarma, and M. Koval. Differential oligomerization of endoplasmic reticulum-retained connexin43/connexin32 chimeras, *Cell Commun. Adhes.* 2003, *10*, 319–322.
36 G. T. Cottrell and J. M. Burt. Functional consequences of heterogeneous gap junction channel formation and its influence in health and disease, *Biochim. Biophys. Acta* 2005, *1711*, 126–141.
37 S. Ahmad, J. A. Diez, C. H. George, and W. H. Evans. Synthesis and assembly of connexins in vitro into homomeric and heteromeric functional gap junction hemichannels, *Biochem. J.* 1999, *339*, 247–253.
38 J. X. Jiang and D. A. Goodenough. Heteromeric connexons in lens gap junction channels, *Proc. Natl. Acad. Sci. USA* 1996, *93*, 1287–1291.
39 J. Das Sarma, R. A. Meyer, F. Wang, V. Abraham, C. W. Lo, and M. Koval. Heteromeric connexin interactions prior to the trans Golgi network, *J. Cell Sci.* 2001, *114*, 4013–4024.
40 M. Koval, S. T. Geist, E. M. Westphale, A. E. Kemendy, R. Civitelli, E. C. Beyer, and T. H. Steinberg. Transfected connexin45 alters gap junction permeability in cells expressing endogenous connexin43, *J. Cell Biol.* 1995, *130*, 987–995.
41 B. M. Altevogt, K. A. Kleopa, F. R. Postma, S. S. Scherer, and D. L. Paul. Connexin29 is uniquely distributed within myelinating glial cells of the

central and peripheral nervous systems, *J. Neurosci.* 2002, *22*, 6458–6470.

42 B. E. Isakson and B. R. Duling. Heterocellular contact at the myoendothelial junction influences gap junction organization, *Circ. Res.* 2005, *97*, 44–51.

43 T. W. White, D. L. Paul, D. A. Goodenough, and R. Bruzzone. Functional analysis of selective interactions among rodent connexins, *Mol. Biol. Cell* 1995, *6*, 459–470.

44 C. Elfgang, R. Eckert, H. Lichtenberg-Frate, A. Butterweck, O. Traub, R. A. Klein, D. F. Hulser, and K. Willecke. Specific permeability and selective formation of gap junction channels in connexin-transfected HeLa cells, *J. Cell Biol.* 1995, *129*, 805–817.

45 G. T. Cottrell, Y. Wu, and J. M. Burt. Cx40 and Cx43 expression ratio influences heteromeric/heterotypic gap junction channel properties, *Am. J. Physiol. Cell Physiol.* 2002, *282*, C1469–C1482.

46 V. Valiunas, R. Weingart, and P. R. Brink. Formation of heterotypic gap junction channels by connexins 40 and 43, *Circ. Res.* 2000, *86*, E42–E49.

47 H. Hennemann, T. Suchyna, H. Lichtenberg-Frate, S. Jungbluth, E. Dahl, J. Schwarz, B. J. Nicholson, and K. Willecke. Molecular cloning and functional expression of mouse connexin40, a second gap junction gene preferentially expressed in lung, *J. Cell Biol.* 1992, *117*, 1299–1310.

48 S. Haubrich, H. J. Schwarz, F. Bukauskas, H. Lichtenberg-Frate, O. Traub, R. Weingart, and K. Willecke. Incompatibility of connexin 40 and 43 hemichannels in gap junctions between mammalian cells is determined by intracellular domains, *Mol. Biol. Cell* 1996, *7*, 1995–2006.

49 N. J. Severs, S. Rothery, E. Dupont, S. R. Coppen, H. I. Yeh, Y. S. Ko, T. Matsushita, R. Kaba, and D. Halliday. Immunocytochemical analysis of connexin expression in the healthy and diseased cardiovascular system, *Microsc. Res. Tech.* 2001, *52*, 301–322.

50 X. F. Figueroa, B. E. Isakson, and B. R. Duling. Connexins: gaps in our knowledge of vascular function, *Physiology (Bethesda)* 2004, *19*, 277–284.

51 B. N. Giepmans. Role of connexin43-interacting proteins at gap junctions, *Adv. Cardiol.* 2006, *42*, 41–56.

52 D. Singh, J. L. Solan, S. M. Taffet, R. Javier, and P. D. Lampe. Connexin 43 interacts with zona occludens-1 and -2 proteins in a cell cycle stage-specific manner, *J. Biol. Chem.* 2005, *280*, 30416–30421.

53 J. G. Laing, M. Koval, and T. H. Steinberg. Association with ZO-1 correlates with plasma membrane partitioning in truncated Connexin45 mutants, *J. Membr. Biol.* 2005, *207*, 45–53.

54 J. G. Laing, B. C. Chou, and T. H. Steinberg. ZO-1 alters the plasma membrane localization and function of Cx43 in osteoblastic cells, *J. Cell Sci.* 2005, *118*, 2167–2176.

55 A. W. Hunter, R. J. Barker, C. Zhu, and R. G. Gourdie. ZO-1 Alters Connexin43 Gap Junction Size and Organization by Influencing Channel Accretion, *Mol. Biol. Cell* 2005, *16*, 5686–5698.

56 T. Kojima, Y. Kokai, H. Chiba, M. Yamamoto, Y. Mochizuki, and N. Sawada. Cx32 but not Cx26 is associated with tight junctions in primary cultures of rat hepatocytes, *Exp. Cell Res.* 2001, *263*, 193–201.

57 B. N. Giepmans and W. H. Moolenaar. The gap junction protein connexin43 interacts with the second PDZ domain of the zona occludens-1 protein, *Curr. Biol.* 1998, *8*, 931–934.

58 B. N. Giepmans, I. Verlaan, T. Hengeveld, H. Janssen, J. Calafat, M. M. Falk, and W. H. Moolenaar. Gap junction protein connexin-43 interacts directly with microtubules, *Curr. Biol.* 2001, *11*, 1364–1368.

59 R. M. Shaw, A. J. Fay, M. A. Puthenveedu, M. von Zastrow, Y. N. Jan, and L. Y. Jan. Microtubule plus-end-tracking proteins target gap junctions directly from the cell interior to adherens junctions, *Cell* 2007, *128*, 547–560.

60 D. W. Laird. Connexin phosphorylation as a regulatory event linked to gap junction internalization and degradation, *Biochim. Biophys. Acta* 2005, *1711*, 172–182.

61. A. P. Moreno. Connexin phosphorylation as a regulatory event linked to channel gating, *Biochim. Biophys. Acta* 2005, *1711*, 164–171.
62. J. L. Solan and P. D. Lampe. Connexin phosphorylation as a regulatory event linked to gap junction channel assembly, *Biochim. Biophys. Acta* 2005, *1711*, 154–163.
63. P. D. Lampe and A. F. Lau. The effects of connexin phosphorylation on gap junctional communication, *Int. J. Biochem. Cell Biol.* 2004, *36*, 1171–1186.
64. T. Toyofuku, Y. Akamatsu, H. Zhang, T. Kuzuya, M. Tada, and M. Hori. c-Src regulates the interaction between connexin-43 and ZO-1 in cardiac myocytes, *J. Biol. Chem.* 2001, *276*, 1780–1788.
65. S. Suarez and K. Ballmer-Hofer. VEGF transiently disrupts gap junctional communication in endothelial cells, *J. Cell Sci.* 2001, *114*, 1229–1235.
66. J. F. Ek-Vitorin, T. J. King, N. S. Heyman, P. D. Lampe, and J. M. Burt. Selectivity of connexin 43 channels is regulated through protein kinase C-dependent phosphorylation, *Circ. Res.* 2006, *98*, 1498–1505.
67. P. D. Lampe. Analyzing phorbol ester effects on gap junctional communication: a dramatic inhibition of assembly, *J. Cell Biol.* 1994, *127*, 1895–1905.
68. E. Leithe, S. Sirnes, Y. Omori, and E. Rivedal. Downregulation of gap junctions in cancer cells, *Crit. Rev. Oncogen.* 2006, *12*, 225–256.
69. P. D. Lampe, C. D. Cooper, T. J. King, and J. M. Burt. Analysis of Connexin43 phosphorylated at S325, S328 and S330 in normoxic and ischemic heart, *J. Cell Sci.* 2006, *119*, 3435–3442.
70. S. Boitano and W. H. Evans. Connexin mimetic peptides reversibly inhibit Ca(2+) signaling through gap junctions in airway cells, *Am. J. Physiol. Lung Cell. Mol. Physiol.* 2000, *279*, L623–L630.
71. Y. Ashino, X. Ying, L. G. Dobbs, and J. Bhattacharya. [Ca2+]i oscillations regulate type II cell exocytosis in the pulmonary alveolus, *Am. J. Physiol.* 2000, *279*, L5–L13.
72. A. S. Patel, D. Reigada, C. H. Mitchell, S. R. Bates, S. S. Margulies, and M. Koval. Paracrine stimulation of surfactant secretion by extracellular ATP in response to mechanical deformation, *Am. J. Physiol. Lung Cell. Mol. Physiol.* 2005, *289*, L489–L496.
73. G. Goldberg, P. D. Lampe, and B. J. Nicholson. Selective transfer of endogenous metabolites through gap junctions composed of different connexins, *Nat. Cell Biol.* 1999, *1*, 457–459.
74. E. B. Trexler, M. V. Bennett, T. A. Bargiello, and V. K. Verselis. Voltage gating and permeation in a gap junction hemichannel, *Proc. Natl. Acad. Sci. USA* 1996, *93*, 5836–5841.
75. X. Bao, Y. Chen, L. Reuss, and G. A. Altenberg. Functional expression in Xenopus oocytes of gap-junctional hemichannels formed by a cysteine-less connexin 43, *J. Biol. Chem.* 2004, *279*, 9689–9692.
76. C. G. Bevans, M. Kordel, S. K. Rhee, and A. L. Harris. Isoform composition of connexin channels determines selectivity among second messengers and uncharged molecules, *J. Biol. Chem.* 1998, *273*, 2808–2816.
77. D. M. Larson, K. H. Seul, V. M. Berthoud, A. F. Lau, G. D. Sagar, and E. C. Beyer. Functional expression and biochemical characterization of an epitope-tagged connexin37, *Mol. Cell. Biol. Res. Commun.* 2000, *3*, 115–121.
78. P. A. Weber, H. C. Chang, K. E. Spaeth, J. M. Nitsche, and B. J. Nicholson. The permeability of gap junction channels to probes of different size is dependent on connexin composition and permeant-pore affinities, *Biophys. J.* 2004, *87*, 958–973.
79. J. M. Burt, A. M. Fletcher, T. D. Steele, Y. Wu, G. T. Cottrell, and D. T. Kurjiaka. Alteration of Cx43:Cx40 expression ratio in A7r5 cells, *Am. J. Physiol. Cell Physiol.* 2001, *280*, C500–C508.
80. D. P. Kelsell, J. Dunlop, and M. B. Hodgins. Human diseases: clues to cracking the connexin code? *Trends Cell Biol.* 2001, *11*, 2–6.
81. V. Krutovskikh and H. Yamasaki. Connexin gene mutations in human

genetic diseases, *Mutat. Res.* 2000, *462*, 197–207.

82 T. Toth, J. Hajdu, T. Marton, B. Nagy, and Z. Papp. connexin43 gene mutations and heterotaxy, *Circulation* 1998, *97*, 117–118.

83 M. Gebbia, J. A. Towbin, and B. Casey. Failure to detect connexin43 mutations in 38 cases of sporadic and familial heterotaxy, *Circulation* 1996, *94*, 1909–1912.

84 V. Krutovskikh, N. Mironov, and H. Yamasaki. Human connexin 37 is polymorphic but not mutated in tumours, *Carcinogenesis* 1996, *17*, 1761–1763.

85 C. W. Wong, T. Christen, I. Roth, C. E. Chadjichristos, J. P. Derouette, B. F. Foglia, M. Chanson, D. A. Goodenough, and B. R. Kwak. Connexin37 protects against atherosclerosis by regulating monocyte adhesion, *Nat. Med.* 2006, *12*, 950–954.

86 D. A. Gerido, A. M. Derosa, G. Richard, and T. W. White. Aberrant hemichannel properties of Cx26 mutations causing skin disease and deafness, *Am. J. Physiol. Cell Physiol.* 2007, *293*, C337–C345.

87 C. K. Abrams, M. V. Bennett, V. K. Verselis, and T. A. Bargiello. Voltage opens unopposed gap junction hemichannels formed by a connexin 32 mutant associated with X-linked Charcot-Marie Tooth disease, *Proc. Natl. Acad. Sci. USA* 2002, *99*, 3980–3984.

88 B. C. Stong, Q. Chang, S. Ahmad, and X. Lin. A novel mechanism for connexin 26 mutation linked deafness: cell death caused by leaky gap junction hemichannels, *Laryngoscope* 2006, *116*, 2205–2210.

89 D. P. Kelsell, J. Dunlop, H. P. Stevens, N. J. Lench, J. N. Liang, G. Parry, R. F. Mueller, and I. M. Leigh. Connexin 26 mutations in hereditary non-syndromic sensorineural deafness, *Nature* 1997, *387*, 80–83.

90 R. L. Snoeckx, P. L. Huygen, D. Feldmann, S. Marlin, F. Denoyelle, J. Waligora, M. Mueller-Malesinska, A. Pollak, R. Ploski, A. Murgia, E. Orzan, P. Castorina, U. Ambrosetti, E. Nowakowska-Szyrwinska, J. Bal, W. Wiszniewski, A. R. Janecke, D. Nekahm-Heis, P. Seeman, O. Bendova, M. A. Kenna, A. Frangulov, H. L. Rehm, M. Tekin, A. Incesulu, H. H. Dahl, D. du Sart, L. Jenkins, D. Lucas, M. Bitner-Glindzicz, K. B. Avraham, Z. Brownstein, I. del Castillo, F. Moreno, N. Blin, M. Pfister, I. Sziklai, T. Toth, P. M. Kelley, E. S. Cohn, L. Van Maldergem, P. Hilbert, A. F. Roux, M. Mondain, L. H. Hoefsloot, C. W. Cremers, T. Lopponen, H. Lopponen, A. Parving, K. Gronskov, I. Schrijver, J. Roberson, F. Gualandi, A. Martini, G. Lina-Granade, N. Pallares-Ruiz, C. Correia, G. Fialho, K. Cryns, N. Hilgert, P. Van de Heyning, C. J. Nishimura, R. J. Smith, and G. Van Camp. GJB2 mutations and degree of hearing loss: a multicenter study, *Am. J. Hum. Genet.* 2005, *77*, 945–957.

91 A. Alvarez, I. del Castillo, M. Villamar, L. A. Aguirre, A. Gonzalez-Neira, A. Lopez-Nevot, M. A. Moreno-Pelayo, and F. Moreno. High prevalence of the W24X mutation in the gene encoding connexin-26 (GJB2) in Spanish Romani (gypsies) with autosomal recessive non-syndromic hearing loss, *Am. J. Med. Genet. A* 2005, *137*, 255–258.

92 C. G. Meyer, G. K. Amedofu, J. M. Brandner, D. PohlAnd, C. Timmann, and R. D. Horstmann. Selection for deafness? *Nat. Med.* 2002, *8*, 1332–1333.

93 G. Richard, L. E. Smith, R. A. Bailey, P. Itin, D. Hohl, E. H. EpsteinJr, ., J. J. DiGiovanna, J. G. Compton, and S. J. Bale. Mutations in the human connexin gene GJB3 cause erythrokeratodermia variabilis [see comments], *Nat. Genet.* 1998, *20*, 366–369.

94 A. R. Janecke, H. C. Hennies, B. Gunther, G. Gansl, J. Smolle, E. M. Messmer, G. Utermann, and O. Rittinger. GJB2 mutations in keratitis-ichthyosis-deafness syndrome including its fatal form, *Am. J. Med. Genet. A* 2005, *133*, 128–131.

95 G. Richard, T. W. White, L. E. Smith, R. A. Bailey, J. G. Compton, D. L. Paul, and S. J. Bale. Functional defects of Cx26 resulting from a heterozygous missense mutation in a family with dominant deaf-mutism and palmoplantar keratoderma, *Hum. Genet.* 1998, *103*, 393–399.

96. D. P. Kelsell, A. L. Wilgoss, G. Richard, H. P. Stevens, C. S. Munro, and I. M. Leigh. Connexin mutations associated with palmoplantar keratoderma and profound deafness in a single family, *Eur. J. Hum. Genet.* 2000, *8*, 469–472.

97. E. Maestrini, B. P. Korge, J. Ocana-Sierra, E. Calzolari, S. Cambiaghi, P. M. Scudder, A. Hovnanian, A. P. Monaco, and C. S. Munro. A missense mutation in connexin26, D66H, causes mutilating keratoderma with sensorineural deafness (Vohwinkel's syndrome) in three unrelated families, *Hum. Mol. Genet.* 1999, *8*, 1237–1243.

98. M. L. Bondeson, A. M. Nystrom, U. Gunnarsson, and A. Vahlquist. Connexin 26 (GJB2) mutations in two Swedish patients with atypical Vohwinkel (mutilating keratoderma plus deafness) and KID syndrome both extensively treated with acitretin, *Acta Dermatol. Venereol.* 2006, *86*, 503–508.

99. F. Alexandrino, E. L. Sartorato, A. P. Marques-de-Faria, and C. E. Steiner. G59S mutation in the GJB2 (connexin 26) gene in a patient with Bart-Pumphrey syndrome, *Am. J. Med. Genet. A* 2005, *136*, 282–284.

100. G. Richard, N. Brown, A. Ishida-Yamamoto, and A. Krol. Expanding the phenotypic spectrum of Cx26 disorders: Bart-Pumphrey syndrome is caused by a novel missense mutation in GJB2, *J. Invest. Dermatol.* 2004, *123*, 856–863.

101. I. del Castillo, M. Villamar, M. A. Moreno-Pelayo, F. J. Del Castillo, A. Alvarez, D. Telleria, I. Menendez, and F. Moreno. A deletion involving the connexin 30 gene in nonsyndromic hearing impairment, *N. Engl. J. Med.* 2002, *346*, 243–249.

102. J. Lamartine, G. Munhoz Essenfelder, Z. Kibar, I. Lannelluc, E. Callouet, D. Laoudj, G. Lemaitre, C. Hand, S. J. Hayflick, J. Zonana, S. Antonarakis, U. Radhakrishna, D. P. Kelsell, A. L. Christianson, A. Pitaval, V. Der Kaloustian, C. Fraser, C. Blanchet-Bardon, G. A. Rouleau, and G. Waksman. Mutations in GJB6 cause hidrotic ectodermal dysplasia, *Nat. Genet.* 2000, *26*, 142–144.

103. S. M. Morley, M. I. White, M. Rogers, D. Wasserman, P. Ratajczak, W. H. McLean, and G. Richard. A new, recurrent mutation of GJB3 (Cx31) in erythrokeratodermia variabilis, *Br. J. Dermatol.* 2005, *152*, 1143–1148.

104. G. Richard, N. Brown, F. Rouan, J. G. Van der Schroeff, E. Bijlsma, L. F. Eichenfield, V. P. Sybert, K. E. Greer, P. Hogan, C. Campanelli, J. G. Compton, S. J. Bale, J. J. DiGiovanna, and J. Uitto. Genetic heterogeneity in erythrokeratodermia variabilis: novel mutations in the connexin gene GJB4 (Cx30.3) and genotype-phenotype correlations, *J. Invest. Dermatol.* 2003, *120*, 601–609.

105. J. H. Xia, C. Y. Liu, B. S. Tang, Q. Pan, L. Huang, H. P. Dai, B. R. Zhang, W. Xie, D. X. Hu, D. Zheng, X. L. Shi, D. A. Wang, K. Xia, K. P. Yu, X. D. Liao, Y. Feng, Y. F. Yang, J. Y. Xiao, D. H. Xie, and J. Z. Huang. Mutations in the gene encoding gap junction protein beta-3 associated with autosomal dominant hearing impairment [see comments], *Nat. Genet.* 1998, *20*, 370–373.

106. G. Richard, N. Brown, L. E. Smith, A. Terrinoni, G. Melino, R. M. Mackie, S. J. Bale, and J. Uitto. The spectrum of mutations in erythrokeratodermias – novel and de novo mutations in GJB3, *Hum. Genet.* 2000, *106*, 321–329.

107. J. Bergoffen, S. S. Scherer, S. Wang, M. O. Scott, L. J. Bone, D. L. Paul, K. Chen, M. W. Lensch, P. F. Chance, and K. H. Fischbeck. Connexin mutations in X-linked Charcot-Marie-Tooth disease, *Science* 1993, *262*, 2039–2042.

108. S. W. Yum, K. A. Kleopa, S. Shumas, and S. S. Scherer. Diverse trafficking abnormalities of connexin32 mutants causing CMTX, *Neurobiol. Dis.* 2002, *11*, 43–52.

109. K. A. Kleopa and S. S. Scherer. Molecular genetics of X-linked Charcot-Marie-Tooth disease, *Neuromolec. Med.* 2006, *8*, 107–122.

110. H. I. Yeh, Y. Chou, H. F. Liu, S. C. Chang, and C. H. Tsai. Connexin37 gene polymorphism and coronary artery disease in Taiwan, *Int. J. Cardiol.* 2001, *81*, 251–255.

111 Y. Yamada, H. Izawa, S. Ichihara, F. Takatsu, H. Ishihara, H. Hirayama, T. Sone, M. Tanaka, and M. Yokota. Prediction of the risk of myocardial infarction from polymorphisms in candidate genes, *N. Engl. J. Med.* 2002, *347*, 1916–1923.

112 M. H. Gollob, D. L. Jones, A. D. Krahn, L. Danis, X. Q. Gong, Q. Shao, X. Liu, J. P. Veinot, A. S. Tang, A. F. Stewart, F. Tesson, G. J. Klein, R. Yee, A. C. Skanes, G. M. Guiraudon, L. Ebihara, and D. Bai. Somatic mutations in the connexin 40 gene (GJA5) in atrial fibrillation, *N. Engl. J. Med.* 2006, *354*, 2677–2688.

113 M. Vreeburg, E. A. de Zwart-Storm, M. I. Schouten, R. G. Nellen, D. Marcus-Soekarman, M. Devies, M. van Geel, and M. A. Van Steensel. Skin changes in oculo-dento-digital dysplasia are correlated with C-terminal truncations of connexin 43, *Am. J. Med. Genet. A* 2007, *143*, 360–363.

114 A. Pizzuti, E. Flex, R. Mingarelli, C. Salpietro, L. Zelante, and B. Dallapiccola. A homozygous GJA1 gene mutation causes a Hallermann-Streiff/ODDD spectrum phenotype, *Hum. Mutat.* 2004, *23*, 286.

115 D. Mackay, A. Ionides, Z. Kibar, G. Rouleau, V. Berry, A. Moore, A. Shiels, and S. Bhattacharya. Connexin46 mutations in autosomal dominant congenital cataract, *Am J Hum. Genet.* 1999, *64*, 1357–1364.

116 M. I. Rees, P. Watts, I. Fenton, A. Clarke, R. G. Snell, M. J. Owen, and J. Gray. Further evidence of autosomal dominant congenital zonular pulverulent cataracts linked to 13q11 (CZP3) and a novel mutation in connexin 46 (GJA3), *Hum. Genet.* 2000, *106*, 206–209.

117 B. Uhlenberg, M. Schuelke, F. Ruschendorf, N. Ruf, A. M. Kaindl, M. Henneke, H. Thiele, G. Stoltenburg-Didinger, F. Aksu, H. Topaloglu, P. Nurnberg, C. Hubner, B. Weschke, and J. Gartner. Mutations in the gene encoding gap junction protein alpha 12 (connexin 46.6) cause Pelizaeus-Merzbacher-like disease, *Am. J. Hum. Genet.* 2004, *75*, 251–260.

118 L. Salviati, E. Trevisson, M. C. Baldoin, I. Toldo, S. Sartori, M. Calderone, R. Tenconi, and A. Laverda. A novel deletion in the GJA12 gene causes Pelizaeus-Merzbacher-like disease, *Neurogenetics* 2007, *8*, 57–60.

119 A. Shiels, D. Mackay, A. Ionides, V. Berry, A. Moore, and S. Bhattacharya. A missense mutation in the human connexin50 gene (GJA8) underlies autosomal dominant "zonular pulverulent" cataract, on chromosome 1q, *Am. J. Hum. Genet.* 1998, *62*, 526–532.

120 J. D. Pal, V. M. Berthoud, E. C. Beyer, D. Mackay, A. Shiels, and L. Ebihara. Molecular mechanism underlying a Cx50-linked congenital cataract, *Am. J. Physiol.* 1999, *276*, C1443–C1446.

121 W. Roscoe, G. I. Veitch, X. Q. Gong, E. Pellegrino, D. Bai, E. McLachlan, Q. Shao, G. M. Kidder, and D. W. Laird. Oculodentodigital dysplasia-causing connexin43 mutants are non-functional and exhibit dominant effects on wild-type connexin43, *J. Biol. Chem.* 2005, *280*, 11458–11466.

122 S. M. Deschenes, J. L. Walcott, T. L. Wexler, S. S. Scherer, and K. H. Fischbeck. Altered trafficking of mutant connexin32, *J. Neurosci.* 1997, *17*, 9077–9084.

123 J. K. VanSlyke, S. M. Deschenes, and L. S. Musil. Intracellular transport, assembly, and degradation of wild-type and disease-linked mutant gap junction proteins, *Mol. Biol. Cell* 2000, *11*, 1933–1946.

124 P. J. Minogue, X. Liu, L. Ebihara, E. C. Beyer, and V. M. Berthoud. An aberrant sequence in a connexin46 mutant underlies congenital cataracts, *J. Biol. Chem.* 2005, *280*, 40788–40795.

125 V. M. Berthoud, P. J. Minogue, J. Guo, E. K. Williamson, X. Xu, L. Ebihara, and E. C. Beyer. Loss of function and impaired degradation of a cataract-associated mutant connexin50, *Eur. J. Cell Biol.* 2003, *82*, 209–221.

126 A. Lai, D. N. Le, W. A. Paznekas, W. D. Gifford, E. W. Jabs, and A. C. Charles. Oculodentodigital dysplasia connexin43 mutations result in non-functional connexin hemichannels and gap

junctions in C6 glioma cells, *J. Cell Sci.* 2006, *119*, 532–541.

127. I. M. Skerrett, W. L. Di, E. M. Kasperek, D. P. Kelsell, and B. J. Nicholson. Aberrant gating, but a normal expression pattern, underlies the recessive phenotype of the deafness mutant Connexin26M34T, *FASEB J.* 2004, *18*, 860–862.

128. F. Rouan, T. W. White, N. Brown, A. M. Taylor, T. W. Lucke, D. L. Paul, C. S. Munro, J. Uitto, M. B. Hodgins, and G. Richard. trans-dominant inhibition of connexin-43 by mutant connexin-26: implications for dominant connexin disorders affecting epidermal differentiation, *J. Cell Sci.* 2001, *114*, 2105–2113.

129. J. Gemel, V. Valiunas, P. R. Brink, and E. C. Beyer. Connexin43 and connexin26 form gap junctions, but not heteromeric channels in co-expressing cells, *J. Cell Sci.* 2004, *117*, 2469–2480.

130. T. Thomas, D. Telford, and D. W. Laird. Functional domain mapping and selective trans-dominant effects exhibited by Cx26 disease-causing mutations, *J. Biol. Chem.* 2004, *279*, 19157–19168.

131. K. Willecke, H. Hennemann, E. Dahl, and S. Jungbluth, The mouse connexin gene family, in: J. E. Hall, G. A. Zampighi, and R. M. Davis (Eds.). *Progress in Cell Research*, Volume 3, 1993, Elsevier Science Publishers: Amsterdam, pp. 33–37.

132. A. Plum, G. Hallas, T. Magin, F. Dombrowski, A. Hagendorff, B. Schumacher, C. Wolpert, J. Kim, W. H. Lamers, M. Evert, P. Meda, O. Traub, and K. Willecke. Unique and shared functions of different connexins in mice, *Curr. Biol.* 2000, *10*, 1083–1091.

133. T. W. White. Unique and redundant connexin contributions to lens development, *Science* 2002, *295*, 319–320.

134. G. Y. Huang, A. Wessels, B. R. Smith, K. K. Linask, J. L. Ewart, and C. W. Lo. Alteration in connexin 43 gap junction gene dosage impairs conotruncal heart development, *Dev. Biol.* 1998, *198*, 32–44.

135. J. Chung, V. M. Berthoud, L. Novak, R. Zoltoski, B. Heilbrunn, P. J. Minogue, X. Liu, L. Ebihara, J. Kuszak, and E. C. Beyer. Transgenic overexpression of connexin50 induces cataracts, *Exp. Eye Res.* 2007, *84*, 513–528.

136. S. Ahmad, W. Tang, Q. Chang, Y. Qu, J. Hibshman, Y. Li, G. Sohl, K. Willecke, P. Chen, and X. Lin. Restoration of connexin26 protein level in the cochlea completely rescues hearing in a mouse model of human connexin30-linked deafness, *Proc. Natl. Acad. Sci. USA* 2007, *104*, 1337–1341.

137. H. D. Gabriel, D. Jung, C. Butzler, A. Temme, O. Traub, E. Winterhager, and K. Willecke. Transplacental uptake of glucose is decreased In embryonic lethal connexin26-deficient mice, *J. Cell Biol.* 1998, *140*, 1453–1461.

138. A. G. Reume, P. A. de Sousa, S. Kulkarni, B. L. Langille, D. Zhu, T. C. Davies, S. C. Juneja, G. M. Kidder, and J. Rossant. Cardiac malformation in neonatal mice lacking connexin43, *Science* 1995, *267*, 1831–1834.

139. M. Cohen-Salmon, T. Ott, V. Michel, J. P. Hardelin, I. Perfettini, M. Eybalin, T. Wu, D. C. Marcus, P. Wangemann, K. Willecke, and C. Petit. Targeted ablation of connexin26 in the inner ear epithelial gap junction network causes hearing impairment and cell death, *Curr. Biol.* 2002, *12*, 1106–1111.

140. M. Theis, C. de Wit, T. M. Schlaeger, D. Eckardt, O. Kruger, B. Doring, W. Risau, U. Deutsch, U. Pohl, and K. Willecke. Endothelium-specific replacement of the connexin43 coding region by a lacZ reporter gene, *Genesis* 2001, *29*, 1–13.

141. Y. Liao, K. H. Day, D. N. Damon, and B. R. Duling. Endothelial cell-specific knockout of connexin 43 causes hypotension and bradycardia in mice, *Proc. Natl. Acad. Sci. USA* 2001, *98*, 9989–9994.

142. A. M. Flenniken, L. R. Osborne, N. Anderson, N. Ciliberti, C. Fleming, J. E. Gittens, X. Q. Gong, L. B. Kelsey, C. Lounsbury, L. Moreno, B. J. Nieman, K. Peterson, D. Qu, W. Roscoe, Q. Shao, D. Tong, G. I. Veitch, I. Voronina, I. Vukobradovic, G. A. Wood, Y. Zhu, R. A. Zirngibl, J. E. Aubin, D. Bai, B. G.

Bruneau, M. Grynpas, J. E. Henderson, R. M. Henkelman, C. McKerlie, J. G. Sled, W. L. Stanford, D. W. Laird, G. M. Kidder, S. L. Adamson, and J. Rossant. A Gja1 missense mutation in a mouse model of oculodentodigital dysplasia, *Development* 2005, *132*, 4375–4386.

143 T. Kudo, S. Kure, K. Ikeda, A. P. Xia, Y. Katori, M. Suzuki, K. Kojima, A. Ichinohe, Y. Suzuki, Y. Aoki, T. Kobayashi, and Y. Matsubara. Transgenic expression of a dominant-negative connexin26 causes degeneration of the organ of Corti and non-syndromic deafness, *Hum. Mol. Genet.* 2003, *12*, 995–1004.

144 Y. Maeda, K. Fukushima, K. Nishizaki, and R. J. Smith. In vitro and in vivo suppression of GJB2 expression by RNA interference, *Hum. Mol. Genet.* 2005, *14*, 1641–1650.

145 K. Parthasarathi, H. Ichimura, E. Monma, J. Lindert, S. Quadri, A. Issekutz, and J. Bhattacharya. Connexin 43 mediates spread of Ca2+-dependent proinflammatory responses in lung capillaries, *J. Clin. Invest.* 2006, *116*, 2193–2200.

146 S. Zahler, A. Hoffmann, T. Gloe, and U. Pohl. Gap-junctional coupling between neutrophils and endothelial cells: a novel modulator of transendothelial migration, *J. Leukoc. Biol.* 2003, *73*, 118–126.

147 E. Oviedo-Orta, T. Hoy, and W. H. Evans. Intercellular communication in the immune system: differential expression of connexin40 and 43, and perturbation of gap junction channel functions in peripheral blood and tonsil human lymphocyte subpopulations, *Immunology* 2000, *99*, 578–590.

148 P. F. Davies, C. Shi, N. Depaola, B. P. Helmke, and D. C. Polacek. Hemodynamics and the focal origin of atherosclerosis: a spatial approach to endothelial structure, gene expression, and function, *Ann. N. Y. Acad. Sci.* 2001, *947*, 7–16; discussion 16–17.

149 J. E. Gabriels and D. L. Paul. Connexin43 is highly localized to sites of disturbed flow in rat aortic endothelium but connexin37 and connexin40 are more uniformly distributed, *Circ. Res.* 1998, *83*, 636–643.

150 D. Polacek, F. Bech, J. F. McKinsey, and P. F. Davies. Connexin43 gene expression in the rabbit arterial wall: effects of hypercholesterolemia, balloon injury and their combination, *J. Vasc. Res.* 1997, *34*, 19–30.

151 H. I. Yeh, F. Lupu, E. Dupont, and N. J. Severs. Upregulation of connexin43 gap junctions between smooth muscle cells after balloon catheter injury in the rat carotid artery, *Arterioscler. Thromb. Vasc. Biol.* 1997, *17*, 3174–3184.

152 B. R. Kwak, F. Mulhaupt, N. Veillard, D. B. Gros, and F. Mach. Altered pattern of vascular connexin expression in atherosclerotic plaques, *Arterioscler. Thromb. Vasc. Biol.* 2002, *22*, 225–230.

153 C. E. Chadjichristos, C. M. Matter, I. Roth, E. Sutter, G. Pelli, T. F. Luscher, M. Chanson, and B. R. Kwak. Reduced connexin43 expression limits neointima formation after balloon distension injury in hypercholesterolemic mice, *Circulation* 2006, *113*, 2835–2843.

154 P. I. Jara, M. P. Boric, and J. C. Saez. Leukocytes express connexin 43 after activation with lipopolysaccharide and appear to form gap junctions with endothelial cells after ischemia-reperfusion, *Proc. Natl. Acad. Sci. USA* 1995, *92*, 7011–7015.

155 D. Polacek, R. Lal, M. V. Volin, and P. F. Davies. Gap junctional communication between vascular cells. Induction of connexin43 messenger RNA in macrophage foam cells of atherosclerotic lesions, *Am. J. Pathol.* 1993, *142*, 593–606.

156 C. W. Wong, T. Christen, A. Pfenniger, R. W. James, and B. R. Kwak. Do allelic variants of the connexin37 1019 gene polymorphism differentially predict for coronary artery disease and myocardial infarction? *Atherosclerosis* 2007, *191*, 355–361.

157 M. Boerma, L. Forsberg, L. Van Zeijl, R. Morgenstern, U. De Faire, C. Lemne, D. Erlinge, T. Thulin, Y. Hong, and I. A. Cotgreave. A genetic polymorphism in connexin 37 as a prognostic marker for atherosclerotic plaque development, *J. Intern. Med.* 1999, *246*, 211–218.

11
Tight Junctions in Simple and Stratified Epithelium

Cara J. Gottardi and Carien M. Niessen

11.1
Introduction

An important feature of epithelial tissues is their ability to form a physical and functional barrier between different compartments. Tight junctions comprise a crucial part of epithelial barriers by forming a molecular pore-like structure that encircles the apex of individual cells, thereby controlling the passage of small molecules between lumenal (apical) and serosal (basolateral) compartments (for a review, see Ref. [1]). The size- and ion-permeability of a particular barrier varies considerably between different epithelia, and depends on the molecular composition of these junctions. Tight junctions are also thought to form a fence between apical and basolateral domains, restricting diffusion of proteins and lipids between these membrane regions. Historically, tight junctions were thought to be functionally restricted to simple epithelia, such as those lining the kidney tubules, gastrointestinal tract, and lung airways. This view has changed, however, as results of recent studies with skin have revealed that tight junction components are also critically required for barrier functions in stratifying epithelia, long considered to rely on other barrier-control mechanisms. An important conceptual advance is that the tight junction is not only crucial for regulating epithelial barriers, but also defines a structural landmark in cells that drives cell polarization by providing spatial cues for signaling complexes and docking sites for vesicle transport. These distinct functions are reflected in the diverse and extensive molecular complexity of tight junctions [2]. In this chapter, an overview is provided of the structure and molecular composition of tight junctions in simple and stratifying epithelia, and how this relates to the functional importance of tight junctions in epithelial physiology and pathology.

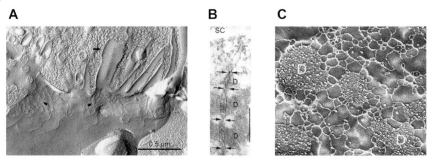

Figure 11.1 Ultrastructural appearance of tight junctions in simple and stratifying epithelia. (A) Freeze-fracture electron microscopy of the apical membrane domain of airway simple epithelia. The arrows indicate the microvilli, whereas the tight junction strands are indicated by arrowheads. (Illustration courtesy of J. Carson, University of North Carolina.) (B) Tight junctions in the stratum granulosum of human epidermis. Tight junctions are often found closely associated with desmosomes (D), and show a similar close apposition of apposing membranes, the so-called "kissing points" (arrows), as observed in simple epithelia. The stratum corneum (SC) overlies the tight junctions in the granulosum layer. (C) Freeze-fracture electron microscopy of the upper stratum granulosum, showing a highly continuous network of bead-like strands closely resembling tight junctional strands in simple epithelia. (Illustrations B and C courtesy of Dr. W. Franke and Elsevier [7].)

11.2
Ultrastructure of Tight Junctions

Tight junctions were first identified as the most apical structure of the terminal bar, a tripartite junctional complex bordering the apico-basolateral membrane in a variety of simple polarized epithelia [3]. Adherens junctions (AJs) and desmosomes form the other two structures of this complex, and these are discussed in Chapters 8 and 12, respectively. Relative to these other junctions, the intercellular membrane space of tight junctions is almost completely obliterated, inspiring their alternative name, zonulae occludens ("kissing points") (Figure 11.1). This transverse ultrastructural view also reveals an electron-dense cytoplasmic plaque that appears closely associated with a circumferential belt of actin filaments. Freeze-fracture replica electron microscopy, which shows an *en face* view of structures between apposing cells (see Figure 11.1), reveals a network of interconnected, anastomosing strands that resemble beaded chains [4, 5]. Since a tight junction strand from one cell appears to associate laterally with a strand from the opposing cell, these strands were long considered to comprise the structural units (i.e., the seal) of the tight junction barrier. In stratifying epithelia, similar strand structures are observed in the uppermost viable layers (Figure 11.1) [6, 7]. The number, length and interconnectivity of these strands vary considerably between different epithelia [8], and the way in which these structural features contribute to barrier function is currently under investigation.

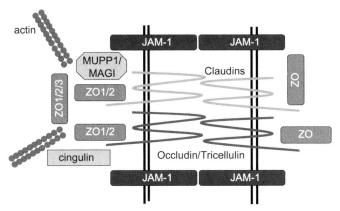

Figure 11.2 A schematic overview of the structural tight junction components.

11.3
Tight Junctions and Epithelial Barrier Function

The first indication that tight junctions were important for epithelial barrier function was obtained from electron-dense tracer studies that revealed a diffusion limit at the ultrastructural tight junction [3]. While limited, this information nonetheless predicted that three functional classes of proteins would be required for barrier function (Figure 11.2): (i) transmembrane proteins that can mediate adhesion ("kissing points"; see Figure 11.1) and form the core of the tight junctional strands; (ii) scaffolding proteins that cluster the transmembrane proteins and form a connection to the cytoskeleton; and (iii) actin and related cytoskeletal proteins.

11.3.1
Transmembrane Components

Occludin was the first identified transmembrane component of tight junctions, and the first component specifically localized to tight junction strands by freeze fracture/immuno-electron microscopy labeling [9]. As with the other tight junction membrane components described below, occludin spans the membrane four times, and its two extracellular loops are important for function. Specifically, peptides corresponding to the sequence of these loops could interfere with barrier function in cells and *Xenopus laevis* embryos [10], perhaps through an ability for these loops to mediate a form of Ca^{2+}-independent adhesion [11]. However, occludin expression is not sufficient to induce the formation of strands, and tight junction structure and function appear unperturbed in cells deficient for occludin [12, 13]. Of interest, occludin-null mice exhibit several different phenotypes, such as growth retardation, mineral deposits in brain, male sterility and gastritis, although how these phenotypes are related to alterations in the barrier or other tight junction functions (see below) is presently unclear [13].

A necessity for occludin in barrier function may be compensated by *tricellulin*, which is a structurally related, four membrane-spanning protein that shows similarity to occludin in the extracellular loops. However, unlike other tight junction proteins, tricellulin appears enriched at, and to enforce contacts, where three cells meet (i.e., at tricellular junctions) [14]. The features of tricellulin that control its targeting to this tight junction subdomain, and how tricellulin uniquely seals spaces between three apposing cells, remains to be defined. Since occludin knockout and tricellulin knockdown phenotypes show relatively mild barrier defects, it is speculated that these proteins compensate for each other, and that double knockouts may reveal essential roles for these related proteins.

The presence of tight junctions in the absence of occludin led Tsukita and colleagues to identify a new class of four membrane-spanning tight junction proteins, named *claudins*. In contrast to occludin, claudins not only mediate Ca^{2+}-independent adhesion but are also essential for strand formation [1, 15]. Specifically, the exogenous expression of claudins in fibroblasts is sufficient to generate tight junction strands that appear similar in morphological terms to fully polarized epithelia; conversely, strands are absent in Sertoli cells of claudin-11-null mice [16]. At present, the means by which claudins co-assemble and drive the formation of these strands is under active investigation.

At least 24 members of the claudin family have now been identified, with each showing a specific organ and tissue distribution [1]. The forced expression of particular claudins, for example in Madin–Darby canine kidney (MDCK) cells, can influence the ion selectivity of the barrier and, remarkably, this selectivity correlates well with the permeability characteristics of epithelia that express these respective claudins *in vivo* [17]. Ion selectivity is determined by the extracellular loop regions of claudins which, in contrast to occludin and tricellulin, are charged and exhibit a wide range of isoelectric points. For example, reversing the key charge residues in the first extracellular loop of claudin-15 is sufficient to change the ion selectivity of the barrier [18]. Thus it is now widely recognized that the variation in ion specificity of tight junction barriers in different epithelia is largely due to the type of claudin(s) expressed [1, 15]. How tight junction pores restrict the size of molecules that pass through the junction is less clear, but this effect may be controlled by some of the same components that regulate ion-selective barrier, as claudin-5-null mice showed size-selective blood–brain barrier defects [19].

The IgG-like family of junctional adhesion molecules (JAMs) is the third group of transmembrane components found at tight junctions. This family consists of the closely related molecules JAM-A, -B and -C, and the more distantly related Coxsackie and adenovirus receptor (CAR), endothelial cell-selective adhesion molecule (ESAM) and JAM-4 [20]. While JAMs can engage in homophilic and heterophilic adhesion, they do not induce the formation of tight junction strands when expressed in fibroblasts. Nevertheless, JAMs appear to be important for tight junctions, as the overexpression of mutant JAM-A proteins can interfere with both the ion and size barrier [21]. JAMs are not exclusively expressed in cells that form tight junctions, but are also found on leukocytes where they mediate transendothelial migration [20]. In addition, Jam-C regulates polarization and the differentiation of spermatids by recruiting polarity protein complexes [22]. Thus, JAMs may be more crucial for regulating cell polarization which may only indirectly affect tight junction barrier function (see below).

11.3.2
Scaffolding Proteins

Similar to what has been shown for integrin- and cadherin-mediated adhesion [23, 24], the incorporation of occludin, claudins and JAMs into tight junction-strands may require the local clustering of these proteins. As no direct interactions have been found between occludin, claudins and JAMs, the cytoplasmic binding partners must fulfill this scaffolding function. One important group of tight junction scaffolding molecules are the zonula occluden proteins ZO-1, ZO-2, and ZO-3. These proteins belong to the MAGUK (membrane-associated guanylate kinase-like homologues) family, and are characterized by three N-terminal PDZ domains, and an SH3 domain followed by the GUK domain. These ZO-proteins can bind directly to claudins via their PDZ1 domain and with occludin via their GUK domain, while their C termini can associate with actin [2, 25]. ZO-1 can also interact with JAMs and form homodimers or heterodimers with either ZO-2 or ZO-3. Thus, ZO proteins are ideal scaffold candidates that can provide a direct link between different transmembrane components and the underlying cytoskeleton. Indeed, a recent study has clearly demonstrated that ZO-1 or ZO-2 are crucial for claudin clustering, strand formation, and barrier function [26].

Several other PDZ-containing scaffolding proteins, such as MUPP1 and MAGI proteins, are associated with the tight junction plaque, and can directly interact with one or more of the tight junction transmembrane components [2]. It is at present unclear if these molecules are directly involved in the formation of the tight junctions, or serve a more regulatory function.

11.3.3
Actin and Related Cytoskeleton Proteins

Early ultrastructural analyses showed that actin is intimately associated with the cytoplasmic face of tight junctions. Not surprisingly, the disruption of actin filaments, for example by cytochalasin D or dominant active/inhibitory forms of Rho family GTPases, can increase paracellular permeability and decrease transepithelial resistance, indicating that actin is important for maintaining the barrier [27, 28]. The means by which actin dynamics are locally modulated to affect tight junction function is of great interest, and recently acquired evidence has indicated that Rho family guanidine exchange factors (GEFs) and GTPAse activating proteins (GAPS) are enriched at tight junctions [29]. For example, ZO-1 binds to the CDC42-specific GEF Tuba, which seems to control the shape of junctions [30]. In addition, the forced activation of actinomyosin contraction by expression of constitutively active myosin light chain kinase can increase epithelial permeability [31]. Cingulin, a tight junction plaque protein, interacts with ZO proteins 1 to 3, JAMs and actin via its head domain, while its central rod domain is required for homodimerization and can interact with myosin and RhoGEF1 [32]. As such, this protein may be an important regulator of tight junctions by coupling transmembrane components to the actinomyosin contraction machinery. As mutant forms of cingulin that cannot interact with myosin fail to affect tight junction function [33], it is presumed that there may be multiple, redundant mechanisms for coupling actin and myosin to tight junctions.

11.4
Regulation of Tight Junction Barrier Function

One of the clearest examples of barrier regulation is during diapedisis, when leukocytes migrate across an endothelial barrier [34]. Importantly, cell rearrangement or migration occurs while preserving the epithelial barrier function, which implies that tight junctions are intrinsically dynamic structures. In this regard, live cell imaging of GFP-claudin reveals the highly dynamic rearrangements of tight junctional strands [35].

The molecular events that drive the dynamic formation and maintenance of tight junctions, either under steady state or under conditions that challenge the barrier function of tight junctions, such as wounding, inflammation or infection, are not well understood A range of mechanisms has been implicated, such as phosphorylation, endocytosis and transcription of key junctional components, as well as actinomyosin contractility, all of which are not necessarily mutually exclusive [36, 37]. Here, the key question that remains is what the initiating event(s) is/are that control association/dissociation of tight junctions.

11.5
Tight Junctions and Epithelial Polarity

A strong interdependent relationship exists between intercellular junction formation and the establishment of apico-basolateral polarity [38]. During recent years, exciting physical interactions between polarity complexes and structural components of the tight junction have been reported, which provide the first molecular evidence that these two processes are interconnected [39]. Studies conducted in *Drosophila* resulted in the identification of two multiprotein complexes, the Par3/Par6/aPKC complex and the Crumbs/Pals/Patj complex, both of which are crucial for the establishment of the apical membrane domain. The discovery that both of these protein complexes are localized to tight junctions in vertebrate epithelia strongly suggested that a tight junction structure may critically contribute to the establishment of epithelial polarity (see below). In contrast, the functional interference of these polarity proteins affects paracellular permeability, thereby indicating their importance in the assembly of functional tight junctions [39–42].

11.5.1
Par3/Par6/aPKC Complex

Atypical protein kinase C (aPKC) belongs to the protein kinase C serine/threonine kinase family but, unlike conventional PKCs, aPKC does not depend on phorbol esters or Ca^{2+} for its activation. This kinase was identified as a key regulator of nearly all forms of polarity in organisms, ranging from *Drosophila* to humans [42]. aPKC function has also been implicated in inflammatory and metabolic pathways [43], which suggests that these kinases can couple structural integrity to the regulation of growth and inflammation. Vertebrates have two highly related proteins, aPKCζ and aPKCι/λ, which are encoded by different genes, although it remains an open question as to whether the two aPKCs share common and/or separate functions.

aPKCs form a complex with Par3 and Par6 – proteins that were initially discovered in the symmetry-breaking event during the first cell division of *Caenorhabditis elegans* embryos. Par6 appears to play a crucial role in the activation of the complex by binding to activated forms of Cdc42/Rac. This results in a conformational change in Par6, thereby releasing aPKC auto-inhibition [44]. Three isoforms exist of Par6, Par6A, Par6B and Par6C, which are encoded by different genes and appear to have distinct effects on tight junctions [45].

Par3 is thought to be crucial for the positioning of the complex by directly interacting with JAMs [20]. The loss of Jam-C *in vivo* results in a mislocalization of key polarity proteins such as Par6 and Cdc42 [22]. Par3 may also regulate tight junctions via its interaction with the Rac-specific exchange factor TIAM1, although conflicting data exist as to whether this stimulates or inhibits barrier formation, and if this requires its interaction with aPKC and Par6 [46, 47]. The inactivation of Par3 affects murine heart development, which may be due to polarity defects associated with a loss of aPKC and Par6 from the membrane [48]. Lastly, there are at least six different splice variants of Par3, with apparently different capacities to associate with tight junctions, thereby underscoring the importance and complexity of Par3 functions [49, 50]. The biological significance of these distinct forms is presently unknown.

11.5.2
Crbs/Patj/Pals

Crumbs (Crb) was initially identified as a protein crucial for the formation of apical membranes in *Drosophila* [51]. It has been shown that three Crbs (1–3) exist in mammals. Whereas Crb2 is largely uncharacterized, Crb1 is mostly expressed in the brain and eyes, and its loss affects photoreceptor polarization [39]. Crb3 is localized at tight junctions, and its expression is sufficient to induce tight junction formation in MCF-10 cells [52]. Crb3 interacts directly with the MAGUK protein PALS1, the mammalian homologue of *Drosophila* Stardust, via its C-terminal PDZ-binding domain. The PALS1 interaction domain, together with the FERM domain, is crucial for Crb3 to regulate tight junction formation [52]. Since the stability of

PALS1 and Crb are mutually dependent, and PALS1 knockdown cells exhibit tight junction defects [53], it is likely that a Crb3/PALS1 complex is required for barrier function.

PALS1 links Crb with PATJ, a multi-PDZ domain protein also found enriched at tight junctions [54, 55]. PATJ may be required for apical targeting of the Crb complex in epithelia, since PATJ downregulation causes Crb3 accumulation in an intracellular compartment and reduces the stability of the PALS/Crb complex [56, 57]. As PATJ directly couples the Crb complex to components required for strand formation (e.g., claudins and ZO-3 [58]), it intimately links tight junction structure to polarity establishment.

These two complexes cooperate functionally to establish polarity, and this is reflected by direct interactions between them. Par6 can bind Crb3 and PALS1 [59, 60], but these interactions appear mutually exclusive from Crb-Pals or Pals-Patj, respectively, due to competing sites. *Drosophila* aPKC can interact with PATJ and Crb, and aPKC-dependent phosphorylation regulates epithelial polarity [61]. However, how these two complexes cooperate functionally to couple the establishment of polarity to the formation of tight junctions remains an open question.

11.5.3
Fence Function

It has been postulated that tight junctions regulate epithelial polarity by forming an intramembranous fence that prevents protein and lipid diffusion between apical and basolateral membrane domains. The best evidence for such a fence function is derived from studies that implicate tight junctions in the regulation of growth by physically separating growth factor receptors and either their coreceptors or ligands to different membrane domains. For example, ErbB2 is often found at the apical membrane in polarized epithelia; this is in contrast to its kinase inactive co-receptor ErbB3, which localizes to AJs at the basolateral membrane domain [62]. Similarly, in polarized airway epithelia the receptors Erb3/4 localize to a basolaterally membrane domain, whereas their ligand is found apically in the overlying airway surface liquid [63]. Only the perturbation of polarity (e.g., by wounding) or selective perturbation of the tight junctional fence/barrier function (e.g., by inflammatory stimuli), allows ligand binding and subsequent activation of the heterodimeric receptor [62].

The existence of a fence function has been recently questioned, as the loss of strands (due to an absence of ZO-1 and -2 proteins) failed to affect the localization of apical markers [26]. This finding was in agreement with the observation that membrane polarity can be observed in single epithelial cells [64]. ZO-1 and/or -2 downregulation did, however, cause a partial loss of polarity, as judged by Par3 and JamA localization [26, 65]. More importantly, E-cadherin leaked into the apical surface, indicating a disturbance of the fence function [65]. Occludin may play a crucial role here, as antibodies to the second extracellular loop disrupt polarity [66], and a loss of polarity during transforming growth factor-β (TGF-β)-induced epithelial to mesenchymal transition appears to require interactions of the TGF-β receptor with the polarity protein Par6 and occludin [67]. Thus, whilst tight junctions may not be crucial for apical membrane identity and apical targeting of proteins, they do appear to be crucial for polarity and the fence function.

11.6
Signaling from Tight Junctions: Coupling Junction Maturation to Transcription-Mediated Differentiation

Non-transformed cells can sense when their surfaces contact other cells, and respond accordingly by limiting their proliferative and motile characteristics, ultimately undergoing terminal differentiation. This phenomenon is known as "contact inhibition", and a loss of this property results in overgrowth and partial epithelial-to-mesenchymal transition. An emerging theme in junction signaling is that some of the structural components of junctions also control the activity of nuclear transcription factors. For example, β-catenin and p120^{ctn} are not only core components of the cadherin adhesive complex central to AJs (see Chapter 8), but β-catenin binds LEF/TCF (lymphoid enhancer factor/T-cell factor)-family transcription factors transcription factors and recruits histone acetylases and chromatin remodeling factors [68], while p120^{ctn} binds and derepresses the transcriptional repressor, Kaiso [69, 70]. Structural components of tight junctions may also directly regulate transcription factor activity. For example, ZO-1 can bind the Y-box transcription factor ZONAB (ZO-1-associated nucleic acid-binding protein) and sequester it at mature tight junctions, thus inhibiting ZONAB target genes, such as cyclin D1 and PCNA [71]. Whilst it has been reported that ZO-1 can localize to nuclei in cells with immature junctions, as during wound healing [72], this observation has not been widely confirmed. Thus, the suggestion that ZO-1 might affect transcription within the nuclear compartment remains speculative. ZO-2, however, has been observed independently to localize to the nucleus, where it may interact with – and potentially inhibit – the AP1 family of transcription factors [73]. Whilst evidence that ZO-proteins affect nuclear transcription functions analogous to catenins remains suggestive, it is clear that a mature tight junction structure can provide a scaffold for the sequestration of transcription factors (e.g., ZONAB). Of course, alternative modes of tight junction signaling will also be important. For example, cingulin can bind the Rho-specific exchange factor GEF-H1, which not only affects paracellular permeability through local modulation of the cytoskeleton, but can also regulate cell cycling through Rho [32]. Thus, as with AJs, tight junctions will most likely utilize similar modes of signaling, from direct binding to transcription factors to the local modulation of tight junction/cytoskeletal structure that initiates a cascade of downstream events.

11.7
Tight Junctions in Stratifying Epithelia

The initial identification of tight junctions in simple epithelia raised the question of whether these structures also exist in stratifying epithelia, where they might contribute to their unique epithelial barrier functions. For keratinizing epithelia, it was originally thought that the secretion and deposition of a crosslinked protein–lipid barrier obviated the need for a tight junction barrier in such tissues. However, electron-dense tracer studies, combined with ultrastructural analysis, revealed that diffusion was restricted to the uppermost viable layer, the stratum granulosum [74, 75]. Although tight junction strands were observed by freeze-fracture, they appeared incomplete and thus the barrier was largely attributed to the intercellular deposition of lamellar bodies [6, 76]. For some time this view remained unchallenged, despite the continued localization/identification of tight junction components in stratifying epithelia [77, 78]. However, a seminal breakthrough was made with the observation that mice lacking the claudin 1 gene displayed massive epidermal water loss due to impaired barrier function of the stratum granulosum [79]. This provided the first functional evidence that a tight junction component was required for barrier function in the epidermis. Subsequently, a dense network of strands resembling tight junctions was shown to be present in the stratum granulosum of human epidermis [7]. It is now widely accepted that tight junctions form a critical part of the barrier in the upper viable layer of stratifying epithelia (Figures 11.1 and 11.3).

Tight junctions and the stratum corneum may cooperate in the formation of a functional barrier in stratifying epithelia. For example, the overexpression of claudin-6 in the upper layers of the epidermis can induce barrier defects that affect both tight junction and stratum corneum barriers [80]. Inactivation of the membrane-anchored serine protease (CAP)1/Prss8 in the epidermis can also disturb both barriers [81]. Although the underlying mechanisms are unknown, they may involve the coordinated regulation of both barriers by signal molecules such as IKK1. Inactivation of IKK1 in the epidermis severely impairs barrier function

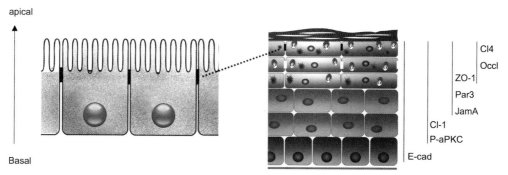

Figure 11.3 A schematic overview of the functional similarity of tight junctions in simple epithelia (left) and stratifying epithelia (right). In each case, tight junctions border a lumen/apical matrix (simple epithelia) or a dead cell layer (stratifying epithelia), indicating functional similarity. In stratifying epithelia, proteins (stellate) and lipid vesicles (circular) are targeted to the stratum corneum, which is equivalent to vesicle transport to the apical domain in simple epithelia. Also indicated is the membrane localization of several tight junction and polarity proteins in the different layers of stratifying epithelia.

associated with improper epidermal lipid processing and changes in tight junction component expression. Epidermal IKK1 function is independent of NFκB signaling, but regulates the expression of retinoic acid receptor target genes, many of which are involved in epidermal barrier function [82]. As formation of the stratum corneum depends on the fusion of lamellar bodies and keratohyalin granules with plasma membranes at the transition between stratum granulosum and stratum corneum layers, it is tempting to speculate that the specific occurrence of tight junctions in the stratum granulosum may target "apical" protein and lipid vesicles directly towards the stratum corneum.

The epidermis is not a classically polarized epithelium like intestine, where the basolateral and apical domains are separated by tight junctions. Instead, the epidermis establishes a form of junctional polarity along the apical to basal axis of the tissue, with the stratum granulosum forming the viable apical boundary (see Figure 11.3). What restricts the assembly of a tight junctional barrier to the uppermost layers of stratifying epithelia is unclear, but as many tight junction components are expressed throughout the epidermal layers it is possible that a local signal in the granular layer triggers tight junction formation. It is speculated that the dead cell/keratin layer may initiate this signal, similar to other simple epithelia, which secrete and are polarized by an apical matrix (e.g., the follicular epithelia in flies that secrete an apical cuticle, or epithelia in thyroid that secrete thyroglobulin colloid). Alternatively, tight junction formation in the lower layers may be actively inhibited by the presence of an overlying viable cell layer. Either mechanism suggests that the restriction of tight junctions to the apical most layer in stratifying epidermis or apical region of simple epithelia may be conserved. Indeed, recent evidence supports this argument, as E-cadherin is required for tight junction formation in both simple and stratifying epithelia. The blocking of E-cadherin function *in vitro* either inhibits or delays tight junctions in simple epithelia [83], whereas a genetic loss of E-cadherin in epidermis perturbs the tight junction barrier, similar to mice lacking claudin-1 [84]. How E-cadherin regulates tight junctions is not yet clear, but this may involve aPKC, as blocking aPKC function in stratifying keratinocytes inhibits *in-vitro* barrier formation [85]. Thus, the junctional and polarity proteins required in simple epithelia also appear to be critical for epidermal barrier function, suggesting that these processes may be mechanistically related.

11.8
Tight Junctions and Disease

The importance of maintaining a regulatory barrier that controls the distribution of small ionic and non-ionic molecules between the lumen and the interstitial compartment of epithelia is best underscored by pathological conditions resulting from a disturbance of the barrier. In certain cases, structural alterations in the barrier are the primary cause of disease, but in other cases it is less clear if tight junction dysfunction is a cause, or a consequence, of the disease. The crucial role of claudins in regulating the specificity of the barrier is emphasized by the observation that mutations in different claudins result in specific human diseases associated with very defined changes in barrier function. For example, mutations in claudin-16 underlie hereditary hypomagnesemia, whereas mutations in claudin-14 cause deafness in humans and in mice [1, 15]. Mutations in tricellulin are also associated with deafness in humans [86], thus providing the first evidence that alterations in occludin-related protein function contribute to barrier-related disease. The importance of tight junctions in stratifying epithelial barrier function is underscored by the finding that claudin-1 mutations were found in neonatal ichthyosis-sclerosing cholangitis (NISCH) syndrome [87, 88]. Many skin barrier diseases are still genetically uncharacterized, and it is tempting to speculate that mutations in other tight junction components could underlie such disorders.

A variety of viruses and bacteria use tight junctions to obtain entry into cells and/or to replicate. Certain viruses or bacteria use tight junction components as receptors, whereas others modulate junctional structure [89]. For example, the reovirus interacts with JAM whereas the *Helicobacter* CagA protein binds ZO-1, Jam-A and the polarity protein Par-1, thereby interfering with epithelial polarity and barrier function [90, 91]. CagA-positive strains are associated with gastritis and gastric adenocarcinoma.

Many inflammatory diseases are associated with barrier dysfunction, both in simple and stratifying epithelia, such as inflammatory bowl disease, Crohn's disease, atopic dermatitis, or psoriasis [37, 92]. By using cell culture systems, it was shown that inflammatory cytokines such as tumor necrosis factor-α or interferon-γ can induce the internalization of tight junction components or change the organization of the actinomyosin cytoskeleton, thereby affecting barrier function [37]. One recently emerging concept is that barrier function is not only affected by inflammatory signals, but that defects in the structural components forming the barrier may be the initiating event for inflammatory diseases. Although no example yet exists for tight junction defects underlying inflammatory diseases, precedents come from findings that mutations in fillagrin, a key structural epidermal barrier protein, cause the inflammatory skin disease atopic dermatitis [93], and that a loss of epidermal p120^{ctn} results in skin inflammation [94].

11.9
Concluding Remarks

During recent years, a number of exciting developments have been recognized, not only in how the key components of tight junctions contribute to the structural formation of a barrier, but also how tight junctions and polarity are coupled functionally. Another important discovery has been that tight junctions contribute functionally to the barrier properties of stratifying epithelia. Hence, an important challenge for the future will be to dissect, on a molecular basis, those pathways that dynamically regulate tight junctions, and to demonstrate how pathological conditions might impinge on tight junctions and thereby contribute to disease.

Acknowledgments

C.J.G. is supported by the American Heart Association (#0530304Z) and NIH-GM076561-01. C.M.N. is supported by DFG grant (NI685/2-1), SFB 589 P7 and a Köln Fortune grant. The authors thank Dr. W. Franke and Dr. J. Carson for providing images, Dr. A. Fanning and C. Van Itallie for providing helpful comments, and Dr. J. Scott for critical reading of the manuscript.

References

1. Van Itallie CM, Anderson JM. 2006. Claudins and epithelial paracellular transport. *Annu. Rev. Physiol.* 68: 403–429.
2. Schneeberger EE, Lynch RD. 2004. The tight junction: a multifunctional complex. *Am. J. Physiol. Cell Physiol.* 286: C1213–C1228.
3. Farquhar MG, Palade GE. 1963. Junctional complexes in various epithelia. *J. Cell Biol.* 17: 375–412.
4. Staehelin LA, Mukherjee TM, Williams AW. 1969. Fine structure of frozen-etched tight junctions. *Naturwissenschaften* 56: 142.
5. Staehelin LA. 1973. Further observations on the fine structure of freeze-cleaved tight junctions. *J. Cell Sci.* 13: 763–786.
6. Elias PM, McNutt NS, Friend DS. 1977. Membrane alterations during cornification of mammalian squamous epithelia: a freeze-fracture, tracer, and thin-section study. *Anat. Rec.* 189: 577–594.
7. Schluter H, Wepf R, Moll I, Franke WW. 2004. Sealing the live part of the skin: the integrated meshwork of desmosomes, tight junctions and curvilinear ridge structures in the cells of the uppermost granular layer of the human epidermis. *Eur. J. Cell Biol.* 83: 655–665.
8. Friend DS, Gilula NB. 1972. Variations in tight and gap junctions in mammalian tissues. *J. Cell Biol.* 53: 758–776.
9. Fujimoto K. 1995. Freeze-fracture replica electron microscopy combined with SDS digestion for cytochemical labeling of integral membrane proteins. Application to the immunogold labeling of intercellular junctional complexes. *J. Cell Sci.* 108 (Pt 11): 3443–3449.
10. Wong V, Gumbiner BM. 1997. A synthetic peptide corresponding to the extracellular domain of occludin perturbs the tight junction permeability barrier. *J. Cell Biol.* 136: 399–409.
11. Van Itallie CM, Anderson JM. 1997. Occludin confers adhesiveness when expressed in fibroblasts. *J. Cell Sci.* 110 (Pt 9): 1113–1121.
12. Saitou M, Fujimoto K, Doi Y, Itoh M, Fujimoto T, et al. 1998. Occludin-deficient embryonic stem cells can differentiate into polarized epithelial cells bearing tight junctions. *J. Cell Biol.* 141: 397–408.
13. Saitou M, Furuse M, Sasaki H, Schulzke JD, Fromm M, et al. 2000. Complex phenotype of mice lacking occludin, a component of tight junction strands. *Mol. Biol. Cell* 11: 4131–4142.
14. Ikenouchi J, Furuse M, Furuse K, Sasaki H, Tsukita S, Tsukita S. 2005. Tricellulin constitutes a novel barrier at tricellular contacts of epithelial cells. *J. Cell Biol.* 171: 939–945.
15. Furuse M, Tsukita S. 2006. Claudins in occluding junctions of humans and flies. *Trends Cell Biol.* 16: 181–188.
16. Gow A, Southwood CM, Li JS, Pariali M, Riordan GP, et al. 1999. CNS myelin and Sertoli cell tight junction strands are absent in Osp/claudin-11 null mice. *Cell* 99: 649–659.
17. Van Itallie C, Rahner C, Anderson JM. 2001. Regulated expression of claudin-4 decreases paracellular conductance through a selective decrease in sodium permeability. *J. Clin. Invest.* 107: 1319–1327.

18. Colegio OR, Van Itallie CM, McCrea HJ, Rahner C, Anderson JM. 2002. Claudins create charge-selective channels in the paracellular pathway between epithelial cells. *Am. J. Physiol. Cell Physiol.* 283: C142–C147.
19. Nitta T, Hata M, Gotoh S, Seo Y, Sasaki H, et al. 2003. Size-selective loosening of the blood-brain barrier in claudin-5-deficient mice. *J. Cell Biol.* 161: 653–660.
20. Ebnet K, Suzuki A, Ohno S, Vestweber D. 2004. Junctional adhesion molecules (JAMs): more molecules with dual functions? *J. Cell Sci.* 117: 19–29.
21. Rehder D, Iden S, Nasdala I, Wegener J, Brickwedde MK, et al. 2006. Junctional adhesion molecule-a participates in the formation of apico-basal polarity through different domains. *Exp. Cell Res.* 312: 3389–3403.
22. Gliki G, Ebnet K, Aurrand-Lions M, Imhof BA, Adams RH. 2004. Spermatid differentiation requires the assembly of a cell polarity complex downstream of junctional adhesion molecule-C. *Nature* 431: 320–324.
23. Wiesner S, Legate KR, Fassler R. 2005. Integrin-actin interactions. *Cell. Mol. Life Sci.* 62: 1081–1099.
24. Gumbiner BM. 2005. Regulation of cadherin-mediated adhesion in morphogenesis. *Nat. Rev. Mol. Cell Biol.* 6: 622–634.
25. Fanning AS, Ma TY, Anderson JM. 2002. Isolation and functional characterization of the actin binding region in the tight junction protein ZO-1. *FASEB J.* 16: 1835–1837.
26. Umeda K, Ikenouchi J, Katahira-Tayama S, Furuse K, Sasaki H, et al. 2006. ZO-1 and ZO-2 independently determine where claudins are polymerized in tight-junction strand formation. *Cell* 126: 741–754.
27. Stevenson BR, Begg DA. 1994. Concentration-dependent effects of cytochalasin D on tight junctions and actin filaments in MDCK epithelial cells. *J. Cell Sci.* 107 (Pt 3): 367–375.
28. Bruewer M, Hopkins AM, Hobert ME, Nusrat A, Madara JL. 2004. RhoA, Rac1, and Cdc42 exert distinct effects on epithelial barrier via selective structural and biochemical modulation of junctional proteins and F-actin. *Am. J. Physiol. Cell Physiol.* 287: C327–C335.
29. Wells CD, Fawcett JP, Traweger A, Yamanaka Y, Goudreault M, et al. 2006. A Rich1/Amot complex regulates the Cdc42 GTPase and apical-polarity proteins in epithelial cells. *Cell* 125: 535–548.
30. Otani T, Ichii T, Aono S, Takeichi M. 2006. Cdc42 GEF Tuba regulates the junctional configuration of simple epithelial cells. *J. Cell Biol.* 175: 135–146.
31. Shen L, Black ED, Witkowski ED, Lencer WI, Guerriero V, et al. 2006. Myosin light chain phosphorylation regulates barrier function by remodeling tight junction structure. *J. Cell Sci.* 119: 2095–2106.
32. Aijaz S, D'Atri F, Citi S, Balda MS, Matter K. 2005. Binding of GEF-H1 to the tight junction-associated adaptor cingulin results in inhibition of Rho signaling and G1/S phase transition. *Dev. Cell* 8: 777–786.
33. Guillemot L, Hammar E, Kaister C, Ritz J, Caille D, et al. 2004. Disruption of the cingulin gene does not prevent tight junction formation but alters gene expression. *J. Cell Sci.* 117: 5245–5256.
34. Vestweber D. 2002. Regulation of endothelial cell contacts during leukocyte extravasation. *Curr. Opin. Cell Biol.* 14: 587–593.
35. Sasaki H, Matsui C, Furuse K, Mimori-Kiyosue Y, Furuse M, Tsukita S. 2003. Dynamic behavior of paired claudin strands within apposing plasma membranes. *Proc. Natl. Acad. Sci. USA* 100: 3971–3976.
36. Ivanov AI, Nusrat A, Parkos CA. 2005. Endocytosis of the apical junctional complex: mechanisms and possible roles in regulation of epithelial barriers. *BioEssays* 27: 356–365.
37. Turner JR. 2006. Molecular basis of epithelial barrier regulation: from basic mechanisms to clinical application. *Am. J. Pathol.* 169: 1901–1909.
38. Nelson WJ. 2003. Adaptation of core mechanisms to generate cell polarity. *Nature* 422: 766–774.
39. Shin K, Fogg VC, Margolis B. 2006. Tight junctions and cell polarity. *Annu. Rev. Cell Dev. Biol.* 22: 207–235.

40. Anderson JM, Van Itallie CM, Fanning AS. 2004. Setting up a selective barrier at the apical junction complex. *Curr. Opin. Cell Biol.* 16: 140–145.
41. Matter K, Balda MS. 2003. Signalling to and from tight junctions. *Nat. Rev. Mol. Cell Biol.* 4: 225–236.
42. Suzuki A, Ohno S. 2006. The PAR-aPKC system: lessons in polarity. *J. Cell Sci.* 119: 979–987.
43. Moscat J, Rennert P, Diaz-Meco MT. 2006. PKCzeta at the crossroad of NF-kappaB and Jak1/Stat6 signaling pathways. *Cell Death Differ.* 13: 702–711.
44. Garrard SM, Capaldo CT, Gao L, Rosen MK, Macara IG, Tomchick DR. 2003. Structure of Cdc42 in a complex with the GTPase-binding domain of the cell polarity protein, Par6. *EMBO J.* 22: 1125–1133.
45. Gao L, Macara IG. 2004. Isoforms of the polarity protein par6 have distinct functions. *J. Biol. Chem.* 279: 41557–41562.
46. Mertens AE, Rygiel TP, Olivo C, van der Kammen R, Collard JG. 2005. The Rac activator Tiam1 controls tight junction biogenesis in keratinocytes through binding to and activation of the Par polarity complex. *J. Cell Biol.* 170: 1029–1037.
47. Chen X, Macara IG. 2005. Par-3 controls tight junction assembly through the Rac exchange factor Tiam1. *Nat. Cell Biol.* 7: 262–269.
48. Hirose T, Karasawa M, Sugitani Y, Fujisawa M, Akimoto K, et al. 2006. PAR3 is essential for cyst-mediated epicardial development by establishing apical cortical domains. *Development* 133: 1389–1398.
49. Gao L, Macara IG, Joberty G. 2002. Multiple splice variants of Par3 and of a novel related gene, Par3L, produce proteins with different binding properties. *Gene* 294: 99–107.
50. Duncan FE, Moss SB, Schultz RM, Williams CJ. 2005. PAR-3 defines a central subdomain of the cortical actin cap in mouse eggs. *Dev. Biol.* 280: 38–47.
51. Tepass U, Theres C, Knust E. 1990. crumbs encodes an EGF-like protein expressed on apical membranes of Drosophila epithelial cells and required for organization of epithelia. *Cell* 61: 787–799.
52. Fogg VC, Liu CJ, Margolis B. 2005. Multiple regions of Crumbs3 are required for tight junction formation in MCF10A cells. *J. Cell Sci.* 118: 2859–2869.
53. Straight SW, Shin K, Fogg VC, Fan S, Liu CJ, et al. 2004. Loss of PALS1 expression leads to tight junction and polarity defects. *Mol. Biol. Cell* 15: 1981–1990.
54. Lemmers C, Medina E, Delgrossi MH, Michel D, Arsanto JP, Le Bivic A. 2002. hINADl/PATJ, a homolog of discs lost, interacts with crumbs and localizes to tight junctions in human epithelial cells. *J. Biol. Chem.* 277: 25408–25415.
55. Roh MH, Makarova O, Liu CJ, Shin K, Lee S, et al. 2002. The Maguk protein, Pals1, functions as an adapter, linking mammalian homologues of Crumbs and Discs Lost. *J. Cell Biol.* 157: 161–172.
56. Shin K, Straight S, Margolis B. 2005. PATJ regulates tight junction formation and polarity in mammalian epithelial cells. *J. Cell Biol.* 168: 705–711.
57. Michel D, Arsanto JP, Massey-Harroche D, Beclin C, Wijnholds J, Le Bivic A. 2005. PATJ connects and stabilizes apical and lateral components of tight junctions in human intestinal cells. *J. Cell Sci.* 118: 4049–4057.
58. Roh MH, Liu CJ, Laurinec S, Margolis B. 2002. The carboxyl terminus of zona occludens-3 binds and recruits a mammalian homologue of discs lost to tight junctions. *J. Biol. Chem.* 277: 27501–27509.
59. Lemmers C, Michel D, Lane-Guermonprez L, Delgrossi MH, Medina E, et al. 2004. CRB3 binds directly to Par6 and regulates the morphogenesis of the tight junctions in mammalian epithelial cells. *Mol. Biol. Cell* 15: 1324–1333.
60. Hurd TW, Gao L, Roh MH, Macara IG, Margolis B. 2003. Direct interaction of two polarity complexes implicated in epithelial tight junction assembly. *Nat. Cell Biol.* 5: 137–142.
61. Sotillos S, Diaz-Meco MT, Caminero E, Moscat J, Campuzano S. 2004. DaPKC-dependent phosphorylation of Crumbs is required for epithelial cell polarity in Drosophila. *J. Cell Biol.* 166: 549–557.

62. Carraway CA, Carraway KL. 2007. Sequestration and segregation of receptor kinases in epithelial cells: implications for ErbB2 oncogenesis. *Sci. STKE 2007, 381*: re3.
63. Vermeer PD, Einwalter LA, Moninger TO, Rokhlina T, Kern JA, et al. 2003. Segregation of receptor and ligand regulates activation of epithelial growth factor receptor. *Nature 422*: 322–326.
64. Baas AF, Smit L, Clevers H. 2004. LKB1 tumor suppressor protein: PARtaker in cell polarity. *Trends Cell Biol. 14*: 312–319.
65. Hernandez S, Chavez Munguia B, Gonzalez-Mariscal L. 2007. ZO-2 silencing in epithelial cells perturbs the gate and fence function of tight junctions and leads to an atypical monolayer architecture. *Exp. Cell Res. 313*: 1533–1547.
66. Tokunaga Y, Kojima T, Osanai M, Murata M, Chiba H, et al. 2007. A novel monoclonal antibody against the second extracellular loop of occludin disrupts epithelial cell polarity. *J. Histochem. Cytochem. 55*: 735–744.
67. Barrios-Rodiles M, Brown KR, Ozdamar B, Bose R, Liu Z, et al. 2005. High-throughput mapping of a dynamic signaling network in mammalian cells. *Science 307*: 1621–1625.
68. Daugherty RL, Gottardi CJ. 2007. Phospho-regulation of β-catenin adhesion and signaling functions. *Physiology (Bethesda) 22*: 303–309.
69. Daniel JM. 2007. Dancing in and out of the nucleus: p120(ctn) and the transcription factor Kaiso. *Biochim. Biophys. Acta 1773*: 59–68.
70. McCrea PD, Park JI. 2007. Developmental functions of the P120-catenin sub-family. *Biochim. Biophys. Acta 1773*: 17–33.
71. Matter K, Balda MS. 2007. Epithelial tight junctions, gene expression and nucleo-junctional interplay. *J. Cell Sci. 120*: 1505–1511.
72. Gottardi CJ, Arpin M, Fanning AS, Louvard D. 1996. The junction-associated protein, zonula occludens-1, localizes to the nucleus before the maturation and during the remodeling of cell-cell contacts. *Proc. Natl. Acad. Sci. USA 93*: 10779–10784.
73. Betanzos A, Huerta M, Lopez-Bayghen E, Azuara E, Amerena J, Gonzalez-Mariscal L. 2004. The tight junction protein ZO-2 associates with Jun, Fos and C/EBP transcription factors in epithelial cells. *Exp. Cell Res. 292*: 51–66.
74. Logan KR, Hopwood D, Milne G. 1978. Cellular junctions in human oesophageal epithelium. *J. Pathol. 126*: 157–163.
75. Elias PM, Friend DS. 1975. The permeability barrier in mammalian epidermis. *J. Cell Biol. 65*: 180–191.
76. Orlando RC, Lacy ER, Tobey NA, Cowart K. 1992. Barriers to paracellular permeability in rabbit esophageal epithelium. *Gastroenterology 102*: 910–923.
77. Brandner JM, Kief S, Grund C, Rendl M, Houdek P, et al. 2002. Organization and formation of the tight junction system in human epidermis and cultured keratinocytes. *Eur. J. Cell Biol. 81*: 253–263.
78. Morita K, Itoh M, Saitou M, Ando-Akatsuka Y, Furuse M, et al. 1998. Subcellular distribution of tight junction-associated proteins (occludin, ZO-1, ZO-2) in rodent skin. *J. Invest. Dermatol. 110*: 862–866.
79. Furuse M, Hata M, Furuse K, Yoshida Y, Haratake A, et al. 2002. Claudin-based tight junctions are crucial for the mammalian epidermal barrier: a lesson from claudin-1-deficient mice. *J. Cell Biol. 156*: 1099–1111.
80. Turksen K, Troy TC. 2002. Permeability barrier dysfunction in transgenic mice overexpressing claudin 6. *Development 129*: 1775–1784.
81. Leyvraz C, Charles RP, Rubera I, Guitard M, Rotman S, et al. 2005. The epidermal barrier function is dependent on the serine protease CAP1/Prss8. *J. Cell Biol. 170*: 487–496.
82. Gareus R, Huth M, Breiden B, Nenci A, Rosch N, et al. 2007. Normal epidermal differentiation but impaired skin-barrier formation upon keratinocyte-restricted IKK1 ablation. *Nat. Cell Biol. 9*: 461–469.
83. Gumbiner B, Stevenson B, Grimaldi A. 1988. The role of the cell adhesion molecule uvomorulin in the formation and

maintenance of the epithelial junctional complex. *J. Cell Biol. 107*: 1575–1587.

84 Tunggal JA, Helfrich I, Schmitz A, Schwarz H, Gunzel D, et al. 2005. E-cadherin is essential for in vivo epidermal barrier function by regulating tight junctions. *EMBO J. 24*: 1146–1156.

85 Helfrich I, Schmitz A, Zigrino P, Michels C, Haase I, et al. 2007. Role of aPKC isoforms and their binding partners Par3 and Par6 in epidermal barrier formation. *J. Invest. Dermatol. 127*: 782–791.

86 Riazuddin S, Ahmed ZM, Fanning AS, Lagziel A, Kitajiri S, et al. 2006. Tricellulin is a tight-junction protein necessary for hearing. *Am. J. Hum. Genet. 79*: 1040–1051.

87 Hadj-Rabia S, Baala L, Vabres P, Hamel-Teillac D, Jacquemin E, et al. 2004. Claudin-1 gene mutations in neonatal sclerosing cholangitis associated with ichthyosis: a tight junction disease. *Gastroenterology 127*: 1386–1390.

88 Feldmeyer L, Huber M, Fellmann F, Beckmann JS, Frenk E, Hohl D. 2006. Confirmation of the origin of NISCH syndrome. *Hum. Mutat. 27*: 408–410.

89 Sousa S, Lecuit M, Cossart P. 2005. Microbial strategies to target, cross or disrupt epithelia. *Curr. Opin. Cell Biol. 17*: 489–498.

90 Krueger S, Hundertmark T, Kuester D, Kalinski T, Peitz U, Roessner A. 2007. *Helicobacter pylori* alters the distribution of ZO-1 and p120ctn in primary human gastric epithelial cells. *Pathol. Res. Pract. 203*: 433–444.

91 Saadat I, Higashi H, Obuse C, Umeda M, Murata-Kamiya N, et al. 2007. Helicobacter pylori CagA targets PAR1/MARK kinase to disrupt epithelial cell polarity. *Nature 447*: 330–333.

92 Laukoetter MG, Bruewer M, Nusrat A. 2006. Regulation of the intestinal epithelial barrier by the apical junctional complex. *Curr. Opin. Gastroenterol. 22*: 85–89.

93 Palmer CN, Irvine AD, Terron-Kwiatkowski A, Zhao Y, Liao H, et al. 2006. Common loss-of-function variants of the epidermal barrier protein filaggrin are a major predisposing factor for atopic dermatitis. *Nat. Genet. 38*: 441–446.

94 Perez-Moreno M, Davis MA, Wong E, Pasolli HA, Reynolds AB, Fuchs E. 2006. p120-catenin mediates inflammatory responses in the skin. *Cell 124*: 631–644.

12
Desmosomes in Development and Disease
Ansgar Schmidt and Peter J. Koch

12.1
Introduction

The formation of specialized cell adhesion structures (junctions) was an essential requirement for the evolution from single eukaryotic cells to complex organisms with cells that differentiate into specialized tissues and organs. Different types of intercellular junctions have developed in multicellular eukaryotes, and have initially been identified and classified using morphological criteria [1]. In addition to the tight junctions (i.e., *zonulae occludentes*) that seal and polarize epithelial sheets [2] and gap junctions that metabolically couple cells [3], adhering junctions are essential for the development of a correct tissue architecture and function during mammalian development. Two different types of adhering junctions have been analyzed in great detail: (i) actin-microfilament anchoring adherens junctions (AJs; *zonulae adherentes*) [4, 5]; and (ii) intermediate filament (IF)-anchoring desmosomes (*maculae adherentes*; see Figure 12.1).

The desmosomes of mammalian cells [6–8] are formed between epithelial cells, as well as between cells of certain non-epithelial tissues such as myocardial cells, Purkinje fiber-cells [9], dendritic reticulum cells of lymph nodes, and cells of the arachnoidea [10]. These junctions are particularly abundant in tissues such as the heart muscle or the skin, which are exposed to considerable mechanical stress. There is substantial experimental evidence in support of the hypothesis that these junctions are absolutely required to maintain tissue cohesion in these organs.

Desmosomes vary in dimension and shape depending on the tissue and cell type examined, but they consistently show a symmetric structure (see Figure 12.1B). The apparent intercellular space of the desmosome is termed the desmoglea; this region contains the extracellular portion of desmosomal transmembrane proteins that establish cell–cell coupling. On the cytoplasmic surface of the plasma membrane at the desmosomal junction, an electron-dense plaque is formed that connects the desmosome to the IF cytoskeleton (Figures 12.1 and 12.2).

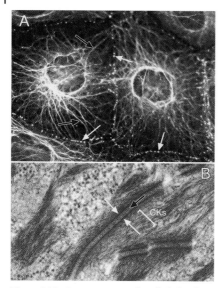

Figure 12.1 Desmosomes of epithelial cells. (A) Human cultured epithelial cells (colon carcinoma cell line CaCo-2) were stained with antibodies against the intermediate filament (IF) protein cytokeratin 18 (open arrow) and the desmosomal plaque protein desmoplakin (white arrows). Desmosomes are aligned along the boundaries of the cells (white arrows) as small dots. Cytokeratin filaments pass through the cytoplasm and terminate in desmosomes at the plasma membrane. (B) Electron micrograph of desmosomes formed between mouse keratinocytes. In the apparent intercellular space between the cells (termed desmoglea), a narrow "midline" is visible (black arrow). The plasma membranes of the two cells that form a desmosome are marked with white arrows. Note the electron-dense plaques on the cytoplasmic surfaces of the plasma membranes that connect the desmosome to the cytokeratin (CKs) intermediate filaments.

12.2
Molecular Composition of Desmosomes, Disease Associations, and Animal Models

12.2.1
Desmosomal Cadherins

Desmosomal adhesion is dependent on transmembrane glycoproteins belonging to the cadherin family of cell adhesion molecules: in humans, both desmogleins (Dsg) and desmocollins (Dsc) form small cadherin subfamilies with four different Dsg isoforms (Dsg1 to 4) and three different Dsc isoforms (Dsc1 to 3). The genes of these cadherins are clustered on chromosome 18q21 [8,11–13]. A special feature of the DSC gene family is that each gene encodes two proteins that differ only with respect to their carboxy-terminal cytoplasmic amino acid sequences (the "a" and "b" forms of Dsc proteins). The mRNAs encoding "a" and "b" forms are generated by differential splicing of the primary transcript. The primary structure of the cadherins can be divided into the extracellular portions consisting of four

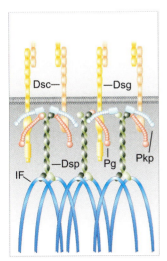

Figure 12.2 Simplified representation of the molecular building blocks of a desmosome. The adhesive transmembrane core of the desmosome is formed by transmembrane glycoproteins that belong to the desmoglein (Dsg1–4 in humans; yellow) and desmocollin (Dsc1–Dsc3 in human; orange) subfamily of desmosomal cadherins. The desmosomal plaque proteins (plakoglobin, Pg red; plakophilins, Pkp, light blue) are recruited to the cytoplasmic domain of the cadherins and engage the linker molecule desmoplakin (Dsp, green). Dsp in turn connects the intermediate filaments (IF; dark blue) to the desmosomes. It appears that the transmembrane protein Perp (not shown) is also essential for the function of desmosomes, at least in stratified epithelia (see text for details).

cadherin domains that mediate intercellular adhesion [13], whereas the intracellular amino acid sequences interact with a multitude of proteins that connect the transmembrane proteins to the IF cytoskeleton (see Figure 12.2). Furthermore, some of these cytoplasmic desmosomal proteins also act as signal transduction molecules (see below).

The expression of both desmosomal cadherin-types is closely related to cellular differentiation [14–16]. DSG2 and DSC2 are broadly expressed and found in all cell types that assemble desmosomes. DSG1 and DSG3 are mainly restricted to stratifying squamous epithelia, and show inverse expression patterns in these tissues (Figure 12.3). For example, while DSG3 is strongly expressed in the lower epidermal layers of the skin, DSG1 expression increases during keratinocyte differentiation, and is thus found mainly in suprabasal keratinocytes of the epidermis. Similarly, DSC1 and DSC3 are mainly found in complex epithelia, with expression patterns similar to those of DSG1 and DSG3. Finally, DSG4 is expressed in highly differentiated keratinocytes and hair follicles [16, 17] (see Figure 12.3).

Several human diseases are caused by an impaired function of desmosomal cadherins; these include cases of autosomal-dominant striate palmoplantar keratoderma (SPPK) [18], a disease characterized by a thickening of the stratum corneum of the soles and palms, the condition having been linked to haplo-

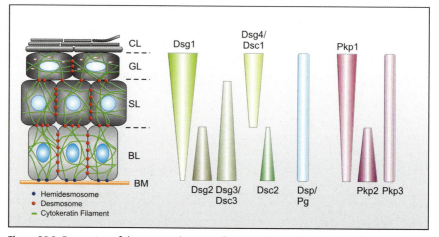

Figure 12.3 Expression of desmosomal proteins in the epidermis. A schematic representation of the cell layers that make up the epidermis is shown on the left side (BL, basal layer; SL, spinous layer; GL, granular layer; CL, cornified layer). The epidermis connects to the underlying dermis via a specialized basement membrane (BM). Desmosomes (red dots) and hemidesmosomes (blue dots) serve as anchorage points for cytokeratin intermediate filaments (green lines) to the plasma membrane. The diagrams on the right-hand side symbolize expression levels in the various layers of the epidermis. Narrow lines indicate low expression levels. Dsg1, for example, is highly expressed in suprabasal keratinocytes layers, whereas Dsg2 is weakly expressed in the basal layer and Dsg3 is present mainly in the basal and first suprabasal keratinocytes layers.

insufficiency in Dsg1 function. The thickened stratum corneum may be due to an increased sensitivity to mechanical stress (i.e., formation of callous) that cannot adequately be absorbed due to defects in desmosomal adhesion. Alternatively, or in addition to this, a change in the ratio of the Dsg isoforms in the upper epidermis may lead to changes in keratinocyte differentiation and abnormal tissue morphogenesis.

Pemphigus is a group of acquired autoimmune diseases in which autoantibodies against Dsg proteins impair desmosome function in certain stratified epithelia, thereby causing intraepithelial blistering [19]. The two main types of pemphigus are pemphigus vulgaris (PV) and pemphigus foliaceus (PF). These differ with respect to the main autoantigen that is targeted and the epidermal cell layers that are affected in the skin. PF is characterized by autoantibodies against Dsg1 and consequently leads to acantholytic blistering in the superficial epidermis where Dsg1 is strongly expressed. In contrast, in PV the autoantibodies against Dsg3 cause blistering of predominantly the oral mucosa, where Dsg3 is strongly expressed. In cases of mucocutaneous PV, autoantibodies against both Dsg1 and Dsg3 are present. In all types of pemphigus diseases, the autoantibodies appear to bind directly to the extracellular domain of Dsg and may thereby directly interfere with heterophilic Dsg/Dsc interactions, consequently inhibiting cell adhesion (steric hindrance hypothesis). Alternatively, these autoantibodies might induce endocytosis followed by lysosomal degradation of Dsg molecules, thereby depleting the cell surface pool of these adhesion molecules [20, 21].

Several recent reports have pointed towards aberrant cell signaling as an additional or alternative pathway to a loss of cell adhesion in keratinocytes exposed to pemphigus IgG [22–24]. It appears that pemphigus IgG can activate the p38 mitogen-activated protein kinase (MAPK) pathway. Inhibitors of the MAPK pathway can prevent autoantibody-induced skin blistering in a mouse model of PV [23]. In this mouse PV model system, the newborn pups are injected with PV IgG, which leads to skin blistering with a histopathology identical to that found in PV patients' skin. Another signaling protein which appears to be involved in the pathogenesis of PV is plakoglobin (Pg) [24, 25]. PV IgG appears to increase the catabolic turnover of Pg, leading to a decrease in the cytoplasmic and nuclear pool of this armadillo protein. Pg is a dual-function protein which is essential for assembling functional desmosomes (see below), although it also acts as a transcription cofactor [26, 27]. It appears that Pg functions as a repressor of the c-Myc gene. Hence, by reducing the nuclear Pg pool, PV IgG antibodies thus relieve Pg-mediated c-Myc suppression, thereby inducing epidermal hyperproliferation. The biological relevance of this signaling cascade for the pathophysiology of PV was demonstrated by the fact that c-Myc inhibitors prevented skin blistering in newborn mice injected with PV IgG.

Blistering skin diseases are not the only disorders associated with impaired desmosomal cadherin function. *Hypotrichiosis* (hair loss) has been linked to mutations in the human DSG4 gene [17], and particular heart defects (e.g., arrhythmogenic right ventricular cardiomyopathy/dysplasia; ARVC/D) have been observed in patients with mutations in the DSC2 and DSG2 genes [28, 29]. This inherited heart disease is characterized by the replacement of cardiomyocytes with fibro-fatty tissue, with resultant heart failure and sudden death. It remains to be seen if, and how, aberrant signaling through functionally impaired desmosomes contributes to these disease phenotypes (see also below).

The analysis of genetically engineered mice with mutations in desmosomal cadherin genes has greatly advanced our understanding of the biological function of these genes. The ablation of murine *Dsg3*, for example, causes intraepithelial blistering of mucous membranes in *Dsg3*-null mice with a histopathology identical to PV [30, 31]. Taken together, these results suggest that a loss of *Dsg3* function in mice leads to a disease phenotype similar to that of human PV.

The loss of *Dsg2* in mice is embryonic lethal, and mutant embryos die shortly after implantation [32]. Although the underlying mechanism is not completely understood, it appears that a loss of cell adhesion is not the cause. Even more surprising was the finding that a loss of *Dsc3* in mice caused an embryonic lethal phenotype, with *Dsc3*-null embryos dying during the early cleavage stages, before embryo implantation occurred [33]. As the embryos died even prior to the formation of desmosomes, it appears that Dsc3 has desmosome-independent functions at this stage of development.

12.2.2
Desmoplakin

Desmoplakin (Dsp) is related to the proteins of the plakin family, which function as linker molecules that connect junctional transmembrane proteins to the cytoskeleton [34]. Dsp is a constitutive component of the desmosomal plaque, and therefore is found ubiquitously in all those tissues which assemble desmosomes [8, 35]. Dsp has also been described as a component of the *complexus adherens*, a particular type of cell adhesion structure formed between endothelial cells [36, 37]. The structure of the protein can be generally divided into three distinct domains: (i) the amino-terminal domain, necessary for the association with desmosomal cadherins via *armadillo*-repeat proteins (Pg. Pkp; see below); (ii) the central rod-domain, responsible for homodimerization; and (iii) the carboxy-terminal domain, required for binding to IFs [38–40].

Insights into the *in-vivo* function of Dsp in the desmosome were obtained from the analysis of genetically engineered *Dsp*-null mice [41–43]. *Dsp*-null embryos develop until the early post-implantation stage, but die around embryonic day (E) 6.5 from adhesion defects in desmosome-rich, extra-embryonic tissues. The IF network of *Dsp*-null cells in extra-embryonic tissues collapses, and the number of desmosomes is greatly reduced. Thus, Dsp seems to be indispensable for the correct organization and stability of desmosomes and the linkage of IFs to the desmosomal cadherins. Further analyses revealed that Dsp also has a vital role in the development and function of other embryonic tissues and organs such as the heart, skin, blood vessels and neuroepithelium [42].

Various inherited human diseases that are caused by the partial loss of DSP function, or by the synthesis of truncated Dsp proteins, have been described. The clinical symptoms of these disorders are evident in skin, skin appendages, and/or the heart [44]. Some autosomal-dominant mutations that occur in the head domain of Dsp lead to haplo-insufficiency for DSP, and some of these patients develop striate palmoplantar keratoderma (SPPK) [45, 46]. Certain autosomal-recessive DSP mutations leading to a protein with a truncated carboxy-terminal domain cause Carvajal syndrome [47, 48], whereby affected individuals suffer from ARVC/D, woolly hair, and epidermolytic palmoplantar keratoderma. Recently, a very severe case of compound DSP mutations that led to a loss of almost the entire carboxy-terminal domain was described [49]. The clinicopathological phenotype caused by these mutations comprised severe fragility of skin and mucous membranes, general alopecia, and nail loss. The affected child died shortly after birth due to a loss of the skin water barrier function. However, the reason why these diverse mutations of DSP lead to such heterogeneous phenotypes remains uncertain.

12.3
Plakoglobin

Like Dsp, plakoglobin (Pg) is a constitutive component of the desmosomal plaque, and is found in all desmosome-forming cells. Pg is closely related to β-catenin, and both proteins belong to the larger family of *armadillo* (or arm-) repeat proteins [50]. These proteins are characterized by a tandem array of repeats of approximately 41 amino acids that comprise the central portion of the proteins. While β-catenin associates directly with the classical cadherins of AJs (e.g., E-cadherin), Pg binds to the classical and desmosomal cadherins (Dsc, Dsg).

Pg mediates the association of cadherins to the cytoskeleton; when Pg is bound to desmosomal cadherins, it recruits Dsp to the desmosomal plaque. Dsp then binds IF proteins, thereby connecting the desmosome to the IF cytoskeleton. When bound to classical cadherins, plakoglobin connects these proteins to the actin microfilament cytoskeleton [51].

The analyses of *Pg*-null mice revealed that this protein is crucial for normal mammalian development [52–54]. *Pg*-mutant mice develop normally until midgestation, at which time the mutant embryos die due to severe defects in heart architecture and function, with the heart walls often bursting and blood leaking into the pericardium. At the ultrastructural level, the adhesive junctions of the intercalated discs between cardiomyocytes are grossly altered, and typical desmosomes are missing. Depending on the genetic backgrounds of the mutant mice studied, a small percentage of *Pg*-deficient embryos develop almost until birth and exhibit severe heart and skin defects, including blistering with subcorneal acantholysis. Surprisingly, β-catenin colocalizes with the desmosomal cadherin in the few desmosomes that form in the basal and spinous cell layers [53, 54]. Thus, appropriate organization and function of the desmosomes is dependent on Pg, and loss of this protein leads to severe tissue fragility in the heart and skin. However, desmosomes of certain other epithelia, such as colon epithelium, appear normal [52]. The molecular basis for this different behavior of skin and, for example, colon epithelial desmosomes is not known.

A profound phenotype was also observed in humans who suffer from the autosomal-recessive inherited Naxos disease, which is caused by *Pg* mutations [55, 56]. Clinical observation of these patients revealed heart defects of the ARVC/D type, which is associated with woolly hair and palmoplantar keratoderma. An analysis of the underlying mutations revealed that the carboxy-terminal domain of Pg is truncated in these patients. As no epidermal acantholysis has yet been described in Naxos patients, it can be assumed that the truncated Pg can maintain cell adhesion in most tissues, except for those that are exposed to tremendous mechanical force, such as the heart. Nevertheless, it remains to be seen whether *Pg* mutations induce disease phenotypes solely via impaired cell adhesion. Pg has also been shown to act as a transcription cofactor, and phenotypes might also be due to changes in the global gene expression profile of affected cells, resulting in altered cellular differentiation [26, 27].

Garcia-Gras and colleagues recently provided evidence indicating that abnormal Pg signaling might be the basis for certain types of cardiomyopathies [57]. These authors showed that a reduced Dsp expression in cardiomyocytes can indirectly increase nuclear Pg levels, which appear to inhibit canonical (β-catenin-dependent) Wnt signaling. This is predicted to increase the expression of adipogenic and fibrogenic genes in cardiomyocytes, thus contributing to the disease phenotype (fibroadipocytic replacement of cardiac myocytes). In essence, a loss of desmosomal function and a signaling defect could contribute to the clinical symptoms of these cardiomyopathies.

An *in-vivo* role of Pg in signaling is also supported by the analysis of transgenic mice that overexpress Pg in the epidermis. These mice showed stunted hair growth, which was attributed to reduced proliferation of epithelial cells and premature exit of hair follicles from the active growth phase (anagen) of the hair growth cycle [58].

12.4
Plakophilins

Plakophilins (Pkp) represent another group of arm-repeat proteins that have been found in the desmosomal plaque [59, 60]. Three classical Pkp isoforms (Pkp1 to 3) have been identified. These proteins comprise a subgroup of the p120^{ctn}-subfamily [61]. p0071, which has been referred to as Pkp4, is more distantly related to the classical Pkp and is found in desmosomes and AJs [62]. Similar to desmosomal cadherins, Pkp show both tissue- and cell-type-specific expression patterns in mammals (Figure 12.3). Pkp2 is the most widespread plakophilin isoform, and is found in all desmosome-forming cell types. Pkp1 is restricted to the desmosomes of stratified and complex epithelia, whereas Pkp3 is integrated into the desmosomes of all stratified epithelia, almost all simple epithelia, and also some non-epithelia tissues. Pkp3 is not synthesized in hepatocytes and cardiomyocytes. Thus, all desmosomes contain at least one Pkp isoform. In addition to their presence in desmosomes, Pkp1 and Pkp2 are also localized in the nucleus of certain cell types [63, 64]. Although Pkp 2 is part of RNA polymerase III holoenzyme complex, its specific function in this protein complex is not known [65]. It is not clear what function Pkp1 exerts in the nucleus, but Pkp3 was recently identified as a component of certain cytoplasmic particles (i.e. stress granules) [66] which regulate post-transcriptional gene activity in cellular stress situations [67].

Insights into the *in-vivo* function of Pkp1 came from the analysis of the inherited human disease ectodermal dysplasia/skin fragility (EDSF) that clinically presents with severe epidermal blistering, dysplasia, and abnormal skin appendages [68, 69]. Null mutations in both alleles of the PKP1 gene were identified in patients suffering from this disease. The severely affected epidermis of these patients is unable to withstand even mild mechanical trauma, which leads to severe blistering. At the ultrastructural level, the epidermis is thickened and the intercellular space between keratinocytes is enlarged. The desmosomes are small and reduced in number, and fail to anchor cytokeratin filaments at the plasma membrane. At the cellular level, the loss of Pkp1 leads to a redistribution of Dsp from the desmosome to the cytoplasm in the patient's keratinocytes. Thus, in combination with data that have been obtained earlier from *in-vitro* studies [70, 71], it has been found that Pkp1 recruits and anchors Dsp to the desmosomal plaque. This is also assumed to be the main function of Pkp2.

Recently, mice with a null mutation in the *Pkp2* gene were generated [72]. The ablation of *Pkp2* in mice is embryonic lethal, and mutant animals die at about E10.5 due to defects in cellular adhesion and abnormal heart architecture that causes rupture of the cardiac walls and blood leakage into the pericardium. Dsp is lost from the intercalated discs in the heart muscle of the mutant mice, leading to desmin IFs retraction from the intercalated discs. Desmosomes found in other epithelia, such as intestinal epithelium, seem to be unaffected by the loss of Pkp2. In those tissues, loss of Pkp2 might be compensated for by other plakophilins such as Pkp3. Inherited heterozygous mutations of the PKP2 gene have been linked to certain types of ARVC/D disease in humans [57, 73].

The main function of Pkp in the desmosome is as a linker between Dsp and desmosomal cadherins. As with the other desmosomal proteins, Pkp also seems to be required for normal epidermal differentiation; this is demonstrated, for example, by an analysis of the EDSF syndrome, where such patients exhibit abnormal differentiation and morphogenesis of skin and skin appendages, such as the hair and nails.

12.5
Accessory Desmosomal Proteins

In addition to the core components that are present in every desmosome, a variety of additional proteins have been located in the desmosomes in particular tissues. Some of these proteins, such as phosphoprotein tyrosine phosphatases [74], kinases [75] and growth factor receptors, are involved in cellular signaling. Perp is a four-pass transmembrane protein related to the larger family of claudin/PMP-22/EMP proteins, and is highly expressed in the skin and heart [76, 77]. The PERP gene is one of the target genes transcriptionally controlled by the transcription factor p63, a key regulator for the development of stratified epithelia [78]. Inactivation of the *Perp* gene in mice leads to acantholysis and the formation of severe blisters in the skin and the oral mucosa which resemble those found in pemphigus vulgaris patients. The majority of mutant mice die within 10 days of birth, most likely due to feeding problems. Although the function of Perp in the desmosome is not known, it is clear that this protein is required for the normal functioning of desmosomes in stratified epithelia. An inherited or acquired human disease has not been linked to abnormal Perp function.

12.6
Desmosomal Proteins in Embryonic Development

The formation of cellular adhesion structures is crucial for the normal development of vertebrate embryos [79]. The formation of AJs, for example, is required for compaction, a developmental process in which the cells of the morula increase cellular adhesion and polarize. These steps are required for the development of the blastula (the developmental stage at which the embryo implants into the uterus). Desmosomes emerge during blastula formation when trophectoderm cells (which are polarized epithelial cells) assemble these junctions [12, 80]. It has been assumed that desmosomes may be required at this stage of development in order to enable trophectoderm cells to withstand the pressure from the fluid-filled blastocoel cavity that is surrounded by these cells. Nevertheless, inconsistent with this hypothesis is the finding that null mutations in either the *Dsg2* or the *Dsp* gene – that is, genes that encode constitutive desmosomal components – do not lead to blastocyst disintegration in mouse embryos. Instead, mutant embryos die shortly before or after implantation [32, 43]. Interestingly, inactivation of the *Dsc3* gene causes embryonic death at the morula stage of development (i.e., before desmosomes are formed in the trophectoderm of the blastula) [33]. Clearly, much needs to be learned about the desmosome-independent function of these proteins early in mammalian development.

The elimination of Pg has no gross effects on the early developmental stages of the mouse embryo [52, 54]. However, mutant embryos begin to die from heart failure at E10.5. Depending on the genetic background of the mouse line used, some animals continue to develop almost until birth and then exhibit additional skin phenotypes, such as intraepidermal blistering and epidermolytic hyperkeratosis [52, 54] (see above). As mentioned above, the appropriate assembly of different junctional complexes which form the intercalated discs of cardiomyocytes is fundamentally disturbed by the elimination of Pg, and desmosomal structures are missing in these cells. The relatively late developmental arrest of mutant Pg animals (as compared to mice carrying mutation in other desmosomal genes) may be due to a partial compensation of Pg deficiency by β-catenin. This notion is supported by the observation that β-catenin is forced into an association with desmosomal cadherins in primary keratinocytes derived from Pg-null embryos [53].

Similarly, the inactivation of *Pkp2* in mice causes a lethal heart phenotype due to impaired cell adhesion between cardiomyocytes [72]. Dsp and IFs are displaced from the intercalated disks in these mice. The relatively late developmental arrest (around E10.5) of *Pkp2*-deficient embryos may be due to the expression of *Pkp3*, which is present in most Pkp2-positive tissues. One of the few tissues where Pkp2 is exclusively expressed is the myocardium, and this explains the severe heart defect observed in *Pkp2*-null mice.

12.7
Concluding Remarks

The results summarized in this chapter demonstrate that desmosomes are indispensable for maintaining tissue integrity in organs exposed to considerable mechanical stress, such as the skin, skin appendages, and the heart. The recent finding that pemphigus autoantibodies induce aberrant signaling by binding to desmogleins also suggests that desmosomes are connected to several cellular signaling cascades, in particular the MAPK pathway and Pg signaling. In the case of Pg, desmosomes could indirectly affect canonical Wnt signaling by either enhancing or inhibiting transcriptional control mediated by β-catenin/Tcf/Lef complexes [57]. Alternatively, Pg could control gene expression independently of Wnt signaling [81]. Clearly, although very little is known about the normal function of desmosomes in cell signaling, the observation that PV autoantibodies can activate signaling cascades suggests that desmosomes might function in transmitting signals from outside the cell to the nucleus – that is, they serve as sensors for the microenvironment. Desmosomes can also actively affect the microenvironment; by changing their molecular composition and their adhesive properties, these junctions can effect both cell migration and cell sorting – processes that are of critical importance for tissue/organ morphogenesis.

It is now clear that desmosomes are not simply static structures needed to connect cells. However, our present understanding of the role of these junctions in development and disease is clearly in its very early stages.

Acknowledgments

These studies were supported by grants from the German Research Foundation (Deutsche Forschungsgemeinschaft; grant Schm 2145/1-1 and Schm 2145/1-2) to A.S., and by grants from the National Institutes of Health (NIH/NIAMS) to P.J.K. (AR47343, AR050439). The authors would like to thank Stephanie Kadel for her assistance in preparing this manuscript.

References

1 Farquhar MG, Palade GE. Junctional complexes in various epithelia. *J. Cell Biol.* 1963; *17*: 375–412; 375–412.

2 Schneeberger EE, Lynch RD. The tight junction: a multifunctional complex. *Am. J. Physiol. Cell Physiol.* 2004; *286*(6): C1213–C1228.

3 Evans WH, Martin PE. Gap junctions: structure and function (Review). *Mol. Membr. Biol.* 2002; *19*(2): 121–136.

4 Borrmann CM, Mertens C, Schmidt A, et al. Molecular diversity of plaques of epithelial-adhering junctions. *Ann. N. Y. Acad. Sci.* 2000; *915*: 144–150.

5 Borrmann CM, Grund C, Kuhn C, et al. The area composita of adhering junctions connecting heart muscle cells of vertebrates. II. Colocalizations of desmosomal and fascia adhaerens molecules in the intercalated disk. *Eur. J. Cell Biol.* 2006; *85*(6): 469–485.

6 Schmidt A, Heid HW, Schafer S, et al. Desmosomes and cytoskeletal architecture in epithelial differentiation: cell type-specific plaque components and intermediate filament anchorage. *Eur. J. Cell Biol.* 1994; *65*(2): 229–245.

7 Kowalczyk AP, Bornslaeger EA, Norvell SM, Palka HL, Green KJ. Desmosomes: intercellular adhesive junctions specialized for attachment of intermediate filaments. *Int. Rev. Cytol.* 1999; *185*: 237–302.

8 Cheng X, Koch PJ. In vivo function of desmosomes. *J. Dermatol.* 2004; *31*(3): 171–187.

9 Franke WW, Moll R, Schiller DL, et al. Desmoplakins of epithelial and myocardial desmosomes are immunologically and

biochemically related. *Differentiation* 1982; 23(2): 115–127.
10 Franke WW, Moll R. Cytoskeletal components of lymphoid organs. I. Synthesis of cytokeratins 8 and 18 and desmin in subpopulations of extrafollicular reticulum cells of human lymph nodes, tonsils, and spleen. *Differentiation* 1987; 36(2): 145–163.
11 Koch PJ, Franke WW. Desmosomal cadherins: another growing multigene family of adhesion molecules. *Curr. Opin. Cell Biol.* 1994; 6(5): 682–687.
12 Cheng X, Den Z, Koch PJ. Desmosomal cell adhesion in mammalian development. *Eur. J. Cell Biol.* 2005; 84(2-3): 215–223.
13 Dusek RL, Godsel LM, Green KJ. Discriminating roles of desmosomal cadherins: beyond desmosomal adhesion. *J. Dermatol. Sci.* 2007; 45(1): 7–21.
14 Schafer S, Koch PJ, Franke WW. Identification of the ubiquitous human desmoglein, Dsg2, and the expression catalogue of the desmoglein subfamily of desmosomal cadherins. *Exp. Cell Res.* 1994; 211(2): 391–399.
15 Nuber UA, Schafer S, Schmidt A, Koch PJ, Franke WW. The widespread human desmocollin Dsc2 and tissue-specific patterns of synthesis of various desmocollin subtypes. *Eur. J. Cell Biol.* 1995; 66(1): 69–74.
16 Bazzi H, Getz A, Mahoney MG, et al. Desmoglein 4 is expressed in highly differentiated keratinocytes and trichocytes in human epidermis and hair follicle. *Differentiation* 2006; 74(2-3): 129–140.
17 Kljuic A, Bazzi H, Sundberg JP, et al. Desmoglein 4 in hair follicle differentiation and epidermal adhesion: evidence from inherited hypotrichosis and acquired pemphigus vulgaris. *Cell* 2003; 113(2): 249–260.
18 Rickman L, Simrak D, Stevens HP, et al. N-terminal deletion in a desmosomal cadherin causes the autosomal dominant skin disease striate palmoplantar keratoderma. *Hum. Mol. Genet.* 1999; 8(6): 971–976.
19 Stanley JR, Amagai M. Pemphigus, bullous impetigo, and the staphylococcal scalded-skin syndrome. *N. Engl. J. Med.* 2006; 355(17): 1800–1810.
20 Calkins CC, Setzer SV, Jennings JM, et al. Desmoglein endocytosis and desmosome disassembly are coordinated responses to pemphigus autoantibodies. *J. Biol. Chem.* 2006; 281(11): 7623–7634.
21 Yamamoto Y, Aoyama Y, Shu E, et al. Anti-desmoglein 3 (Dsg3) monoclonal antibodies deplete desmosomes of Dsg3 and differ in their Dsg3-depleting activities related to pathogenicity. *J. Biol. Chem.* 2007; 282(24): 17866–17876.
22 Waschke J, Spindler V, Bruggeman P, et al. Inhibition of Rho A activity causes pemphigus skin blistering. *J. Cell Biol.* 2006; 175(5): 721–727.
23 Berkowitz P, Hu P, Liu Z, et al. Desmosome signaling. Inhibition of p38MAPK prevents pemphigus vulgaris IgG-induced cytoskeleton reorganization. *J. Biol. Chem.* 2005; 280(25): 23778–23784.
24 Williamson L, Raess NA, Caldelari R, et al. Pemphigus vulgaris identifies plakoglobin as key suppressor of c-Myc in the skin. *EMBO J.* 2006; 25(14): 3298–3309.
25 de Bruin A, Caldelari R, Williamson L, et al. Plakoglobin-dependent disruption of the desmosomal plaque in pemphigus vulgaris. *Exp. Dermatol.* 2007; 16(6): 468–475.
26 Kolly C, Zakher A, Strauss C, Suter MM, Muller EJ. Keratinocyte transcriptional regulation of the human c-Myc promoter occurs via a novel Lef/Tcf binding element distinct from neoplastic cells. *FEBS Lett.* 2007; 581(10): 1969–1976.
27 Maeda O, Usami N, Kondo M, et al. Plakoglobin (gamma-catenin) has TCF/LEF family-dependent transcriptional activity in beta-catenin-deficient cell line. *Oncogene* 2004; 23(4): 964–972.
28 Pilichou K, Nava A, Basso C, et al. Mutations in desmoglein-2 gene are associated with arrhythmogenic right ventricular cardiomyopathy. *Circulation* 2006; 113(9): 1171–1179.
29 Heuser A, Plovie ER, Ellinor PT, et al. Mutant desmocollin-2 causes arrhythmogenic right ventricular cardiomyopathy. *Am. J. Hum. Genet.* 2006; 79(6): 1081–1088.
30 Koch PJ, Mahoney MG, Ishikawa H, et al. Targeted disruption of the pemphigus

vulgaris antigen (desmoglein 3) gene in mice causes loss of keratinocyte cell adhesion with a phenotype similar to pemphigus vulgaris. *J. Cell Biol*. 1997; *137*(*5*): 1091–1102.
31. Koch PJ, Mahoney MG, Cotsarelis G, et al. Desmoglein 3 anchors telogen hair in the follicle. *J. Cell Sci*. 1998; *111*(*Pt 17*): 2529–2537.
32. Eshkind L, Tian Q, Schmidt A, et al. Loss of desmoglein 2 suggests essential functions for early embryonic development and proliferation of embryonal stem cells. *Eur. J. Cell Biol*. 2002; *81*(*11*): 592–598.
33. Den Z, Cheng X, Merched-Sauvage M, Koch PJ. Desmocollin 3 is required for pre-implantation development of the mouse embryo. *J. Cell Sci*. 2006; *119*(*Pt 3*): 482–489.
34. Leung CL, Green KJ, Liem RK. Plakins: a family of versatile cytolinker proteins. *Trends Cell Biol*. 2002; *12*(*1*): 37–45.
35. Kottke MD, Delva E, Kowalczyk AP. The desmosome: cell science lessons from human diseases. *J. Cell Sci*. 2006; *119*(*Pt 5*): 797–806.
36. Franke WW, Koch PJ, Schafer S, et al. The desmosome and the syndesmos: cell junctions in normal development and in malignancy. *Princess Takamatsu Symp*. 1994; *24*: 14–27.
37. Hammerling B, Grund C, Boda-Heggemann J, Moll R, Franke WW. The complexus adhaerens of mammalian lymphatic endothelia revisited: a junction even more complex than hitherto thought. *Cell Tissue Res*. 2006; *324*(*1*): 55–67.
38. Bornslaeger EA, Corcoran CM, Stappenbeck TS, Green KJ. Breaking the connection: displacement of the desmosomal plaque protein desmoplakin from cell-cell interfaces disrupts anchorage of intermediate filament bundles and alters intercellular junction assembly. *J. Cell Biol*. 1996; *134*(*4*): 985–1001.
39. Godsel LM, Huen AC, Amargo EV, et al. Functional and dynamic analyses reveal desmoplakin's roles in epithelial integrity and desmosome remodeling in living cells. *Mol. Biol. Cell* 2002; *13*: 286A.
40. Kowalczyk AP, Bornslaeger EA, Borgwardt JE, et al. The amino-terminal domain of desmoplakin binds to plakoglobin and clusters desmosomal cadherin-plakoglobin complexes. *J. Cell Biol*. 1997; *139*(*3*): 773–784.
41. Vasioukhin V, Bowers E, Bauer C, Degenstein L, Fuchs E. Desmoplakin is essential in epidermal sheet formation. *Nat. Cell Biol*. 2001; *3*(*12*): 1076–1085.
42. Gallicano GI, Bauer C, Fuchs E. Rescuing desmoplakin function in extra-embryonic ectoderm reveals the importance of this protein in embryonic heart, neuroepithelium, skin and vasculature. *Development* 2001; *128*(*6*): 929–941.
43. Gallicano GI, Kouklis P, Bauer C, et al. Desmoplakin is required early in development for assembly of desmosomes and cytoskeletal linkage. *J. Cell Biol*. 1998; *143*(*7*): 2009–2022.
44. Lai Cheong JE, Wessagowit V, McGrath JA. Molecular abnormalities of the desmosomal protein desmoplakin in human disease. *Clin. Exp. Dermatol*. 2005; *30*(*3*): 261–266.
45. Armstrong DK, McKenna KE, Purkis PE, et al. Haploinsufficiency of desmoplakin causes a striate subtype of palmoplantar keratoderma. *Hum. Mol. Genet*. 1999; *8*(*1*): 143–148.
46. Whittock NV, Ashton GH, Dopping-Hepenstal PJ, et al. Striate palmoplantar keratoderma resulting from desmoplakin haploinsufficiency. *J. Invest. Dermatol*. 1999; *113*(*6*): 940–946.
47. Norgett EE, Hatsell SJ, Carvajal-Huerta L, et al. Recessive mutation in desmoplakin disrupts desmoplakin-intermediate filament interactions and causes dilated cardiomyopathy, woolly hair and keratoderma. *Hum. Mol. Genet*. 2000; *9*(*18*): 2761–2766.
48. Carvajal-Huerta L. Epidermolytic palmoplantar keratoderma with woolly hair and dilated cardiomyopathy. *J. Am. Acad. Dermatol*. 1998; *39*(*3*): 418–421.
49. Jonkman MF, Pasmooij AM, Pasmans SG, et al. Loss of desmoplakin tail causes lethal acantholytic epidermolysis bullosa. *Am. J. Hum. Genet*. 2005; *77*(*4*): 653–660.
50. Hatzfeld M. The armadillo family of structural proteins. *Int. Rev. Cytol*. 1999; *186*: 179–224.

51 Gates J, Peifer M. Can 1000 reviews be wrong? Actin, alpha-catenin, and adherens junctions. *Cell* 2005; *123*(5): 769–772.

52 Ruiz P, Brinkmann V, Ledermann B, et al. Targeted mutation of plakoglobin in mice reveals essential functions of desmosomes in the embryonic heart. *J. Cell Biol.* 1996; *135*(1): 215–225.

53 Bierkamp C, Schwarz H, Huber O, Kemler R. Desmosomal localization of beta-catenin in the skin of plakoglobin null-mutant mice. *Development* 1999; *126*(2): 371–381.

54 Bierkamp C, Mclaughlin KJ, Schwarz H, Huber O, Kemler R. Embryonic heart and skin defects in mice lacking plakoglobin. *Dev. Biol.* 1996; *180*(2): 780–785.

55 McKoy G, Protonotarios N, Crosby A, et al. Identification of a deletion in plakoglobin in arrhythmogenic right ventricular cardiomyopathy with palmoplantar keratoderma and woolly hair (Naxos disease). *Lancet* 2000; *355*(9221): 2119–2124.

56 Protonotarios N, Tsatsopoulou A. Naxos disease: cardiocutaneous syndrome due to cell adhesion defect. *Orphanet. J. Rare Dis.* 2006; *1*: 4–9.

57 Garcia-Gras E, Lombardi R, Giocondo MJ, et al. Suppression of canonical Wnt/beta-catenin signaling by nuclear plakoglobin recapitulates phenotype of arrhythmogenic right ventricular cardiomyopathy. *J. Clin. Invest.* 2006; *116*(7): 2012–2021.

58 Charpentier E, Lavker RM, Acquista E, Cowin P. Plakoglobin suppresses epithelial proliferation and hair growth in vivo. *J. Cell Biol.* 2000; *149*(2): 503–520.

59 Schmidt A, Jager S. Plakophilins – hard work in the desmosome, recreation in the nucleus? *Eur. J. Cell Biol.* 2005; *84*(2-3): 189–204.

60 Hatzfeld M. Plakophilins: Multifunctional proteins or just regulators of desmosomal adhesion? *Biochim. Biophys. Acta* 2007; *1773*(1): 69–77.

61 Hatzfeld M. The p120 family of cell adhesion molecules. *Eur. J. Cell Biol.* 2005; *84*(2-3): 205–214.

62 Hatzfeld M, Nachtsheim C. Cloning and characterization of a new armadillo family member, p0071, associated with the junctional plaque: evidence for a subfamily of closely related proteins. *J. Cell Sci.* 1996; *109*(Pt 11): 2767–2778.

63 Mertens C, Kuhn C, Franke WW. Plakophilins 2a and 2b: constitutive proteins of dual location in the karyoplasm and the desmosomal plaque. *J. Cell Biol.* 1996; *135*(4): 1009–1025.

64 Schmidt A, Langbein L, Rode M, et al. Plakophilins 1a and 1b: widespread nuclear proteins recruited in specific epithelial cells as desmosomal plaque components. *Cell Tissue Res.* 1997; *290*(3): 481–499.

65 Mertens C, Hofmann I, Wang Z, et al. Nuclear particles containing RNA polymerase III complexes associated with the junctional plaque protein plakophilin 2. *Proc. Natl. Acad. Sci. USA* 2001; *98*(14): 7795–7800.

66 Anderson P, Kedersha N. RNA granules. *J. Cell Biol.* 2006; *172*(6): 803–808.

67 Hofmann I, Casella M, Schnolzer M, et al. Identification of the junctional plaque protein plakophilin 3 in cytoplasmic particles containing RNA-binding proteins and the recruitment of plakophilins 1 and 3 to stress granules. *Mol. Biol. Cell* 2006; *17*(3): 1388–1398.

68 McGrath JA, McMillan JR, Shemanko CS, et al. Mutations in the plakophilin 1 gene result in ectodermal dysplasia/skin fragility syndrome. *Nat. Genet.* 1997; *17*(2): 240–244.

69 South AP. Plakophilin 1: an important stabilizer of desmosomes. *Clin. Exp. Dermatol.* 2004; *29*(2): 161–167.

70 Bornslaeger EA, Godsel LM, Corcoran CM, et al. Plakophilin 1 interferes with plakoglobin binding to desmoplakin, yet together with plakoglobin promotes clustering of desmosomal plaque complexes at cell-cell borders. *J. Cell Sci.* 2001; *114*(Pt 4): 727–738.

71 Kowalczyk AP, Hatzfeld M, Bornslaeger EA, et al. The head domain of plakophilin-1 binds to desmoplakin and enhances its recruitment to desmosomes. Implications for cutaneous disease. *J. Biol. Chem.* 1999; *274*(26): 18145–18148.

72. Grossmann KS, Grund C, Huelsken J, et al. Requirement of plakophilin 2 for heart morphogenesis and cardiac junction formation. *J. Cell Biol*. 2004; *167*(*1*): 149–160.
73. Gerull B, Heuser A, Wichter T, et al. Mutations in the desmosomal protein plakophilin-2 are common in arrhythmogenic right ventricular cardiomyopathy. *Nat. Genet*. 2004; *36*(*11*): 1162–1164.
74. Aicher B, Lerch MM, Muller T, Schilling J, Ullrich A. Cellular redistribution of protein tyrosine phosphatases LAR and PTPsigma by inducible proteolytic processing. *J. Cell Biol*. 1997; *138*(*3*): 681–696.
75. Calautti E, Grossi M, Mammucari C, et al. Fyn tyrosine kinase is a downstream mediator of Rho/PRK2 function in keratinocyte cell-cell adhesion. *J. Cell Biol*. 2002; *156*(*1*): 137–148.
76. Ihrie RA, Marques MR, Nguyen BT, et al. Perp is a p63-regulated gene essential for epithelial integrity. *Cell* 2005; *120*(*6*): 843–856.
77. Marques MR, Ihrie RA, Horner JS, Attardi LD. The requirement for perp in postnatal viability and epithelial integrity reflects an intrinsic role in stratified epithelia. *J. Invest. Dermatol*. 2006; *126*(*1*): 69–73.
78. Koster MI, Roop DR. Mechanisms Regulating Epithelial Stratification. *Annu. Rev. Cell Dev. Biol*. 2007; *23*: 93–113.
79. Bloor JW, Kiehart DP. Drosophila RhoA regulates the cytoskeleton and cell-cell adhesion in the developing epidermis. *Development* 2002; *129*(*13*): 3173–3183.
80. Jackson BW, Grund C, Schmid E, et al. Formation of cytoskeletal elements during mouse embryogenesis. Intermediate filaments of the cytokeratin type and desmosomes in preimplantation embryos. *Differentiation* 1980; *17*(*3*): 161–179.
81. Teuliere J, Faraldo MM, Shtutman M, et al. beta-catenin-dependent and -independent effects of DeltaN-plakoglobin on epidermal growth and differentiation. *Mol. Cell. Biol*. 2004; *24*(*19*): 8649–8661.

13
Cadherin Trafficking and Junction Dynamics
Christine M. Chiasson and Andrew P. Kowalczyk

13.1
Introduction

Intercellular junctions are crucial for the formation and maintenance of tissues and organs within multicellular organisms. Interactions between cells impact not only tissue architecture, but also cellular function [1, 2]. Junctional adhesion receptors have long been known to carry out a fundamental structural role as a cellular "glue" that mediates tight interactions between cells. However, these receptors are becoming increasingly appreciated for their function as signaling receptors capable of mediating changes in cellular behavior [3, 4]. The dynamic behavior of these receptors serves to modulate cell–cell adhesion during morphogenesis, cell growth and differentiation, and cell communication.

Cell–cell adhesion is mediated by distinct junctional complexes: the apical tight junction; the subapical adherens junction (AJ); and the basolateral desmosome. The AJ, with cadherins as the central adhesive component of the complex, is the primary junction responsible for homotypic cell–cell adhesion [5, 6]. Cadherins are commonly referred to as the master regulators of cell adhesion, as they provide the initial cell contact and cell signaling that is required for the establishment of other junctional complexes, including tight junctions and desmosomes [7]. The cadherin superfamily consists of over 50 members that vary widely in their tissue distribution and function [8]. Cadherins are single-span transmembrane receptors that promote homophilic interactions between cells in a calcium-dependent manner. Classical cadherins, including E-, N-, P-, and VE-cadherin, are found primarily in AJs, and will form the focal point of this chapter [9]. The cadherin extracellular domain functions to selectively ligate cadherins on neighboring cells. The cytoplasmic tail is highly conserved between family members and interacts with members of the armadillo family of proteins, including β-catenin and p120-catenin (p120). β-Catenin binds to the catenin-binding domain at the distal region of the cadherin cytoplasmic tail and functions in linking the cadherin to the actin cytoskeleton. p120 interacts with the cadherin juxtamembrane domain (JMD), and has been identified as a key regulator of cadherin metabolic stability and clustering [10–12].

The diverse range of functions modulated by cell-adhesion receptors requires that cellular adhesion be highly adaptable to the cellular microenvironment [13]. Cell–cell adhesion must be dynamically regulated in response to changing developmental cues and physiological conditions [14]. During recent years, an important role has been established for membrane trafficking of adhesion receptors in the rapid and precise modulation of changes in cellular adhesive status [15, 16]. Exocytic and endocytic membrane trafficking pathways provide a tightly regulated system for controlling the abundance of cell-surface receptors in a spatial and temporal manner. In this chapter, the recent advancements made in understanding the dynamic regulation of cell-adhesion receptors through membrane trafficking will be reviewed.

13.2
Exocytosis and Polarized Sorting of Adherens Junction Proteins

The establishment and maintenance of a polarized epithelium consisting of distinct apical and basolateral domains is required for the formation of an organized barrier between the internal and external compartments in multicellular organisms. The correct sorting and targeted delivery of exocytic transport vesicles from the trans Golgi network (TGN) to the correct membrane domain is a fundamental mechanism of cell polarization [17, 18]. The targeted delivery of E-cadherin to the basolateral membrane domain is crucial not only for the formation and establishment of adherens junctions, but also for the establishment of a polarized epithelium [19].

The sorting of newly synthesized E-cadherin occurs first in the TGN, where a highly conserved dileucine motif in the E-cadherin cytoplasmic tail has been shown to be required for targeting of E-cadherin to the basolateral membrane [20] (Figure 13.1). The use of chimeric proteins containing the cytoplasmic tail of E-cadherin fused to the extracellular domain of the interleukin-2 receptor α subunit (TAC) revealed that the cytoplasmic tail of E-cadherin is sufficient for proper sorting to the basolateral membrane. The mutation of a dileucine motif at amino acids 587–588 of E-cadherin causes missorting of E-cadherin to the apical surface, and results in the disruption of cell polarity and morphology [20]. One of the earliest regulators of post-Golgi trafficking of E-cadherin is golgin-97, a GRIP domain protein that functions in membrane matrix tethering within the Golgi stacks. siRNA studies support a specific role for golgin-97 in the exocytic trafficking of E-cadherin at the point of exit from the TGN [21]. Basolateral sorting of newly synthesized E-cadherin at or near the TGN is further regulated by members of the Rho family of small GTPases, including Rac1 and Cdc42. The expression of dominant-negative Rac1 or Cdc2 mutants causes E-cadherin to accumulate in a perinuclear vesicular compartment overlapping with, but extending beyond, the classical Golgi markers. This regulation is specific to the basolateral transport of E-cadherin, as the delivery of a missorted E-cadherin mutant is not affected [22]. Following sorting in the TGN, E-cadherin localizes to Rab11-positive recycling endosomes, an intermediate compartment that is involved in exocytic biosynthetic transport [23]. In both polarized and non-polarized cells, a dominant-negative Rab11 mutant disrupted the proper sorting of E-cadherin to the basolateral surface, resulting instead in missorting to the apical membrane and morphological and functional transformation of Madin–Darby canine kidney (MDCK) cells to an invasive phenotype [23] (see Figure 13.1).

Generally, sorting signals function by interacting with adaptor complexes at specific subcellular organelles [24]. These adaptor complexes recruit cargo into clathrin-coated vesicles for delivery to the target membrane. An adaptor interaction with the E-cadherin dileucine motif has not been identified, but it has been suggested that these motifs interact with the β subunit of AP1B [25]. A recent study identified an indirect association of E-cadherin with the μ subunit of the AP1B adaptor complex through type 1γ phosphatidylinositol phosphate kinase (PIPK1γ). The interaction of PIPK1γ with AP1B is functionally important for E-cadherin trafficking. In MDCK cells expressing the isoform of PIPK1γ lacking the AP1B

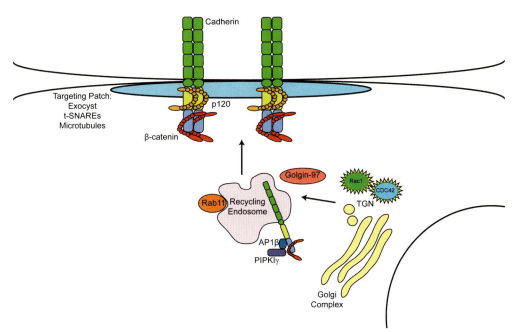

Figure 13.1 Newly synthesized cadherin undergoes active sorting through membrane trafficking pathways en route to the plasma membrane, where it is incorporated into adherens junctions and serves as a spatial cue for the polarized delivery of basolateral proteins. Several points of regulation for the exocytic trafficking of E-cadherin have been identified. At the trans Golgi network (TGN), golgin-97 and the small GTPases Rac1 and Cdc42 are required for basolateral targeting of E-cadherin. A dileucine motif in the E-cadherin cytoplasmic tail also regulates sorting at the TGN. Following exit from the TGN, E-cadherin localizes to Rab11-positive endosomes, where it interacts indirectly with the adaptor protein AP1B through an association with the type 1-gamma phosphatidylinositol kinase.

binding site, E-cadherin is sequestered in a cytosolic compartment and its internalization and recycling are inhibited. Interestingly, a naturally occurring germline mutation in E-cadherin, found in hereditary diffuse gastric cancer, contains a mutation (V832M) in the proposed PIPK1γ binding site and has impaired PIPK1γ binding and accumulates in a cytosolic compartment, thus highlighting the importance of efficient E-cadherin trafficking. The authors propose a model in which PIPK1γ serves a dual function as both scaffold and signaling molecule. PIPK1γ acts as a cargo adaptor to enable the recruitment of AP1B to E-cadherin and also regulates the localized generation of $PI4,5P_2$, which drives the function of the trafficking machinery [26]. As AP1B is preferentially localized to the recycling endosome rather than to the TGN, it is possible that PIPK1γ functions as an additional sorting step to the dileucine motif. Additionally, the dileucine motif may also interact with AP1B, perhaps with the β subunit, to stabilize the interaction [25].

Further evidence suggests that E-cadherin acts as a spatial cue to define the apical junctional complex [18, 27, 28]. Yeaman et al. found that E-cadherin, along with the calcium-independent adhesion molecule nectin, was required for the efficient recruitment of the exocyst (Sec 6/8 complex) to the membrane [29]. Vesicles transported through the post-Golgi sorting machinery to the plasma membrane are tethered to the basolateral membrane through interactions with the exocyst, as well as t-SNARES [30, 31]. This suggests a mechanism in which E-cadherin-based cell contact, along with the exocyst and SNARES, function as a "targeting patch" for the polarized delivery of basolateral proteins (Figure 13.1). Nejsum and Nelson tested this hypothesis by comparing the trafficking of basolaterally targeted aquoporin (AQP)-3 to the apically targeted AQP-5. These authors found that AQP-3 – but not AQP-5 – accumulates at sites of cell contact, and colocalizes with E-cadherin in a manner dependent on the exocyst, t-SNARES, and microtubules. Thus, the targeted basolateral delivery of E-cadherin is important not only for the establishment and maturation of cell contacts, but also for general basolateral protein sorting and membrane domain organization [32]. A similar mechanism has been identified for the delivery of the gap junction protein Cx43 to cell borders [33]. These findings suggest a bidirectional system of regulation between cadherin-based adhesion and protein sorting machinery.

At AJs, E-cadherin forms a complex with catenin proteins. The status of E-cadherin complex assembly is dynamic during E-cadherin trafficking, and is an important factor in the proper delivery of E-cadherin to the cell surface and the establishment of nascent cell junctions. E-cadherin is sorted to the basolateral surface in a complex with β-catenin [33, 34]. Furthermore, formation of the E-cadherin/β-catenin complex correlates with an efficient exit from the endoplasmic reticulum (ER) [35]. Chimeric proteins containing the E-cadherin cytoplasmic tail fused to GP-2, which is normally attached to the membrane via a GPI anchor, were efficiently targeted to the basolateral membrane, again demonstrating that the cytoplasmic domain of E-cadherin contains information required for basolateral sorting. By using a series of E-cadherin mutants unable to bind β-catenin, Chen et al. showed that E-cadherin mutants unable to bind to β-catenin are retained in the ER, which suggests that β-catenin may act as a chaperone for E-cadherin during delivery to the basolateral membrane. Further evidence suggests that, in the absence of proper E-cadherin targeting, β-catenin is also mislocalized [36]. Other components of the adherens junctional complex, including p120 and α-catenin, are only recruited to the complex at the basolateral surface. p120 does not colocalize with E-cadherin on intracellular membranes as does β-catenin, and p120's junctional localization is not disrupted in cells expressing E-cadherin mutants that are improperly sorted. α-Catenin also joins the E-cadherin complex at the basolateral membrane, in conjunction with the arrival of the E-cadherin–β-catenin complex at the plasma membrane [36]. This spatial and temporal regulation of the AJ complex assembly is consistent with the established roles of these proteins at intercellular junctions. α-Catenin assembly with the cadherin complex may coincide with incorporation of the junctional complex into the actin cytoskeleton.

Interestingly, the situation may occur differently for N-cadherin complex formation. Evidence suggests a model in which p120 binds to immature proN-cadherin immediately following synthesis. Later, following phosphorylation of the procadherin, β-catenin becomes associated with the cadherin, possibly in complex with α-catenin. The proregion of N-cadherin is then removed by furin proteases, and the mature cadherin is transported to the plasma membrane in a complex with p120, β-catenin, and α-catenin [37]. The early association of p120 with N-cadherin suggests a potential role for p120 in N-cadherin delivery to the plasma membrane. Mary et al. showed that N-cadherin transport to the plasma membrane occurs in a microtubule-dependent, kinesin-driven manner [38]. p120 also interacts with microtubules through kinesin and plays a role in N-cadherin transport to the plasma membrane [39, 40]. Chen et al. showed that disruption of the interaction of p120 with either N-cadherin or kinesin slows the accumulation of N-cadherin at cell–cell contacts [41]. Further studies are required to elucidate the apparent differences in the role of p120 during the trafficking of various cadherins to sites of adherens junction assembly.

13.3
Endocytosis of Adherens Junction Proteins

Cadherin function at cellular junctions is not static, but rather requires the dynamic modulation of adhesive strength in response to changing developmental or environmental cues [27]. Adherens junctions are highly dynamic and undergo continual remodeling, as evidenced by live cell imaging of GFP-tagged E-cadherin [42]. The molecular mechanisms underlying the dynamic nature of cadherin-based cell adhesion are not fully understood, and have remained a matter of great controversy within the field. The traditional model of the dynamic nature of cadherin-mediated adhesion is based on low-affinity initial cadherin–cadherin interactions that are subsequently strengthened by lateral cadherin clustering and cytoskeletal anchoring [42, 43]. An alternative model is that cadherin-adhesive dimers are formed by strong, but dynamic, interactions that are rapidly assembled and disassembled [44, 45].

Over the past decade, research investigations have increasingly indicated that the expression and function of cadherins at AJs is largely influenced by a balance between exocytic and endocytic transport mechanisms. The regulated uptake of cadherins from the cell surface has been documented during both development and post-natal life, when large-scale cellular rearrangements are required. For example, in gastrulating sea urchin embryos, cells undergoing epithelial–mesenchymal transition (EMT) exhibit AJ disassembly, which correlates with changes in cadherin localization from junctions to intracellular organelles [46]. Dynamic changes in cadherin expression have also been observed during EMT in invasive tumors, angiogenesis, and wound healing [47–50].

Cadherin recycling and degradation occurs both constitutively and under conditions in which cell adhesion is compromised. Membrane trafficking of cadherins to and from the cell surface serves as a crucial determinant of cellular adhesive strength, and acts as a way to dynamically modulate cadherin expression levels within cells. The degree of endocytosis and the fate of the cadherin are dependent on the degree of cell–cell contact and the cellular signaling environment. Under normal physiological conditions, the majority of cadherin is located on the cell surface at intercellular junctions; however, a certain pool is endocytosed into intracellular vesicular compartments. In the case of E-cadherin in MDCK cells, the internalized pool of cadherin is rapidly recycled to the cell surface [51]. However, in endothelial cells, VE-cadherin undergoes a certain level of constitutive lysosomal degradation, as revealed by treatment with chloroquine to inhibit lysosomal degradation [52]. Basal levels of constitutive cadherin endocytosis could involve predominantly non-cytoskeletal associated cadherins that are able to flow within the plane of the plasma membrane and enter endocytic routes. Alternatively, a recent report suggests that endocytosis functions as the driving mechanism for the disassembly of adhesive cadherin dimers. By using ATP depletion or hypertonic sucrose treatment to inhibit endocytosis, Troyanovsky et al. showed that E-cadherin adhesive dimers are stabilized in the absence of endocytosis. This blockage of dimer dissociation resulted in a dramatic increase in the amount of adhesive dimers, along with a parallel decrease in the pool of cadherin monomers [53]. In both of these models, cadherin endocytosis plays a fundamental role in regulating the dynamics and plasticity of AJs.

The levels of endocytosis are greatly increased in cells lacking stable cell–cell contacts, such as in preconfluent monolayers, or following the disruption of cell junctions by Ca^{2+} depletion. Upon the replenishment of extracellular Ca^{2+}, recycling of surface E-cadherin is sufficient, and necessary, to restore the epithelial monolayer in the absence of protein synthesis [51]. Under conditions in which cadherin-based adhesion is disrupted, cadherin expression is often downregulated. For instance, EMT correlates with a loss of E-cadherin, which is often attributed to transcriptional repression or genetic mutations [54]. However, in some model systems, cadherin expression is greatly reduced despite normal gene expression levels [16, 49, 55]. During EMT induced by v-src activation, E-cadherin is degraded in the lysosome following endocytosis. The expression of v-Src leads to the phosphorylation-dependent ubiquitination of E-cadherin. Ubiquitinated E-cadherin is trafficked to the lysosome for degradation through a pathway that requires hepatocyte growth factor (HGF)-regulated tyrosine kinase substrate (Hrs) and the GTPases Rab5 and Rab7 [56]. Another study identified Hakai as a c-Cbl-like E3 ubiquitin ligase that promotes the monoubiquitination and endocytosis of E-cadherin upon phosphorylation of the E-cadherin JMD by Src [57]. At present, it is not fully understood whether Hakai functions at the plasma membrane, or at endocytic compartments.

The mechanisms that regulate cadherin endocytosis and degradation are central to an understanding of how membrane trafficking pathways contribute to the dynamic regulation of cell adhesion. Numerous studies have indicated that cadherin endocytosis can be mediated by both clathrin-dependent and clathrin-independent pathways, and that the cellular machinery used by cadherins is most likely dependent on the cellular context. The recycling of E-cadherin in MDCK cells – both constitutive and in the absence of extracellular Ca^{2+} – occurs in a clathrin-dependent manner [51]. In T84 cells, Ivanov et al. observed a coordinated internalization of the entire apical junctional into an intracellular compartment enriched in syntaxin-4 upon Ca^{2+} depletion [58]. Whilst AJ and tight junction proteins appeared to be segregated into distinct populations within the syntaxin-4 vesicles, endocytosis of both types of junctions was blocked by inhibitors of clathrin-dependent endocytosis, but not by the disruption of cavaolae/lipid rafts or macropinocytosis [58]. In endothelial cells, VE-cadherin is also internalized in a clathrin-dependent manner, and is trafficked through early and late endosomes to the lysosome, where it undergoes degradation [59, 60].

Clathrin-independent pathways have also been implicated in cadherin endocytosis. In response to epidermal growth factor (EGF) treatment, E-cadherin in A431 cells is internalized through a pathway that requires caveolin-1 [61]. A similar mechanism may be involved in E-cadherin endocytosis in keratinocytes upon Rac activation, as cadherin containing intracellular vesicles exhibited colocalization with caveolin [62]. The desmosomal cadherin Desmoglein-3 (Dsg3) also undergoes clathrin-independent endocytosis. The epidermal blistering disease, pemphigus vulgaris (PV) is an autoimmune disease which is characterized by the loss of epidermal cell adhesion in response to autoantibodies generated against Dsg3. In a cell culture model, the treatment of keratinocytes with PV IgG induces the loss of cell adhesion associated with the endocytosis and degradation of Dsg3 in the lysosome [63]. Further characterization of the endocytic machinery involved in Dsg3 endocytosis indicates that Dsg3 is internalized through a cholesterol-dependent, but clathrin- and dynamin-independent pathway (E. Delva and A. P. Kowalczyk, unpublished results). The multitude of endocytic pathways utilized by cadherins for the modulation of cellular adhesive contacts points towards a fine-tuned system for the regulation of cadherin levels that is highly dependent on the cellular context.

13.4
Catenin Regulation of Cadherin Endocytosis

In addition to components of the endosomal trafficking machinery, cadherin endocytosis may also be regulated by other components of the AJ complex. p120 in particular, has been identified as a key regulator of cadherin expression and function through its ability to stabilize the cadherin at the plasma membrane [64, 65]. The first evidence for a role for p120 in cadherin-based adhesion came from the identification of a tumor cell line with cell-adhesion deficiencies due to mutations in p120. The absence of p120 in these cells results in a corresponding decrease in E-cadherin stability, even though mRNA levels are not affected [66]. Cell adhesion and E-cadherin expression can be rescued by exogenous expression of either p120 or E-cadherin, suggesting that p120 functions in regulating E-cadherin turnover [66].

Further studies have helped to elucidate the mechanism by which p120 acts to regulate E-cadherin expression and function. siRNA knockdown and overexpression of p120 in mammalian cell culture revealed that p120 functions as a set point, or rheostat, of cadherin expression levels in multiple cell types and for multiple cadherins, including E-cadherin, N-cadherin and VE-cadherin [52, 67]. The loss of p120 by siRNA leads to a concomitant reduction in cadherin levels, in a dose-dependent manner, as well as decreases in cell adhesion. Conversely, the expression of exogenous p120 causes increased cadherin expression. In cells depleted of p120 by siRNA, unbound E-cadherin is properly delivered to the cell surface [67], but cannot be retained and therefore is immediately targeted for endocytosis and degradation in the lysosome [52, 59].

In-vivo studies in mice using p120 conditional knockout models in a variety of tissues have substantiated these cell-culture findings. The loss of p120 in several epithelial tissues, including the salivary gland, results in a decreased E-cadherin expression, ultimately leading to hyperproliferation or inflammation [68, 69]. Additionally, mice with a conditional knockout of p120 in endothelial cells die embryonically and display severe defects in endothelial barrier function and vessel patterning (K. Xiao and A. P. Kowalczyk, unpublished observation). These defects are associated with a decrease in VE-cadherin levels in endothelial cells *in vivo*. Nonetheless, a key issue to resolve is which *in-vivo* phenotypes reflect p120 functions that are independent of cadherin.

These findings also provided an important insight governing the mechanism of action of the dominant-negative cadherin chimeras which, as described above, lead to the endocytosis and degradation of endogenous cadherins when expressed in cells at high levels. Several studies have demonstrated this phenomenon upon expression of cadherin molecules consisting of a cadherin cytoplasmic domain tethered to a non-adhesive extracellular domain. These cadherin mutants act in a dominant fashion, inducing the downregulation of endogenous cadherin [70–72]. In endothelial cells, the expression of an interleukin (IL)-2R–VE-cadherin chimera results in the internalization of endogenous VE-cadherin that is processed and degraded through an endolysosomal pathway [52, 59]. Xiao et al. found that an

IL-2R–VE-cadherin chimera with a mutation in the p120 binding site does not cause cadherin downregulation. Furthermore, VE-cadherin levels could be restored upon the exogenous expression of p120, which suggests that the dominant-negative cadherin mutants function by competing with endogenous cadherins for interactions with p120 [52]. Thus, p120 acts as a limiting factor in stabilizing cadherins at the cell surface. In cells expressing dominant-negative cadherins, p120 is sequestered by the mutant cadherin and is therefore unable to bind to endogenous cadherin, resulting in its endocytosis and degradation. The downregulation of endogenous cadherins by the expression of dominant-negative chimeric cadherins is reminiscent of the cadherin switching that occurs during EMT [73]. During EMT, E-cadherin expression is transcriptionally repressed, while non-epithelial cadherins such as N-cadherin, R-cadherin or cadherin 11 are upregulated, a change that is required for the increased motility associated with EMT, especially in metastasis of breast cancer cells [74–76]. When R-cadherin is expressed in A431 epithelial cells (which normally express E- and P-cadherin), the endogenous cadherin is shifted from a recycling pathway to a lysosomal degradation pathway. Experiments utilizing an R-cadherin mutant unable to bind to p120 showed that R-cadherin expression causes degradation of E-cadherin by competing for the availability of p120 [77]. The ability of non-epithelial cadherins to downregulate E-cadherin expression is cell-type-specific, as evidenced by the fact that N-cadherin expression in oral squamous epithelial cells reduces levels of E-cadherin but has no effect on E-cadherin in some breast epithelial cells [70, 78]. These findings may be explained by the relative abundance of p120 in the cell, based on p120's function as a set point for overall cadherin levels.

In endothelial cells, the expression of exogenous p120 prevents VE-cadherin from undergoing clathrin-dependent endocytosis through a mechanism that requires the direct interaction between p120 and the VE-cadherin JMD [60]. Additionally, p120 does not colocalize with the endosomal pool of VE-cadherin following internalization, indicating that p120 dissociates from the VE-cadherin tail during endocytosis. During clathrin-mediated endocytosis, transmembrane receptors are recruited into clathrin-coated pits by adaptor proteins that interact with short tyrosine or dileucine sorting motifs contained within the cytoplasmic tail of the cargo protein [79]. Interestingly, gain-of-function experiments with the IL-2R–VE-cadherin chimera demonstrated that the VE-cadherin cytoplasmic domain harbors information that positively mediates endocytosis [60]. A sequence analysis of the VE-cadherin cytoplasmic domain revealed the presence of several putative tyrosine and dileucine motifs, suggesting a potential mechanism for the recruitment of VE-cadherin into a clathrin-mediated endocytic pathway. In further support of this model, an interaction has been identified between the VE-cadherin cytoplasmic tail and the clathrin adaptor complex AP-2 (C. M. Chiasson and A. P. Kowalczyk, unpublished observation). Taken together, these findings lend support for a model in which p120 stabilizes cadherins at the cell surface by competing with components of the clathrin endocytic machinery for interactions with the cadherin tail. p120 may therefore function as a cap to prevent clathrin adaptor molecules from binding to sorting motifs and recruiting the cadherin into clathrin-coated vesicles (Figure 13.2A). Interestingly, a recent report proposed that the dileucine motif located at amino acids 587–588 in the E-cadherin JMD is required for E-cadherin endocytosis in MDCK cells, providing further support for a model in which p120 functions as a cap on the cadherin tail [80]. This dileucine motif is the same as that found to be required for proper basolateral targeting of E-cadherin [36], thus raising the possibility that the same sorting motif may regulate multiple aspects of cadherin trafficking within the cell. Importantly, while this dileucine motif is conserved in several classical cadherins, it is absent in VE-cadherin, indicating that alternative motifs mediate VE-cadherin internalization.

The Takai laboratory recently proposed a similar, yet indirect, role for p120 in the stabilization of cadherin at intercellular AJs [81]. In a cell-free assay system,

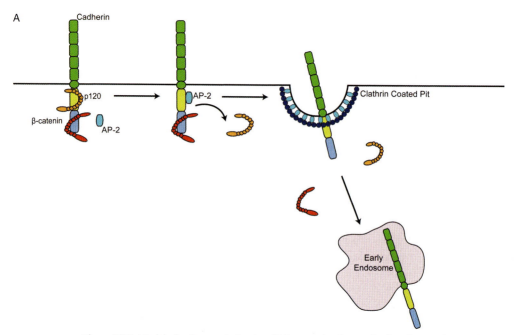

Figure 13.2 Models for the regulation by p120 of cadherin endocytosis. p120 stabilizes cadherin expression at the plasma membrane by preventing the cadherin from entering a clathrin-dependent endocytic pathway. (A) p120 may function as a "cap" on the cadherin cytoplasmic tail. In this model, the binding of p120 to the cadherin juxtamembrane domain (JMD) prevents the interaction of clathrin adaptor proteins, such as AP-2, with sorting motifs in the cadherin tail. Upon dissociation of p120 from the cadherin JMD, the cadherin would be recruited into clathrin-coated pits and internalized from the plasma membrane. (B) p120's regulation of RhoGTPase activity may also be involved in the regulation of cadherin endocytosis. Upon activation of Rac1 by PDGF signaling, p120 inhibits RhoA GTPase activity at cadherin complexes through an interaction with p190RhoGAP. In the absence of p120 or p190RhoGAP, activation of RhoA dramatically increases, resulting in mislocalization of cadherin from cell junctions. RhoA has been shown to be involved in endocytosis and cadherin-based adhesion, suggesting a potential relationship between the p120-mediated stabilization of cadherin and the localized inhibition of Rho activity by p120 and p190RhoGAP.

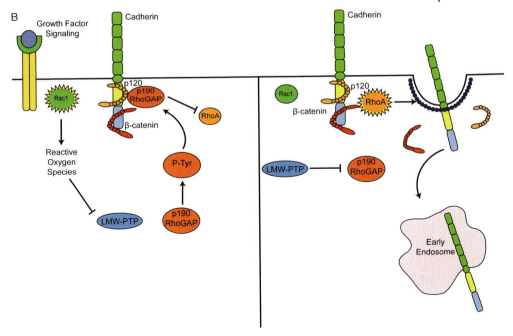

Figure 13.2 *Continued*

non-*trans*-interacting E-cadherin was constitutively endocytosed through a clathrin-dependent pathway [82]. The adhesion molecule nectin, which has been shown to recruit cadherins into AJs, interacts with afadin to link nectin to the actin cytoskeleton. Hoshino et al. found that *trans*-interacting nectin formed a complex with afadin and the small GTPase Rap1, and that this association strengthened p120's interaction with the E-cadherin cytoplasmic tail, thereby preventing the endocytosis of non-*trans*-interacting E-cadherin [81]. The stabilization of non-*trans*-interacting E-cadherin allows for the accumulation and formation of E-cadherin *trans*-interactions, ultimately leading to strong adhesion at AJs.

13.5
Rho GTPase Regulation of Cadherin Endocytosis

Rho GTPases, including RhoA and Rac, have previously been implicated in cadherin endocytosis, either through regulation of the cadherin/catenin complex, or through general regulation of intracellular trafficking steps. The activation of Rac1 has been reported to have both stimulatory [62, 83] and inhibitory [82, 84, 85] effects on cadherin endocytosis, likely depending on the adhesive status and cell type. Braga et al. demonstrated that the requirement for Rho and Rac activity in cadherin-dependent adhesion varied greatly among different cadherin receptors and cell types, and was also dependent on cell confluency. For example, in endothelial cells, VE-cadherin is refractory to regulation by Rho or Rac activity; however, when expressed in Chinese hamster ovary (CHO) cells, the inhibition of Rho or Rac caused disruption of VE-cadherin from junctions [86]. On the other hand, E-cadherin and P-cadherin, both of which are expressed in keratinocytes and form independent complexes, are both lost from cell junctions and degraded upon inhibition of Rho or Rac activity [86]. Interestingly, when E-cadherin is expressed in fibroblast L cells, its function is regulated by Rho but not Rac, again suggesting that Rho GTPase activity and its role in regulating cadherin-based adhesion and trafficking is highly sensitive to cellular context [86].

The model of p120 directly acting to prevent cadherin endocytosis by competing for interactions with the cadherin tail offers a simple and direct explanation for p120's ability to stabilize cadherins at the cell surface. However, an alternative possibility is that p120 exerts its influence over cadherin in a more indirect manner. A recent study has identified a potential mechanism by which p120 may regulate cadherin function by integrating diverse signaling systems through an interaction with p190RhoGAP. Using platelet-derived growth factor (PDGF)-activated actin remodeling as a model system, Wildenberg et al. observed a dramatic increase in actin stress fibers and the loss of actin remodeling, accompanied by cell transformation and serum-free proliferation in NIH3T3 cells depleted of p120 by siRNA. These changes in actin cytoskeleton dynamics and cell growth result from a dramatic increase in Rho GTPase activity upon loss of p120 [87]. By dissecting the molecular pathway of PDGF-mediated actin remodeling, it was found that activation of Rac by PDGF causes transient translocation of p190RhoGAP to cadherin complexes through an interaction with p120. Through this interaction, p120 mediates antagonism between Rac and Rho signaling by locally inhibiting Rho at cadherin complexes. In the absence of p190RhoGAP, p120 and N-cadherin are mislocalized, suggesting a role for p190RhoGAP in AJ stability [87] (Figure 13.2B). Taken together, these findings support a model whereby p120 regulates the stability of cadherin complexes by acting as a nexus to integrate extracellular signals into a coordinated cellular response to environmental cues. Ongoing studies in the present authors' laboratory, and of others, are designed to clarify the roles of p120 in Rho GTPases regulation and clathrin adaptor interactions with cadherins.

13.6
Co-Regulation of Cadherin and Receptor Tyrosine Kinase Function by Endocytosis

The multitude of endocytic pathways utilized by cadherins for the modulation of cellular adhesive contacts points towards a fine-tuned system for the regulation of cadherin levels by differences in the local signaling environment. Evidence suggests that even the same signaling molecules can stimulate cadherin internalization through different pathways, depending on the cellular context. For instance, short-term treatment of tumor cells overexpressing epidermal growth factor receptor (EGFR) with EGF leads to cell contact disruption and EMT as a result of E-cadherin endocytosis through a cholesterol-dependent pathway that may involve caveolae [87]. In another system, Rac1 activation downstream of EGF signaling induces macropinocytosis of E-cadherin, along with its associated catenins, p120 and β-catenin, into a recycling endosome. Macropinocytosis occurs from regions of the cell lacking cell contacts, but is recycled back to cell contacts through a pathway requiring sorting nexin 1 (SNX1). This pathway could therefore be useful in the remodeling of existing cell contacts, without the loss of adhesion [88]. Taken together, these results suggest that endocytosis of junctional proteins is tightly regulated and that cells utilize several different membrane trafficking pathways to regulate cadherin function, depending on the signaling cues in the cellular environment.

The functional interplay between cell-adhesion molecules and receptor tyrosine kinases (RTKs) has recently been identified as a novel mechanism for the co-regulation of adhesion and signaling activity [89, 90]. Several examples of bidirectional regulation of cadherin-RTK function by endocytic trafficking machinery have recently been reported, suggesting a synergistic mechanism for the regulation of cell growth, proliferation, and migration. In MDCK cells, HGF activation results in AJ disruption and the endocytosis of E-cadherin. The HGF/SF receptor c-met colocalizes with E-cadherin at cell junctions and is endocytosed into a perinuclear vesicular compartment along with E-cadherin upon HGF treatment or Ca^{2+} depletion. HGF-induced endocytosis of E-cadherin and c-met is specifically dependent on GTPase activity. The expression of DA Rho, DA Rac1, or DN Rab5 mutants prevented E-cadherin and c-met endocytosis upon HGF treatment, but had no effect on internalization following Ca^{2+} depletion [84]. This suggests an indirect mechanism for the GTPase regulation of cadherin and receptor trafficking in this system, and again underscores the importance of cellular context in the regulation of cadherin function. Activation of the small GTPase ARF6, which lies downstream of the HGF receptor (c-met), has also been identified as a positive regulator of E-cadherin endocytosis in response to HGF. ARF6 promotes both clathrin- and dynamin-dependent endocytosis of E-cadherin [85, 91]. Paterson et al. also found ARF6 activity to be required for the endocytosis of unbound pools of E-cadherin in MCF-7 cells lacking cell–cell contacts. Interestingly, however, in these cells endocytosis occurred through a dynamin-dependent, but clathrin-independent, pathway resembling macropinocytosis [92].

Fibroblast growth factor (FGF) signaling is involved in embryonic patterning during development and epithelial cell morphogenesis through its role in EMT. Activation of the fibroblast growth factor receptor (FGFR) by FGF stimulates the co-endocytosis of FGFR1 and E-cadherin and the eventual nuclear translocation of FGFR1 [93]. The retention of E-cadherin at cell–cell contacts, either by overexpression of E-cadherin or p120, directly impacts FGFR function by preventing its endocytosis, signaling to the MAPK pathway, and nuclear translocation [93]. The parallel regulation of E-cadherin and FGF function through the modulation of intracellular trafficking suggests an elegant mechanism for maintaining the balance between epithelial and mesenchymal phenotypes during changes in the cellular environment. Interestingly, the association of N-cadherin with FGFR1 at the cell surface also prevents FGFR internalization, but inhibition of FGFR endocytosis by N-cadherin leads to the prolonged activation of MAPK signaling downstream of FGF activation, thereby promoting the invasiveness of tumor cells [94].

In endothelial cells, VE-cadherin mediates the contact inhibition of endothelial cell growth through a mechanism involving attenuation of VEGF-induced signaling. In confluent cell monolayers, VE-cadherin forms a complex with VEGF receptor (VEGFR)-2 upon VEGF stimulation. The association of VE-cadherin with VEGFR-2 results in the downregulation of VEGFR-2 phosphorylation and downstream MAPK signaling through a β-catenin-dependent mechanism involving the phosphotyrosine phosphatase density-enhanced phosphatase-1 (DEP-1) [95]. In endothelial cells, the phenomenon of cadherin-mediated attenuation of RTK signaling also requires the endocytosis of growth factor receptors. Analogous to the situation in epithelial cells, an association of VE-cadherin with VEGFR-2 in endothelial cells prevents the internalization of VEGFR-2 through a clathrin-dependent pathway. In VE-cadherin-null cells, VEGFR-2 is rapidly internalized into early endosomes, where it retains its signaling abilities [96]. In contrast to the co-endocytosis of FGFR and E-cadherin, VE-cadherin does not colocalize with internalized VEGFR-2, suggesting that the proteins dissociate prior to internalization. This mechanism of VE-cadherin-mediated regulation of VEGF-induced signaling helps to explain the persistence of proliferative signaling in the absence of VE-cadherin. This suggests a model whereby VE-cadherin modulates cell growth through the inhibition of VEGFR-2 signaling by retaining it at the cell surface, where it is maintained in an inactive state by DEP phosphatase activity [96].

VEGF signaling through VEGFR-2 can also promote VE-cadherin endocytosis via a novel signaling pathway that appears to regulate endothelial monolayer permeability [83]. In this system, VE-cadherin is internalized into an intracellular compartment and colocalizes with markers of clathrin-mediated endocytosis, but not with caveolin1. The treatment of endothelial cells with VEGF resulted in an activation of Rac in a Src- and Vav2-dependent manner. The VEGF-stimulated endocytosis of VE-cadherin is blocked by the expression of a dominant-negative Rac mutant, while a constitutively active form of Rac leads to VE-cadherin endocytosis, even in the absence of VEGF signaling. Gavard and Gutkind identified a serine-threonine cluster (Ser665) unique to VE-cadherin, but highly conserved among vertebrate species, downstream of the p120 binding site. The phosphorylation of serine 665 is required for VEGF-mediated endocytosis of VE-cadherin, due to its ability to interact with β-arrestin-2, which is best known for its role in G-protein-coupled receptor (GPCR) ligand-dependent endocytosis. Furthermore, the knockdown of β-arrestin-2 by shRNA prevents VEGF-mediated internalization of VE-cadherin, in addition to a loss of endothelial barrier function associated with VEGF signaling [83]. Further studies must be conducted to determine if this pathway is also used as a general mechanism of VE-cadherin endocytosis, or is specific to VEGF signaling.

13.7
Conclusions

Adherens junctions are central to the establishment and maintenance of cell polarity and tissue architecture, as well as for the regulation of cellular morphogenetic processes. The rapid and dynamic modulation of these cell contacts is required for both the structural and signaling functions of adherens junctional components. In this chapter, the accumulating evidence for a fundamental role of membrane trafficking in the dynamic regulation of cell adhesion through the movement of AJ proteins to and from the plasma membrane has been presented.

Exocytic and endocytic trafficking pathways provide a rapid and highly regulated mechanism to control the abundance of cadherin receptors and their associated proteins at the plasma membrane. Cadherin trafficking is regulated on many levels by diverse cellular processes in order to create a finely tuned system for the regulation of cell adhesion throughout the life of multicellular organisms. During AJ assembly, cadherin delivery to the basolateral membrane is tightly regulated by post-Golgi trafficking machinery, a process that is crucial for the establishment of cell junctions and cell polarity (see Figure 13.1). A newly emerging concept that will require further exploration is the idea that E-cadherin-based cell contacts function as a "targeting patch" for the sorting and delivery of basolateral proteins. Once established, AJs undergo continual remodeling in response to cellular cues that require changes in the cellular adhesive status. The cadherin-associated catenin, p120 has emerged as a primary regulator of cadherin expression at the plasma membrane by preventing cadherin endocytosis. Further studies are required to fully understand the mechanism of p120's function as a set point of cadherin expression. One possible model is that p120 functions as a "cap" by competing with components of the endocytic machinery for interactions with the cadherin cytoplasmic domain (see Figure 13.2A). In order to thoroughly test this model, however, a more complete understanding of the molecular pathways governing cadherin endocytosis, as well as the regulation of p120's interaction with the cadherin JMD is necessary. Alternatively, p120's role in regulating Rho GTPase activity may be tied to the regulation of cadherin stability, as suggested by Wildenberg et al. [87] (Figure 13.2B). It is not yet clear if the cadherin mislocalization observed is caused by increased endocytosis, or by other mechanisms. Studies of Rho GTPases in cadherin endocytosis indicate that the mechanisms of regulation are highly variable, depending on the cellular context and the signaling environment, which makes it difficult to develop a generalized view of the pathways regulating cadherin endocytosis. Nonetheless, these issues will need to be addressed.

Cadherin-based AJs are increasingly recognized for their role as centers of cell signaling, in addition to their structural role as sites of cell adhesion. Recent studies have shown that cadherin endocytosis functions as an additional means of regulating the signaling molecules that complex with AJs. The ability of cadherins to regulate the signaling activity of RTKs by the regulation of endocytosis indicates a central role for cadherin expression itself in modulating cellular signaling. An understanding of the subtleties of the relationship between cadherin trafficking and RTK signaling will be important for the development of therapeutics aimed at regulating cadherin function. Overall, there is a need to better understand the functional interplay between membrane trafficking pathways and cadherins in the context of development and human disease.

Acknowledgments

The authors are grateful to Dr. Kathleen J. Green for insightful comments and suggestions. The studies conducted in the authors' laboratory were supported by grants from the NIH/NIAMS (R01AR050501 and R01AR048266). C.M.C. is supported by a predoctoral fellowship from the American Heart Association.

References

1 Halbleib JM, Nelson WJ: Cadherins in development: cell adhesion, sorting, and tissue morphogenesis. *Genes Dev.* 2006, *20*(23): 3199–3214.
2 Lien WH, Klezovitch O, Vasioukhin V: Cadherin-catenin proteins in vertebrate development. *Curr. Opin. Cell Biol.* 2006, *18*(5): 499–506.
3 Braga VM: Cell-cell adhesion and signalling. *Curr. Opin. Cell Biol.* 2002, *14*(5): 546–556.
4 Wheelock MJ, Johnson KR: Cadherin-mediated cellular signaling. *Curr. Opin. Cell Biol.* 2003, *15*(5): 509–514.
5 Wheelock MJ, Johnson KR: Cadherins as modulators of cellular phenotype. *Annu. Rev. Cell Dev. Biol.* 2003, *19*: 207–235.
6 Yap AS, Brieher WM, Gumbiner BM: Molecular and functional analysis of cadherin-based adherens junctions. *Annu. Rev. Cell Dev. Biol.* 1997, *13*: 119–146.
7 Braga VM, Yap AS: The challenges of abundance: epithelial junctions and small GTPase signalling. *Curr. Opin. Cell Biol.* 2005, *17*(5): 466–474.
8 Nollet F, Kools P, van Roy F: Phylogenetic analysis of the cadherin superfamily allows identification of six major subfamilies besides several solitary members. *J. Mol. Biol.* 2000, *299*(3): 551–572.
9 Angst BD, Marcozzi C, Magee AI: The cadherin superfamily: diversity in form and function. *J. Cell Sci.* 2001, *114*(Pt 4): 629–641.
10 Anastasiadis PZ, Reynolds AB: The p120 catenin family: complex roles in adhesion, signaling and cancer. *J. Cell Sci.* 2000, *113*(Pt 8): 1319–1334.
11 Thoreson MA, Anastasiadis PZ, Daniel JM, Ireton RC, Wheelock MJ, Johnson KR, Hummingbird DK, Reynolds AB: Selective uncoupling of p120(ctn) from E-cadherin disrupts strong adhesion. *J. Cell Biol.* 2000, *148*(1): 189–202.
12 Yap AS, Niessen CM, Gumbiner BM: The juxtamembrane region of the cadherin cytoplasmic tail supports lateral clustering, adhesive strengthening, and interaction with p120ctn. *J. Cell Biol.* 1998, *141*(3): 779–789.
13 Gumbiner BM: Regulation of cadherin adhesive activity. *J. Cell Biol.* 2000, *148*(3): 399–404.
14 Takeichi M, Nakagawa S, Aono S, Usui T, Uemura T: Patterning of cell assemblies regulated by adhesion receptors of the cadherin superfamily. *Philos. Trans. R. Soc. Lond. B Biol. Sci.* 2000, *355*(1399): 885–890.
15 Bryant DM, Stow JL: The ins and outs of E-cadherin trafficking. *Trends Cell Biol.* 2004, *14*(8): 427–434.
16 D'Souza-Schorey C: Disassembling adherens junctions: breaking up is hard to do. *Trends Cell Biol.* 2005, *15*(1): 19–26.
17 Rodriguez-Boulan E, Musch A, Le Bivic A: Epithelial trafficking: new routes to familiar places. *Curr. Opin. Cell Biol.* 2004, *16*(4): 436–442.
18 Hsu SC, TerBush D, Abraham M, Guo W: The exocyst complex in polarized exocytosis. *Int. Rev. Cytol.* 2004, *233*: 243–265.
19 Nelson WJ: Adaptation of core mechanisms to generate cell polarity. *Nature* 2003, *422*(6933): 766–774.
20 Miranda KC, Khromykh T, Christy P, Le TL, Gottardi CJ, Yap AS, Stow JL, Teasdale

RD: A dileucine motif targets E-cadherin to the basolateral cell surface in Madin-Darby canine kidney and LLC-PK1 epithelial cells. *J. Biol. Chem.* 2001, 276(25): 22565–22572.
21. Lock JG, Hammond LA, Houghton F, Gleeson PA, Stow JL: E-cadherin transport from the trans-Golgi network in tubulovesicular carriers is selectively regulated by golgin-97. *Traffic* 2005, 6(12): 1142–1156.
22. Wang B, Wylie FG, Teasdale RD, Stow JL: Polarized trafficking of E-cadherin is regulated by Rac1 and Cdc42 in Madin-Darby canine kidney cells. *Am. J. Physiol. Cell Physiol.* 2005, 288(6): C1411–C1419.
23. Lock JG, Stow JL: Rab11 in recycling endosomes regulates the sorting and basolateral transport of E-cadherin. *Mol. Biol. Cell* 2005, 16(4): 1744–1755.
24. Bonifacino JS, Glick BS: The mechanisms of vesicle budding and fusion. *Cell* 2004, 116(2): 153–166.
25. Rapoport I, Chen YC, Cupers P, Shoelson SE, Kirchhausen T: Dileucine-based sorting signals bind to the beta chain of AP-1 at a site distinct and regulated differently from the tyrosine-based motif-binding site. *EMBO J.* 1998, 17(8): 2148–2155.
26. Ling K, Bairstow SF, Carbonara C, Turbin DA, Huntsman DG, Anderson RA: Type Igamma phosphatidylinositol phosphate kinase modulates adherens junction and E-cadherin trafficking via a direct interaction with mu 1B adaptin. *J. Cell Biol.* 2007, 176(3): 343–353.
27. Emery G, Knoblich JA: Endosome dynamics during development. *Curr. Opin. Cell Biol.* 2006, 18(4): 407–415.
28. Grindstaff KK, Yeaman C, Anandasabapathy N, Hsu SC, Rodriguez-Boulan E, Scheller RH, Nelson WJ: Sec6/8 complex is recruited to cell-cell contacts and specifies transport vesicle delivery to the basal-lateral membrane in epithelial cells. *Cell* 1998, 93(5): 731–740.
29. Yeaman C, Grindstaff KK, Nelson WJ: Mechanism of recruiting Sec6/8 (exocyst) complex to the apical junctional complex during polarization of epithelial cells. *J. Cell Sci.* 2004, 117(Pt 4): 559–570.
30. Guo W, Novick P: The exocyst meets the translocon: a regulatory circuit for secretion and protein synthesis? *Trends Cell Biol.* 2004, 14(2): 61–63.
31. Chen YA, Scheller RH: SNARE-mediated membrane fusion. *Nat. Rev. Mol. Cell. Biol.* 2001, 2(2): 98–106.
32. Nejsum LN, Nelson WJ: A molecular mechanism directly linking E-cadherin adhesion to initiation of epithelial cell surface polarity. *J. Cell Biol.* 2007, 178(2): 323–335.
33. Shaw RM, Fay AJ, Puthenveedu MA, von Zastrow M, Jan YN, Jan LY: Microtubule plus-end-tracking proteins target gap junctions directly from the cell interior to adherens junctions. *Cell* 2007, 128(3): 547–560.
34. Hinck L, Nathke IS, Papkoff J, Nelson WJ: Dynamics of cadherin/catenin complex formation: novel protein interactions and pathways of complex assembly. *J. Cell Biol.* 1994, 125(6): 1327–1340.
35. Chen YT, Stewart DB, Nelson WJ: Coupling assembly of the E-cadherin/beta-catenin complex to efficient endoplasmic reticulum exit and basal-lateral membrane targeting of E-cadherin in polarized MDCK cells. *J. Cell Biol.* 1999, 144(4): 687–699.
36. Miranda KC, Joseph SR, Yap AS, Teasdale RD, Stow JL: Contextual binding of p120ctn to E-cadherin at the basolateral plasma membrane in polarized epithelia. *J. Biol. Chem.* 2003, 278(44): 43480–43488.
37. Wahl JK, 3rd, , Kim YJ, Cullen JM, Johnson KR, Wheelock MJ: N-cadherin-catenin complexes form prior to cleavage of the proregion and transport to the plasma membrane. *J. Biol. Chem.* 2003, 278(19): 17269–17276.
38. Mary S, Charrasse S, Meriane M, Comunale F, Travo P, Blangy A, Gauthier-Rouviere C: Biogenesis of N-cadherin-dependent cell-cell contacts in living fibroblasts is a microtubule-dependent kinesin-driven mechanism. *Mol. Biol. Cell* 2002, 13(1): 285–301.
39. Chen X, Kojima S, Borisy GG, Green KJ: p120 catenin associates with kinesin and facilitates the transport of cadherin-catenin complexes to intercellular junctions. *J. Cell Biol.* 2003, 163(3): 547–557.
40. Yanagisawa M, Kaverina IN, Wang A, Fujita Y, Reynolds AB, Anastasiadis PZ: A novel interaction between kinesin and p120 modulates p120 localization and

function. *J. Biol. Chem.* 2004, *279*(10): 9512–9521.

41 Chen HC, Chu RY, Hsu PN, Hsu PI, Lu JY, Lai KH, Tseng HH, Chou NH, Huang MS, Tseng CJ, et al: Loss of E-cadherin expression correlates with poor differentiation and invasion into adjacent organs in gastric adenocarcinomas. *Cancer Lett.* 2003, *201*(1): 97–106.

42 Adams CL, Chen YT, Smith SJ, Nelson WJ: Mechanisms of epithelial cell-cell adhesion and cell compaction revealed by high-resolution tracking of E-cadherin-green fluorescent protein. *J. Cell Biol.* 1998, *142*(4): 1105–1119.

43 Kusumi A, Suzuki K, Koyasako K: Mobility and cytoskeletal interactions of cell adhesion receptors. *Curr. Opin. Cell Biol.* 1999, *11*(5): 582–590.

44 Troyanovsky S: Cadherin dimers in cell-cell adhesion. *Eur. J. Cell Biol.* 2005, *84*(2-3): 225–233.

45 Klingelhofer J, Laur OY, Troyanovsky RB, Troyanovsky SM: Dynamic interplay between adhesive and lateral E-cadherin dimers. *Mol. Cell. Biol.* 2002, *22*(21): 7449–7458.

46 Miller JR, McClay DR: Characterization of the role of cadherin in regulating cell adhesion during sea urchin development. *Dev. Biol.* 1997, *192*(2): 323–339.

47 Alexander JS, Jackson SA, Chaney E, Kevil CG, Haselton FR: The role of cadherin endocytosis in endothelial barrier regulation: involvement of protein kinase C and actin-cadherin interactions. *Inflammation* 1998, *22*(4): 419–433.

48 Detmar M: Tumor angiogenesis. *J. Invest. Dermatol. Symp. Proc.* 2000, *5*(1): 20–23.

49 Thiery JP: Epithelial-mesenchymal transitions in tumour progression. *Nat. Rev. Cancer* 2002, *2*(6): 442–454.

50 Tonnesen MG, Feng X, Clark RA: Angiogenesis in wound healing. *J. Invest. Dermatol. Symp. Proc.* 2000, *5*(1): 40–46.

51 Le TL, Yap AS, Stow JL: Recycling of E-cadherin: a potential mechanism for regulating cadherin dynamics. *J. Cell Biol.* 1999, *146*(1): 219–232.

52 Xiao K, Allison DF, Buckley KM, Kottke MD, Vincent PA, Faundez V, Kowalczyk AP: Cellular levels of p120 catenin function as a set point for cadherin expression levels in microvascular endothelial cells. *J. Cell Biol.* 2003, *163*(3): 535–545.

53 Troyanovsky RB, Sokolov EP, Troyanovsky SM: Endocytosis of cadherin from intracellular junctions is the driving force for cadherin adhesive dimer disassembly. *Mol. Biol. Cell* 2006, *17*(8): 3484–3493.

54 Cavallaro U, Christofori G: Cell adhesion in tumor invasion and metastasis: loss of the glue is not enough. *Biochim. Biophys. Acta* 2001, *1552*(1): 39–45.

55 Janda E, Nevolo M, Lehmann K, Downward J, Beug H, Grieco M: Raf plus TGFbeta-dependent EMT is initiated by endocytosis and lysosomal degradation of E-cadherin. *Oncogene* 2006, *25*(54): 7117–7130.

56 Palacios F, Tushir JS, Fujita Y, D'Souza-Schorey C: Lysosomal targeting of E-cadherin: a unique mechanism for the down-regulation of cell-cell adhesion during epithelial to mesenchymal transitions. *Mol. Cell. Biol.* 2005, *25*(1): 389–402.

57 Fujita Y, Krause G, Scheffner M, Zechner D, Leddy HE, Behrens J, Sommer T, Birchmeier W: Hakai, a c-Cbl-like protein, ubiquitinates and induces endocytosis of the E-cadherin complex. *Nat. Cell Biol.* 2002, *4*(3): 222–231.

58 Ivanov AI, Nusrat A, Parkos CA: Endocytosis of epithelial apical junctional proteins by a clathrin-mediated pathway into a unique storage compartment. *Mol. Biol. Cell* 2004, *15*(1): 176–188.

59 Xiao K, Allison DF, Kottke MD, Summers S, Sorescu GP, Faundez V, Kowalczyk AP: Mechanisms of VE-cadherin processing and degradation in microvascular endothelial cells. *J. Biol. Chem.* 2003, *278*(21): 19199–19208.

60 Xiao K, Garner J, Buckley KM, Vincent PA, Chiasson CM, Dejana E, Faundez V, Kowalczyk AP: p120-Catenin regulates clathrin-dependent endocytosis of VE-cadherin. *Mol. Biol. Cell* 2005, *16*(11): 5141–5151.

61 Lu Z, Ghosh S, Wang Z, Hunter T: Downregulation of caveolin-1 function by EGF leads to the loss of E-cadherin, increased transcriptional activity of beta-catenin, and enhanced tumor cell invasion. *Cancer Cell* 2003, *4*(6): 499–515.

62. Akhtar N, Hotchin NA: RAC1 regulates adherens junctions through endocytosis of E-cadherin. *Mol. Biol. Cell* 2001, *12*(4): 847–862.

63. Calkins CC, Setzer SV, Jennings JM, Summers S, Tsunoda K, Amagai M, Kowalczyk AP: Desmoglein endocytosis and desmosome disassembly are coordinated responses to pemphigus autoantibodies. *J. Biol. Chem.* 2006, *281*(11): 7623–7634.

64. Kowalczyk AP, Reynolds AB: Protecting your tail: regulation of cadherin degradation by p120-catenin. *Curr. Opin. Cell Biol.* 2004, *16*(5): 522–527.

65. Xiao K, Oas RG, Chiasson CM, Kowalczyk AP: Role of p120-catenin in cadherin trafficking. *Biochim. Biophys. Acta* 2007, *1773*(1): 8–16.

66. Ireton RC, Davis MA, van Hengel J, Mariner DJ, Barnes K, Thoreson MA, Anastasiadis PZ, Matrisian L, Bundy LM, Sealy L, et al: A novel role for p120 catenin in E-cadherin function. *J. Cell Biol.* 2002, *159*(3): 465–476.

67. Davis MA, Ireton RC, Reynolds AB: A core function for p120-catenin in cadherin turnover. *J. Cell Biol.* 2003, *163*(3): 525–534.

68. Davis MA, Reynolds AB: Blocked acinar development, E-cadherin reduction, and intraepithelial neoplasia upon ablation of p120-catenin in the mouse salivary gland. *Dev. Cell* 2006, *10*(1): 21–31.

69. Perez-Moreno M, Davis MA, Wong E, Pasolli HA, Reynolds AB, Fuchs E: p120-catenin mediates inflammatory responses in the skin. *Cell* 2006, *124*(3): 631–644.

70. Nieman MT, Kim JB, Johnson KR, Wheelock MJ: Mechanism of extracellular domain-deleted dominant negative cadherins. *J. Cell Sci.* 1999, *112*(Pt 10): 1621–1632.

71. Norvell SM, Green KJ: Contributions of extracellular and intracellular domains of full length and chimeric cadherin molecules to junction assembly in epithelial cells. *J. Cell Sci.* 1998, *111*(Pt 9): 1305–1318.

72. Troxell ML, Chen YT, Cobb N, Nelson WJ, Marrs JA: Cadherin function in junctional complex rearrangement and posttranslational control of cadherin expression. *Am. J. Physiol.* 1999, *276*(2 Pt 1): C404–C418.

73. Cowin P, Rowlands TM, Hatsell SJ: Cadherins and catenins in breast cancer. *Curr. Opin. Cell Biol.* 2005, *17*(5): 499–508.

74. Maeda M, Johnson KR, Wheelock MJ: Cadherin switching: essential for behavioral but not morphological changes during an epithelium-to-mesenchyme transition. *J. Cell Sci.* 2005, *118*(Pt 5): 873–887.

75. Hazan RB, Phillips GR, Qiao RF, Norton L, Aaronson SA: Exogenous expression of N-cadherin in breast cancer cells induces cell migration, invasion, and metastasis. *J. Cell Biol.* 2000, *148*(4): 779–790.

76. Pishvaian MJ, Feltes CM, Thompson P, Bussemakers MJ, Schalken JA, Byers SW: Cadherin-11 is expressed in invasive breast cancer cell lines. *Cancer Res.* 1999, *59*(4): 947–952.

77. Maeda M, Johnson E, Mandal SH, Lawson KR, Keim SA, Svoboda RA, Caplan S, Wahl JK3rd, , Wheelock MJ, Johnson KR: Expression of inappropriate cadherins by epithelial tumor cells promotes endocytosis and degradation of E-cadherin via competition for p120(ctn). *Oncogene* 2006, *25*(33): 4595–4604.

78. Islam S, Carey TE, Wolf GT, Wheelock MJ, Johnson KR: Expression of N-cadherin by human squamous carcinoma cells induces a scattered fibroblastic phenotype with disrupted cell-cell adhesion. *J. Cell Biol.* 1996, *135*(6 Pt 1): 1643–1654.

79. Bonifacino JS, Traub LM: Signals for sorting of transmembrane proteins to endosomes and lysosomes. *Annu. Rev. Biochem.* 2003, *72*: 395–447.

80. Miyashita Y, Ozawa M: Increased internalization of p120-uncoupled E-cadherin and a requirement for a dileucine motif in the cytoplasmic domain for endocytosis of the protein. *J. Biol. Chem.* 2007, *282*(15): 11540–11548.

81. Hoshino T, Sakisaka T, Baba T, Yamada T, Kimura T, Takai Y: Regulation of E-cadherin endocytosis by nectin through afadin, Rap1, and p120ctn. *J. Biol. Chem.* 2005, *280*(25): 24095–24103.

82. Izumi G, Sakisaka T, Baba T, Tanaka S, Morimoto K, Takai Y: Endocytosis of E-cadherin regulated by Rac and Cdc42 small G proteins through IQGAP1 and

actin filaments. *J. Cell Biol.* 2004, *166*(2): 237–248.

83 Gavard J, Gutkind JS: VEGF controls endothelial-cell permeability by promoting the beta-arrestin-dependent endocytosis of VE-cadherin. *Nat. Cell Biol.* 2006, *8*(11): 1223–1234.

84 Kamei T, Matozaki T, Sakisaka T, Kodama A, Yokoyama S, Peng YF, Nakano K, Takaishi K, Takai Y: Coendocytosis of cadherin and c-Met coupled to disruption of cell-cell adhesion in MDCK cells – regulation by Rho, Rac and Rab small G proteins. *Oncogene* 1999, *18*(48): 6776–6784.

85 Palacios F, Schweitzer JK, Boshans RL, D'Souza-Schorey C: ARF6-GTP recruits Nm23-H1 to facilitate dynamin-mediated endocytosis during adherens junctions disassembly. *Nat. Cell Biol.* 2002, *4*(12): 929–936.

86 Braga VM, Del Maschio A, Machesky L, Dejana E: Regulation of cadherin function by Rho and Rac: modulation by junction maturation and cellular context. *Mol. Biol Cell.* 1999, *10*(1): 9–22.

87 Wildenberg GA, Dohn MR, Carnahan RH, Davis MA, Lobdell NA, Settleman J, Reynolds AB: p120-catenin and p190RhoGAP regulate cell-cell adhesion by coordinating antagonism between Rac and Rho. *Cell* 2006, *127*(5): 1027–1039.

88 Bryant DM, Kerr MC, Hammond LA, Joseph SR, Mostov KE, Teasdale RD, Stow JL: EGF induces macropinocytosis and SNX1-modulated recycling of E-cadherin. *J. Cell Sci.* 2007, *120*(Pt 10): 1818–1828.

89 Conacci-Sorrell M, Zhurinsky J, Ben-Ze'ev A: The cadherin-catenin adhesion system in signaling and cancer. *J. Clin. Invest.* 2002, *109*(8): 987–991.

90 Qian X, Karpova T, Sheppard AM, McNally J, Lowy DR: E-cadherin-mediated adhesion inhibits ligand-dependent activation of diverse receptor tyrosine kinases. *EMBO J.* 2004, *23*(8): 1739–1748.

91 Palacios F, Price L, Schweitzer J, Collard JG, D'Souza-Schorey C: An essential role for ARF6-regulated membrane traffic in adherens junction turnover and epithelial cell migration. *EMBO J.* 2001, *20*(17): 4973–4986.

92 Paterson AD, Parton RG, Ferguson C, Stow JL, Yap AS: Characterization of E-cadherin endocytosis in isolated MCF-7 and Chinese hamster ovary cells: the initial fate of unbound E-cadherin. *J. Biol. Chem.* 2003, *278*(23): 21050–21057.

93 Bryant DM, Wylie FG, Stow JL: Regulation of endocytosis, nuclear translocation, and signaling of fibroblast growth factor receptor 1 by E-cadherin. *Mol. Biol. Cell* 2005, *16*(1): 14–23.

94 Suyama K, Shapiro I, Guttman M, Hazan RB: A signaling pathway leading to metastasis is controlled by N-cadherin and the FGF receptor. *Cancer Cell* 2002, *2*(4): 301–314.

95 Grazia Lampugnani M, Zanetti A, Corada M, Takahashi T, Balconi G, Breviario F, Orsenigo F, Cattelino A, Kemler R, Daniel TO et al: Contact inhibition of VEGF-induced proliferation requires vascular endothelial cadherin, beta-catenin, and the phosphatase DEP-1/CD148. *J. Cell Biol.* 2003, *161*(4). 793–804.

96 Lampugnani MG, Orsenigo F, Gagliani MC, Tacchetti C, Dejana E: Vascular endothelial cadherin controls VEGFR-2 internalization and signaling from intracellular compartments. *J. Cell Biol.* 2006, *174*(4): 593–604.

Part Three   Cell–Matrix and Cell–Cell Crosstalk

14
Crosstalk Between Cell–Cell and Cell–Matrix Adhesion
C. Michael DiPersio

14.1
Introduction

14.1.1
Coordinate Regulation of Cell–Cell and Cell–Matrix Adhesion

The individual cells that comprise epithelial, endothelial, neuronal, and many other tissues adhere simultaneously to both neighboring cells and to the extracellular matrix (ECM). While a wide variety of cell–cell and cell–matrix adhesion mechanisms have evolved to mediate intercellular cohesion and/or communication in both invertebrate and vertebrate organisms, many fundamental aspects of these mechanisms have been conserved in metazoan evolution [1]. Changes in cell–cell and cell–matrix adhesion are important during both normal and pathological processes, including developmental morphogenesis, differentiation, inflammatory response, angiogenesis, wound healing, tumor progression, and metastasis. Indeed, the adhesion properties of individual cells that comprise a tissue are highly regulated, and abnormal changes in cell–cell or cell–matrix adhesion can contribute to tissue dysfunction or pathological tissue remodeling.

  In general, spatial and temporal coordination of cell–cell and cell–matrix adhesion is critical for maintaining the control of migration, proliferation, survival, and/or differentiation of the individual cells within a tissue. However, the functional interrelationships between cell–cell and cell–matrix adhesion can differ greatly between different tissue remodeling processes. In some processes, the disassembly of cell–cell adhesions is coupled with increased cell–matrix adhesion. For example, during epithelial-to-mesenchymal transition (EMT), reduced cell–cell adhesion occurs simultaneously with changes in cell–matrix adhesion that promote individual cell migration and scattering [2–4]. This process occurs in a highly controlled manner during embryonic development in order to prevent the unregulated dissemination of cells to inappropriate tissues, whilst it occurs in an uncontrolled manner during malignant carcinoma progression to promote metastasis. In other processes, cell–cell adhesion is maintained during the collective migration of cohesive cells, such as during embryonic tissue morphogenesis or the re-epithelialization of cutaneous wounds. It follows from these examples that the overall spatial and temporal coordination of cell–matrix or cell–cell adhesion must involve considerable crosstalk between cell-surface receptors that mediate these adhesions.

The regulation of cell adhesion has been an active area of investigation for decades. Indeed, a wide variety of cell-surface receptors have been identified that mediate different types of cell–matrix adhesions (i.e., focal adhesions, hemidesmosomes) or cell–cell adhesions (i.e., adherens junctions, tight junctions, desmosomes, gap junctions), and numerous studies in various cell types and model organisms have elucidated molecular mechanisms whereby specific receptors for cell–cell or cell–matrix adhesion regulate cell function, or by which they are themselves regulated [1, 2, 5–8]. While the notion of crosstalk between distinct adhesion receptors is not new, most early studies focused on individual receptors without directly addressing crosstalk mechanisms between different adhesion receptor families. However, more recent studies have identified clear examples of crosstalk between specific cell–cell and cell–matrix receptors (for recent reviews on this subject, see Refs. [4, 9, 10]). In this chapter, the mechanisms of crosstalk between cell–matrix adhesions and cell–cell adhesions that are mediated by integrins and cadherins, respectively will be reviewed.

14.1.2
Crosstalk Between Integrins and Cadherins

Integrins are the major receptors for cell adhesion to the ECM [6]. The integrin family consists of at least 24 distinct members, all of which are obligate, transmembrane heterodimers consisting of an α and a β subunit. While the cytoplasmic domains of most integrins are relatively small (usually less than 50 amino acids), as a group these receptors can interact either directly or indirectly with a variety of intracellular proteins in order to regulate cytoskeletal connections and signal transduction [6, 7, 11]. However, these cytoplasmic interactions, as well as binding to ECM ligands, are highly integrin-specific [6].

Cadherins mediate the calcium-dependent formation of adherens junctions (AJs), which are essential for stable epithelial and endothelial cell–cell adhesion, and dynamic changes in AJ assembly/disassembly are an important point of regulation in many tissue remodeling processes [2, 5]. Cadherins consist of a relatively large extracellular domain, a single transmembrane domain, and a short cytoplasmic domain. Homophilic interactions between cadherin extracellular domains mediate cell–cell adhesion, while cadherin cytoplasmic domains interact directly or indirectly with β-catenin, α-catenin, p120 catenin (p120^{ctn}), and other cytoplasmic proteins that mediate cytoskeletal connections and/or signal transduction.

Integrins and cadherins can each signal bidirectionally across the cell membrane and it is clear that, under certain circumstances, there is considerable crosstalk between these two receptor families [4]. Indeed, numerous examples have been reported where integrin-mediated adhesion to ECM, or integrin activation, leads to changes in cell–cell adhesion or cadherin function [12–25]. There are relatively fewer examples where cadherins have been shown to regulate integrin function [26, 27], but this may be due to the fact that crosstalk in this direction has been explored less extensively. Indeed, the signaling protein Rap1 was recently identified as an important mediator of cadherin-to-integrin crosstalk [9, 28] (see Section 14.2.5.3). Fundamental details regarding the structure and function of integrins or cadherins can be found in other chapters of this book. This chapter will focus on recent findings towards molecular mechanisms of crosstalk between integrins and cadherins, and on some of the molecules and pathways that have emerged as strong candidates for mediating this crosstalk.

14.2
Mechanisms of Integrin–Cadherin Crosstalk

For purposes of discussion, the potential mechanisms of integrin–cadherin crosstalk can be grouped into at least five different categories: (1) extracellular proteolysis; (2) cross-regulation of gene expression; (3) changes in intracellular force generation; (4) integrin–cadherin interactions at cell adhesion sites; and (5) intracellular signal transduction pathways. Obviously, these mechanisms are not mutually exclusive, and in many cases they probably occur simultaneously and influence one another. Nevertheless, it is instructive to consider each mechanism individually.

14.2.1
Extracellular Proteolysis

Cells of actively remodeling tissues produce an array of extracellular proteases that collectively cleave a wide range of substrates that includes ECM proteins, cell-adhesion receptors, and other extracellular or cell-surface proteins, thereby leading to alterations in cell–cell and/or cell–matrix interactions with profound effects on cell function [29]. Members of the matrix metalloproteinase (MMP) family of extracellular proteases play a prominent role in many normal and pathological processes, including wound repair, angiogenesis, and tumor invasion and metastasis [30]. As described in a recent review, there are many clear examples where either integrins or cadherins regulate MMP-mediated proteolysis [31]. Indeed, integrins or integrin-associated proteins have been shown to regulate MMPs at the levels of gene transcription [32, 33], post-transcriptional mRNA stability [34], endocytosis [35], and extracellular localization and function [36–38]. Similarly, E-cadherin-mediated cell–cell adhesion can suppress MMP expression and ECM proteolysis in some models [39–41]. In addition, both cadherins and integrins, as well as many of their associated proteins or ligands, can be cleaved by MMPs or other extracellular proteases [30, 31, 42, 43]. Taken together, these observations place extracellular proteolysis both upstream and downstream of either cell–cell or cell–matrix adhesion, and suggest that it plays an important role in regulating crosstalk between integrins and cadherins (Figure 14.1).

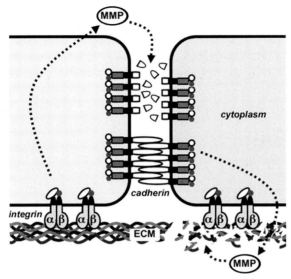

Figure 14.1 Crosstalk mediated by extracellular proteases. Some integrins regulate the production or activities of matrix metalloproteinases (MMP) or other extracellular proteases that can cleave cadherins or other adherens junction proteins, thereby altering cell–cell adhesion, as shown in the cell on the left. Similarly, cadherins can regulate the expression of MMPs that can cleave receptors or ligands at sites of cell adhesion to extracellular matrix (ECM), thereby altering cell–matrix adhesion, as shown in the cell on the right. Integrins and cadherins are indicated at sites of cell–matrix and cell–cell adhesion, respectively.

MMP-mediated crosstalk in the direction of integrins to cadherins is readily illustrated by a recent study of ovarian carcinoma cells, in which a functional link was identified between integrin-mediated cell adhesion to collagen and MMP-9-dependent proteolysis of E-cadherin [44]. In this system, the integrin-mediated induction of MMP-9 expression and subsequent cleavage of E-cadherin resulted in the shedding of an E-cadherin ectodomain fragment that disrupts AJs. These findings suggest a novel mechanism whereby the induction of MMP-9 by cell–matrix adhesion contributes to tumor cell metastasis by causing downregulation of E-cadherin-mediated cell–cell adhesion [44].

Considerable evidence also exists to support MMP-mediated crosstalk in the direction of cadherins to integrins [31]. Indeed, several studies have shown that a loss of E-cadherin function leads to enhanced MMP gene expression in some carcinoma cells, which in turn leads to increased ECM proteolysis and invasion [39–41]. Although the mechanistic details remain unclear, a likely mechanism of MMP gene induction in response to E-cadherin involves functional changes in β-catenin, a structural component of AJs. As discussed below (see Section 14.2.2), the downregulation of E-cadherin can lead to translocation of β-catenin into the nucleus, where it can bind to T-cell factor/lymphoid enhancer factor (Tcf/Lef) transcription factors and activate transcription of certain MMP genes [45]. However, some MMP genes that are suppressed by E-cadherin do not contain Tcf/Lef binding elements, indicating that β-catenin-independent pathways of MMP gene induction also exist [31]. Other mechanisms of cadherin-mediated MMP induction have also been described, such as synergistic interactions between N-cadherin and the fibroblast growth factor (FGF) receptor that induce MMP-9 expression in breast cancer cells [46, 47].

14.2.2
Cross-Regulation of Gene Expression

Transcription factors that promote EMT, such as Twist, Snail, and Slug, probably do so in part by regulating the expression of genes involved in cell–cell and cell–matrix adhesion [4, 48]. In addition, there are examples where signaling pathways that can be controlled by adhesion receptors can regulate the nuclear translocation of transcription factors/cofactors, or of kinases that activate transcription factors [49, 50]. Taken together, these findings implicate the cross-regulation of gene expression as a potentially important mechanism of crosstalk between cell adhesion receptors (Figure 14.2).

One of the most straightforward examples of integrin-mediated induction of cadherin gene expression comes from a recent study in prostate carcinoma cells, which showed that β1 integrin-mediated adhesion leads to nuclear translocation of Twist, which subsequently binds to a regulatory element within the N-cadherin gene to activate its transcription [51] (see Figure 14.2). In another example, integrin-linked kinase (ILK) was implicated in the suppression of E-cadherin gene

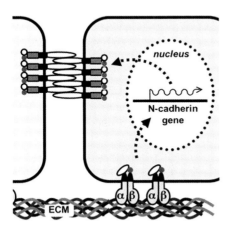

Figure 14.2 Crosstalk mediated by cross-regulation of gene expression. Some adhesion receptors can regulate the expression of genes that encode other adhesion receptors or their ligands. For example, β1 integrin-mediated adhesion of prostate cancer cells to matrix promotes transcription of the N-cadherin gene by inducing nuclear localization of the transcription factor Twist [51]. The nucleus is depicted as a dashed circle; the other symbols are as in Figure 14.1.

transcription in colon carcinoma cells through regulation of β-catenin/Tcf and snail transcription factors [52].

It seems likely that the cross-regulation of gene expression also occurs in the direction of cadherin to integrin, since a loss of E-cadherin binding to β-catenin can lead to enhanced nuclear transport and transcriptional activation function of β-catenin [45, 53, 54]. Briefly, β-catenin is normally a component of AJs, where it is found complexed with the cytoplasmic domain of E-cadherin [2]. A loss of E-cadherin causes the release of β-catenin into the cytoplasm, where it is phosphorylated by glycogen synthase kinase 3 (GSK3) in a complex that includes adenomatous polyposis coli (APC), and then undergoes degradation through the ubiquitin pathway. However, under conditions where β-catenin degradation is inhibited (i.e., when Wnt or ILK signaling is activated, or APC is mutated), β-catenin accumulates in the cytoplasm and translocates into the nucleus where it binds Tcf/Lef transcription factors and activates gene transcription, including genes involved in integrin-mediated ECM adhesion and invasion [45, 53]. Similar roles have been suggested for the AJ protein p120^{ctn} which, upon the loss of interaction with E-cadherin, may translocate to the nucleus and regulate the transcriptional repressor Kaiso [55, 56].

14.2.3
Changes in Intracellular Force Generation

Cell–matrix or cell–cell contacts can be weakened or even lost as a result of increased intracellular tension at sites of cell adhesion, which suggests that changes in cytoskeletal force generation can mediate integrin–cadherin crosstalk (Figure 14.3). While it seems likely that cadherins can influence integrin-mediated adhesion to ECM through changes in force generation, this mechanism has been best characterized in the direction of integrin to cadherin, and there are good examples where integrin-mediated adhesion leads to a separation of cell–cell contacts without

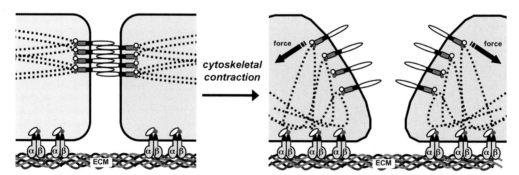

Figure 14.3 Crosstalk mediated by intracellular force generation. Increased intracellular tension on the actin cytoskeleton that is generated through changes in cell–matrix adhesion can lead to the physical separation of cell–cell contacts, often without changes in the expression or intrinsic adhesive function of cadherins. The cytoskeleton is depicted by dashed lines; the other symbols are as in the previous figures.

concurrent loss of E-cadherin expression, intrinsic adhesive function, or interactions with catenins [4, 5]. Consistently, some invasive tumor cells that lose cell–cell adhesion retain E-cadherin expression [3, 46], and studies in breast or colon cancer cells have indicated that loss of E-cadherin-mediated cell–cell adhesion can be caused by increased myosin activity and cytoskeletal contractility [57, 58].

A recent study effectively illustrates the potential importance of actin cytoskeletal contraction as a mediator of integrin-to-cadherin crosstalk in epithelial cells [23]. This study of Madin–Darby canine kidney (MDCK) cells showed that a loss of cell–cell adhesion and subsequent cell scattering in response to hepatocyte growth factor (HGF) involved an enhancement of integrin-mediated adhesion and mysosin light chain phosphorylation, which subsequently increased the actomyosin traction forces applied to cell–cell junctions and pulled them apart. Importantly, HGF did not alter the intrinsic adhesive function of E-cadherin, indicating that the loss of cell–cell adhesion was due primarily to the increased tension imposed upon the cell–cell contacts. Interestingly, the loss of cell–cell adhesion and subsequent cell scattering was dependent on ECM composition and compliance, indicating that this mechanism is specific to certain integrins [23]. Integrin-to-cadherin crosstalk through cytoskeletal force generation is also likely to extend to many other cell types. For example, cytoskeletal contraction is also important in endothelial cells for the downregulation of VE-cadherin and subsequent changes in cell–cell adhesion, which regulates vascular permeability or transendothelial migration of leukocytes and tumor cells [59, 60].

Integrin–cadherin crosstalk through intracellular force generation is likely to involve members of the Rho family of guanosine triphosphatases (GTPases), such as Rho, Rac, and Cdc42, which are known to play critical roles in regulating cytoskeletal dynamics and contractility both upstream and downstream of integrins and cadherins [61, 62]. Similarly, Rap1, a Ras family GTPase, has been linked to actin cytoskeletal dynamics [10, 63]. The potential mechanisms of integrin–cadherin crosstalk that involve Rho or Ras family GTPase signaling pathways are discussed further in Section 14.2.5.

14.2.4
Integrin–Cadherin Associations at Sites of Cell Adhesion

Integrins have traditionally been considered to function primarily as receptors for cell–matrix adhesion, with some exceptions that include members of the $\beta2$ integrin subfamily [6]. However, early studies in epidermal keratinocytes suggested that certain $\beta1$ integrins, namely $\alpha3\beta1$ and $\alpha2\beta1$, may also have adhesive roles at sites of cell–cell contact [64–66]. Consistently, later studies with $\alpha3$-null epithelial cells provided evidence that an absence of $\alpha3\beta1$ leads to defects in AJ integrity, in the organization of the cortical actin cytoskeleton, and/or cell–cell adhesion [67–69]. However, it now appears that functions of $\alpha3\beta1$ at cell–cell adhesions may not require direct binding to a ligand on the surface of an adjacent cell, as occurs for other integrins such as $\alpha4\beta7$ or $\beta2$ [6]. Rather, the results of recent studies suggest that $\alpha3\beta1$ can associate laterally with E-cadherin and other AJ-associated proteins,

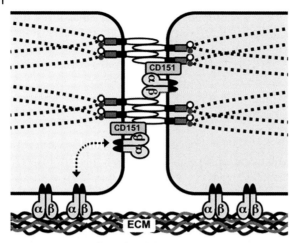

Figure 14.4 Crosstalk mediated by integrin–cadherin associations. Some integrins may associate with cadherins at adherens junctions, perhaps through interactions with tetraspanins (i.e., CD151) or other proteins, as has been described for α3β1 in kidney epithelial cells [70]. Although these interactions probably occur independently of integrin binding to extracellular matrix (ECM) ligands at the basal cell surface, there is potential for crosstalk between cell–cell and cell–matrix adhesions through integrin shuttling between these distinct adhesion sites, as depicted by the dotted arrow. The other symbols are as in the previous figures.

suggesting another mechanism whereby some integrins may regulate AJ function [70] (Figure 14.4). α3β1 forms a very stable complex on the epithelial cell surface with CD151 [71], a tetraspanin protein that can influence E-cadherin-mediated cell–cell adhesion through PKC-dependent and Cdc42-dependent effects on the actin cytoskeleton [72]. Studies using α3-null kidney epithelial cells identified a novel role for the α3β1–CD151 interaction as part of a multi-molecular, AJ-associated complex that appears required for stabilizing E-cadherin association with the actin cytoskeleton and maintaining cell–cell adhesion [70]. While the direct association of α3β1 with CD151 has been well characterized, the fact that other tetraspanin family members also localize to sites of cell–cell adhesion raises the possibility that this regulation extends to other integrin–tetraspanin interactions [73].

Interestingly, the pool of α3β1 that is associated with AJs appears distinct from that involved in cell adhesion to the ECM [70], suggesting that roles for α3β1 in cell–cell adhesion may occur independently of its well-known functions in cell–matrix adhesion. However, an alternative possibility is that α3β1 translocates to some extent between cell–matrix adhesions and cell–cell adhesions, depending on the availability of ECM ligands (see Figure 14.4). In this case, an increased binding of α3β1 to cell–matrix adhesions under conditions where epithelial cells secrete abundant laminin-332 into the ECM (i.e., during wound healing or carcinoma cell invasion) [74, 75] could conceivably influence cell–cell adhesion by titrating α3β1 from the AJs. α3β1 interactions with other cell-surface proteins may also influence cell–cell adhesion. For example, it has been reported that α3β1 binds to the uro-kinase receptor uPAR through a site on the integrin that is distinct from the laminin-binding site, and that this interaction leads to downregulation of E-cadherin-mediated cell–cell adhesion in a Src- dependent manner [76].

14.2.5
Intracellular Signal Transduction Pathways

One of the most obvious mechanisms of integrin–cadherin crosstalk – and perhaps the most prevalent – is the activation of intracellular signal transduction pathways by one type of adhesion receptor to regulate the adhesive function of another receptor. Indeed, a wide variety of structural and signaling proteins have been implicated in the regulation of both cell–cell and cell–matrix adhesions, including kinases, phosphatases, GTPases, and various adaptors/structural proteins. While many of these proteins may have dual, but largely separable roles in regulating cell–cell or cell–matrix adhesion, some of them have emerged in recent years as *bona fide* mediators of integrin–cadherin crosstalk [4, 9]. Obviously, such crosstalk through signal transduction pathways may be involved in any of the other crosstalk mechanisms discussed above (see Sections 14.2.1–14.2.4).

There are a number of different ways whereby a signaling protein might regulate integrin–cadherin crosstalk. First, some proteins may be activated from within one type of adhesion site, leading to induction of a signaling pathway that ultimately regulates adhesion at another site (Figure 14.5A). Second, some signaling proteins or adaptors may physically translocate between cell–cell and cell–matrix adhesions (Figure 14.5B). Third, some proteins may occupy a key position at the hub of a

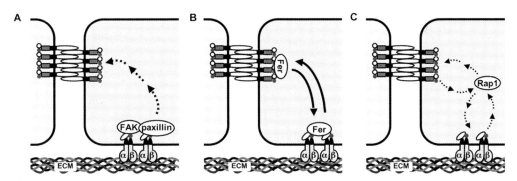

Figure 14.5 Crosstalk mediated by intracellular signal transduction. Signaling pathways that are activated at one type of adhesion junction may regulate the functions of receptors at another type of adhesion junction. Three basic strategies are depicted in (A–C) and discussed in detail in the text. (A) In some cells, FAK and paxillin can be activated at cell–matrix adhesions to initiate pathways that ultimately regulate cell–cell adhesion. (B) Some signaling proteins, such as Fer, can translocate between cell–cell and cell–matrix adhesions. (C) Some signaling proteins, such as the Rap1 GTPase, appear to occupy key positions at the intersections of signaling pathways that are controlled by either cell–cell or cell–matrix adhesions. The symbols are as in the previous figures.

signaling network, thereby serving as a nexus to connect individual pathways that control cell–cell and cell–matrix adhesions (Figure 14.5C). A few specific examples of each of these mechanisms are discussed below, but a more comprehensive coverage of these and other signaling pathways that may regulate integrin–cadherin crosstalk in different systems can be found in several excellent reviews [4, 9, 77, 78].

14.2.5.1 FAK: An Integrin Effector that Signals to Cadherins

Focal adhesion kinase (FAK) has emerged in recent years as an important mediator of crosstalk between integrins and cadherins [77], and it provides an interesting example of a signaling protein that can be activated at a site of integrin–ECM adhesion to regulate the assembly of distal, cadherin-based cell–cell adhesions (Figure 14.5A). The integrin-mediated activation of FAK appears to promote either the assembly or disassembly of cell–cell adhesions, depending on the cell type or physiological circumstances [77]. For example, Src-mediated activation of FAK leads to the disassembly of E-cadherin-based cell–cell adhesions in colon cancer cells, suggesting a role in reduced cell–cell adhesion during EMT [18]. On the other hand, FAK activation promotes cell–cell adhesion and barrier strengthening in endothelial cells [79]. Similarly, integrin-mediated signaling through FAK and paxillin promotes cadherin-mediated cell–cell adhesions in some epithelial cells that migrate as a cohesive sheet [21]. In the latter study, siRNA-mediated knockdown of either FAK or paxillin impaired the assembly of N-cadherin-based cell–cell adhesions in HeLa cells, leading to cell–cell detachment and individual cell migration out of the epithelial sheet. Experiments using a fluorescence resonance energy transfer (FRET)-based biosensor to assess the localized activity of Rac1 (a Rho family GTPase) indicated that a paxillin-dependent recruitment of FAK to cell–matrix adhesion sites leads to localized Rac1 repression at the cell periphery, thereby permitting the assembly of N-cadherin-based cell–cell adhesions [21]. The mechanisms and downstream FAK/paxillin effectors involved in this regulation remain unclear, but they are presumably distinct from integrin/FAK signaling pathways that appear to activate Rac1 in some cells [80–82].

14.2.5.2 Fer: A Tyrosine Kinase that Translocates between Cadherins and Integrins

Some key regulatory proteins may be exchanged directly between the cytoplasmic domains of cadherins and integrins (Figure 14.5B). Fer is a non-receptor tyrosine kinase that interacts with cadherins indirectly through $p120^{ctn}$ and stabilizes the cadherin–catenin complex at cell–cell junctions [83, 84]. In one study, the treatment of neural retina explants with a cell-permeable peptide that mimics the juxtamembrane region of N-cadherin caused the release of Fer from the cadherin complex, which allowed it to translocate and accumulate in integrin–ECM adhesions [85]. Fer translocation was accompanied by the downregulation of both N-cadherin and integrins, leading to a disruption of neurite outgrowth, and suggesting that Fer mediates crosstalk between N-cadherin and β1 integrins by shuttling between their respective sites of cell adhesion [85]. It has been suggested that the proteoglycan neurocan may induce Fer translocation from N-cadherin to integrins in the developing central nervous system, and that this may be important for coordinating the inactivation of both cadherin-mediated adhesion and integrin-mediated adhesion during regulated axon guidance [86].

14.2.5.3 Rap1: A Central Regulator of Integrin–Cadherin Crosstalk Pathways?

Several Ras or Rho family GTPases are implicated at key intersections of signaling pathways that control communication between cell–cell and cell–matrix adhesions. For example, the E-cadherin-mediated sequestration of p120^{ctn} to AJs suppresses its ability to promote cell motility through pathways that involve Rho family GTPases, providing a mechanism whereby the disassembly of cell–cell adhesions may enhance integrin-mediated cell motility [87]. The regulation of Rho GTPases by integrins is also well known, and Rho and Rac have been implicated in the establishment of cell–cell adhesion through control of cytoskeletal assembly and membrane protrusion [62, 88].

Rap1, a member of the Ras family of GTPases [89], has emerged in recent years as a particularly intriguing candidate for coordinating the spatial and temporal regulation of cadherin-mediated AJ formation and integrin-mediated cell–matrix adhesion, as discussed in several excellent reviews [9, 10, 90]. As with other GTPases, Rap1 is activated by specific guanine nucleotide exchange factors (GEFs) that convert the inactive GDP-bound form to the active GTP-bound form, while it is inactivated by specific GTPase-activating proteins (GAPs) that induce GTP hydrolysis. As discussed below, Rap1 signaling can occur both upstream and downstream of cadherins [9]. Rap1 can also regulate the inside-out activation of integrins [9, 10], and a recent report has indicated a prominent role for Rap1 in the control of cell adhesion and motility through the regulation of myosin and actin cytoskeletal dynamics [63]. Rap1 and certain Rap1 GEFs have been linked, both directly and indirectly, to junctional proteins at sites of both cell–cell and cell–matrix adhesion, potentially placing Rap1 at an important crossroads between cadherin and integrin signaling pathways (Figure 14.5C) [9, 10, 90].

Rap1 Signaling in Cadherin-Mediated Cell–Cell Adhesion Although Rap1 has been implicated in the regulation of several types of cell–cell junction, the discussion here will focus on roles for Rap1 in AJ formation. Early clues that Rap1 directly regulates cadherin function were provided by genetic studies demonstrating that it is required for cadherin localization and cell–cell cohesion during *Drosophila* morphogenesis [91]. Later studies confirmed that Rap1 signaling also promotes AJ formation in mammalian epithelial and endothelial cells [92–95].

Rap1 has been implicated both upstream and downstream of cadherin function in the regulation of AJ assembly and maturation. For example, Rap1 is activated in epithelial cells by homophilic E-cadherin interactions at sites of nascent cell–cell contact through a mechanism that involves the binding of C3G (a Rap1 GEF) to the E-cadherin cytoplasmic domain. Newly activated Rap1 then leads to the recruitment of more E-cadherin and the maturation of AJs [92]. Rap1-mediated AJ maturation may also involve the recruitment of GEFs for Rac and Cdc42 [90, 92, 96, 97], as well as of other junctional proteins, through mutual interactions with adaptor proteins such as AF6/afadin (see below). Rap1 is downregulated as AJs mature [90], which is consistent with reports that E-cadherin-mediated cell–cell adhesion suppresses Rap1 activity [28], and provides further evidence that Rap1 is downstream of cadherins in some pathways. On the other hand, Rap1 has also been shown to induce AJ formation by regulating E-cadherin cell-surface expression [93], placing it upstream of cadherins in some pathways.

Rap1 Signaling in Integrin-Mediated Cell–Matrix Adhesion Rap1 also has roles in the regulation of integrin-mediated cell–matrix adhesion. Although some reports have suggested a role for Rap1 in outside-in integrin signaling, in some systems Rap1 activity was not influenced by integrin signaling [28], and so the general importance of Rap1 for outside-in integrin signaling is somewhat ambiguous [9]. However, numerous studies have supported a clear role for Rap1 in regulating inside-out integrin signaling through mechanisms that impact integrin affinity (i.e., activation), integrin avidity (i.e., clustering), or integrin recycling [9, 10, 28, 98]. Potential mechanisms for Rap1-mediated effects on integrin function include the spatial control of Rac GEFs, association with integrin complexes, effects on cytoskeleton, and the recruitment of integrins or integrin-associated proteins to sites of cell–matrix adhesion.

An excellent example of linkage between Rap1 and cell–matrix adhesion comes from studies of Rap1 interactions with Tiam-1 and Vav2, two RacGEFs with important roles in the regulation of cell–matrix adhesion and cell spreading [99]. This study in HeLa cells showed that Rap1 binds to both Tiam1 and Vav2, and targets their localization to the edges of cell protrusions, which promotes Rac-mediated cell spreading. Consistently, Rap1 activation can lead to Rac activation in some systems [100].

Roles of Rap1 in Cadherin–Integrin Crosstalk The involvement of Rap1 in both cell–matrix and cell–cell adhesion suggests that this GTPase plays a pivotal role in crosstalk between integrins and cadherins. Many potential Rap1 effectors have been identified, and several of them are implicated in the regulation of cell–cell adhesion and/or cell–matrix adhesion [10]. Many of these effectors exert their regulatory effects by binding directly to Rap1 through their Ras-association (RA) domains and/or other domains. Some of these effectors are discussed in brief below, but have been reviewed in more detail elsewhere [10].

Some Rap1 effectors may function by linking Rap1 to cadherins or other AJ proteins. For example, as mentioned above, AF-6/afadin is an AJ-associated adaptor protein that binds to Rap1 through its RA domain. AF-6/afadin may negatively regulate Rap1 in AJs through the co-recruitment of Spa1, a Rap1 GAP [101]. Other effectors appear to link Rap1 to sites of integrin-mediated cell adhesion. For example, RapL associates with the cytoplasmic domain of the αL integrin subunit and also binds to Rap1 through its RA domain, and this interaction appears important for Rap1-induced, $\alpha L \beta 2$-dependent cell adhesion [102]. Still other effectors appear to link Rap1 function to integrins less directly through the regulation of actin cytoskeletal dynamics. For example, Riam is an RA domain-containing cytoskeletal regulator that functions as a Rap1 effector during integrin-mediated adhesion and cell spreading [103]. Similarly, Arap3 is an RA domain-containing RhoGAP that is activated by binding to Rap1, leading to downregulation of RhoA [104].

A distinct and intriguing mechanism whereby Rap1 may regulate cadherin–integrin crosstalk is by linking E-cadherin endocytosis to the recruitment of integrins or other proteins to sites of cell–matrix adhesion [28]. Results from the latter study showed that, following AJ disassembly, the internalization and subsequent trafficking of E-cadherin along the endocytic pathway triggers Rap1 activation through a Src-dependent mechanism. Importantly, this cadherin-dependent activation of Rap1 promoted the formation of integrin-mediated focal adhesions [28], indicating a novel Rap1-mediated mechanism of cadherin–integrin crosstalk.

14.2.5.4 The Epidermis as a Model for Investigating Cadherin–Integrin Crosstalk

During cutaneous wound healing, the coordinated regulation of cell–matrix adhesion and cell–cell cohesion in the epidermis is required for migration of the epithelial tongue and correct re-epithelialization of the wound. On the other hand, the malignant conversion of epidermal tumors involves alterations in cadherin-mediated cell–cell adhesion and cell–matrix adhesion that promote tumor cell invasion and metastasis. Both of these processes involve integrin–cadherin crosstalk [4], and the epidermis may provide a useful and accessible physiological context within which to test involvement of candidate signaling pathways. Indeed, the roles of cadherins and integrins in the regulation of keratinocyte adhesion and motility have been well characterized [74, 105–107], and some of the specific adhesion receptors and signaling molecules that are implicated in integrin–cadherin crosstalk through *in vitro* studies (as discussed in this chapter) have also been shown to be important for regulating certain aspects of wound healing, epidermal tumorigenesis, or other epidermal functions *in vivo* [108–116].

Integrin-mediated Rac1 activation, and the RacGEF Tiam1, have important roles in regulating keratinocyte migration both *in vitro* and *in vivo* [69, 115], and a requirement for Rac1 signaling *in vivo* has been established in models of both wound healing and epidermal tumorigenesis [111, 115–117]. Indeed, Rac activation in keratinocytes is triggered by α3β1 integrin-mediated adhesion to laminin-332 in the ECM and promotes cell polarization and migration [69, 82, 118], while Tiam1 appears necessary for this regulation *in vivo* [115]. Tiam1/Rac signaling can also regulate cadherin-mediated cell–cell adhesion, where its effects are dependent on the ECM substrate, indicating that Tiam1/Rac interactions may mediate integrin–cadherin crosstalk in the epidermis [119, 120].

It is interesting to speculate that α3β1 integrin-mediated adhesion to laminins in the cutaneous ECM may activate Tiam1/Rac1 signaling pathways that promote stable cell–cell adhesions. This model would be consistent with reports that reduced levels of active Rac in α3β1-deficient keratinocytes is accompanied by reduced cell–cell adhesion and increased cell scattering [69], and that the formation of epithelial cell–cell adhesions is driven by localized Rac1 activity in peripheral membrane zones of expanding cell contact [62]. As discussed above (see Section 14.2.5.3), Rap1 might be a good candidate for mediating such crosstalk, as it has been implicated upstream of both Tiam1-mediated activation of Rac [99] and E-cadherin-mediated cell–cell adhesion [92, 93].

In a separate – but not mutually exclusive – model, α3β1 may promote Rac1 signaling from within AJs through its effects on E-cadherin function (see Section 14.2.4) [70], since E-cadherin has been shown to activate Rac through Rap1 [97]. In addition, cadherin-mediated activation of Rap1 could conceivably feed back to α3β1 and promote adhesion to its ECM ligands, since Rap1 signaling has been shown to promote the α3β1-mediated adhesion of epithelial cells to laminin-332 [121]. Clearly, crosstalk between integrins and cadherins is likely to be very complex in the epidermis, and may involve several of the mechanisms discussed in this chapter, with varying levels of importance in distinct processes such as skin development, wound healing, and epidermal carcinogenesis. Adding to this complexity, keratinocyte adhesion and motility is regulated by several integrins that bind to diverse ECM proteins, include laminins, collagens, and fibronectin [7, 106, 122, 123]. Therefore, distinct integrins may simultaneously regulate crosstalk with cadherins using the mechanisms described in Sections 14.2.1 to 14.2.5. With the continued development of excellent genetic models and cell culture systems to pursue the molecular dissection of individual integrin–cadherin crosstalk pathways within the epidermis, the relative importance of these mechanisms – and of specific adhesion receptors that are involved in them – should begin to be elucidated.

14.3
Future Prospects

In summary, several diverse mechanisms have been identified whereby cells can coordinately regulate cell–cell and cell–matrix adhesion through integrin–cadherin crosstalk. Ongoing studies in cell-culture systems (including many excellent studies not described here) continue to identify useful paradigms for investigating the detailed mechanisms that govern integrin–cadherin crosstalk and regulate cell behavior during normal tissue development and remodeling, as well as during pathological processes such as carcinoma progression. The relative importance *in vivo* of the different mechanisms described here remains to be seen, and the results of some early *in-vivo* studies indicate that crosstalk between cell–cell and cell–matrix adhesion is likely to be even more complex than has been indicated from the findings of *in-vitro* studies with cultured cells. For example, while Rac1 was required for cadherin-mediated cell–cell adhesion in cultured keratinocytes [88], studies in conditional Rac1 knockout mice have suggested that this GTPase is not essential to maintain cell–cell adhesion in the epidermis [124], possibly indicating the existence of compensatory or overlapping mechanisms *in vivo*. Future *in-vivo* studies of integrin–cadherin crosstalk within resting or remodeling tissues will continue to benefit from new *in-vivo* approaches and compound transgenic/knockout models that will allow the manipulation of relevant receptors and signaling proteins with temporal and spatial precision. In addition, organotypic models–as well as models that exploit embryo or tissue explants–hold much promise for elucidating the molecular details of coordinated cell–cell and cell–matrix adhesion during tissue morphogenesis and remodeling. This is because such systems allow for the tightly controlled manipulation of candidate adhesion genes/proteins, through genetic, physical, or chemical means, within the context of relatively intact tissues [19, 27, 40, 125].

Acknowledgments

The author thanks the members of his laboratory and colleagues at Albany Medical College for valuable discussions. Thanks are expressed in particular to Dr. Peter Vincent for his critical reading of the manuscript. Apologies are also offered to the many authors whose relevant data were not cited for reasons of space constraint.

References

1. Hynes, R. O., and Zhao, Q. (2000). The evolution of cell adhesion. *J. Cell Biol.* 150, F89–F95.
2. Wheelock, M. J., and Johnson, K. R. (2003). Cadherins as modulators of cellular phenotype. *Annu. Rev. Cell Dev. Biol.* 19, 207–235.
3. Hazan, R. B., Qiao, R., Keren, R., Badano, I., and Suyama, K. (2004). Cadherin switch in tumor progression. *Ann. N. Y. Acad. Sci.* 1014, 155–163.
4. Chen, X., and Gumbiner, B. M. (2006). Crosstalk between different adhesion molecules. *Curr. Opin. Cell Biol.* 18, 572–578.
5. Gumbiner, B. M. (2000). Regulation of cadherin adhesive activity. *J. Cell Biol.* 148, 399–404.
6. Hynes, R. O. (2002). Integrins: bidirectional, allosteric signaling machines. *Cell* 110, 673–687.
7. Litjens, S. H., de Pereda, J. M., and Sonnenberg, A. (2006). Current insights into the formation and breakdown of hemidesmosomes. *Trends Cell Biol.* 16, 376–383.
8. Delon, I., and Brown, N. H. (2007). Integrins and the actin cytoskeleton. *Curr. Opin. Cell Biol.* 19, 43–50.
9. Retta, S. F., Balzac, F., and Avolio, M. (2006). Rap1: a turnabout for the crosstalk between cadherins and integrins. *Eur. J. Cell Biol.* 85, 283–293.
10. Bos, J. L. (2005). Linking Rap to cell adhesion. *Curr. Opin. Cell Biol.* 17, 123–128.
11. Liu, S., Calderwood, D. A., and Ginsberg, M. H. (2000). Integrin cytoplasmic domain-binding proteins. *J. Cell Sci.* 113, 3563–3571.
12. Hodivala, K. J., and Watt, F. M. (1994). Evidence that cadherins play a role in the downregulation of integrin expression that occurs during keratinocyte terminal differentiation. *J. Cell Biol.* 124, 589–600.
13. Monier-Gavelle, F., and Duband, J. L. (1997). Cross talk between adhesion molecules: control of N-cadherin activity by intracellular signals elicited by beta1 and beta3 integrins in migrating neural crest cells. *J. Cell Biol.* 137, 1663–1681.
14. Huttenlocher, A., Lakonishok, M., Kinder, M., Wu, S., Truong, T., Knudsen, K. A., and Horwitz, A. F. (1998). Integrin and cadherin synergy regulates contact inhibition of migration and motile activity. *J. Cell Biol.* 141, 515–526.
15. Gimond, C., van Der Flier, A., Van Delft, S., Brakebusch, C., Kuikman, I., Collard, J. G., Fassler, R., and Sonnenberg, A. (1999). Induction of cell scattering by expression of beta1 integrins in beta1-

deficient epithelial cells requires activation of members of the rho family of GTPases and downregulation of cadherin and catenin function. *J. Cell Biol.* 147, 1325–1340.

16 Menke, A., Philippi, C., Vogelmann, R., Seidel, B., Lutz, M. P., Adler, G., and Wedlich, D. (2001). Down-regulation of E-cadherin gene expression by collagen type I and type III in pancreatic cancer cell lines. *Cancer Res.* 61, 3508–3517.

17 Ojakian, G. K., Ratcliffe, D. R., and Schwimmer, R. (2001). Integrin regulation of cell-cell adhesion during epithelial tubule formation. *J. Cell Sci.* 114, 941–952.

18 Avizienyte, E., Wyke, A. W., Jones, R. J., McLean, G. W., Westhoff, M. A., Brunton, V. G., and Frame, M. C. (2002). Src-induced de-regulation of E-cadherin in colon cancer cells requires integrin signalling. *Nat. Cell Biol.* 4, 632–638.

19 Marsden, M., and DeSimone, D. W. (2003). Integrin-ECM interactions regulate cadherin-dependent cell adhesion and are required for convergent extension in Xenopus. *Curr. Biol.* 13, 1182–1191.

20 Sakai, T., Larsen, M., and Yamada, K. M. (2003). Fibronectin requirement in branching morphogenesis. *Nature* 423, 876–881.

21 Yano, H., Mazaki, Y., Kurokawa, K., Hanks, S. K., Matsuda, M., and Sabe, H. (2004). Roles played by a subset of integrin signaling molecules in cadherin-based cell-cell adhesion. *J. Cell Biol.* 166, 283–295.

22 Hintermann, E., Yang, N., O'Sullivan, D., Higgins, J. M., and Quaranta, V. (2005). Integrin alpha6beta4-erbB2 complex inhibits haptotaxis by up-regulating E-cadherin cell-cell junctions in keratinocytes. *J. Biol. Chem.* 280, 8004–8015.

23 de Rooij, J., Kerstens, A., Danuser, G., Schwartz, M. A., and Waterman-Storer, C. M. (2005). Integrin-dependent actomyosin contraction regulates epithelial cell scattering. *J. Cell Biol.* 171, 153–164.

24 Wang, Y., Jin, G., Miao, H., Li, J. Y., Usami, S., and Chien, S. (2006). Integrins regulate VE-cadherin and catenins: dependence of this regulation on Src, but not on Ras. *Proc. Natl. Acad. Sci. USA* 103, 1774–1779.

25 Koenig, A., Mueller, C., Hasel, C., Adler, G., and Menke, A. (2006). Collagen type I induces disruption of E-cadherin-mediated cell-cell contacts and promotes proliferation of pancreatic carcinoma cells. *Cancer Res.* 66, 4662–4671.

26 Tzima, E., Irani-Tehrani, M., Kiosses, W. B., Dejana, E., Schultz, D. A., Engelhardt, B., Cao, G., DeLisser, H., and Schwartz, M. A. (2005). A mechanosensory complex that mediates the endothelial cell response to fluid shear stress. *Nature* 437, 426–431.

27 Zhang, W., Alt-Holland, A., Margulis, A., Shamis, Y., Fusenig, N. E., Rodeck, U., and Garlick, J. A. (2006). E-cadherin loss promotes the initiation of squamous cell carcinoma invasion through modulation of integrin-mediated adhesion. *J. Cell Sci.* 119, 283–291.

28 Balzac, F., Avolio, M., Degani, S., Kaverina, I., Torti, M., Silengo, L., Small, J. V., and Retta, S. F. (2005). E-cadherin endocytosis regulates the activity of Rap1: a traffic light GTPase at the crossroads between cadherin and integrin function. *J. Cell Sci.* 118, 4765–4783.

29 Werb, Z. (1997). ECM and cell surface proteolysis: regulating cellular ecology. *Cell* 91, 439–442.

30 McCawley, L. J., and Matrisian, L. M. (2001). Matrix metalloproteinases: they're not just for matrix anymore! *Curr. Opin. Cell Biol.* 13, 534–540.

31 Munshi, H. G., and Stack, M. S. (2006). Reciprocal interactions between adhesion receptor signaling and MMP regulation. *Cancer Metastasis Rev.* 25, 45–56.

32 Vo, H. P., Lee, M. K., and Crowe, D. L. (1998). alpha2beta1 integrin signaling via the mitogen activated protein kinase pathway modulates retinoic acid-dependent tumor cell invasion and transcriptional downregulation of matrix metalloproteinase 9 activity. *Int. J. Oncol.* 13, 1127–1134.

33 Troussard, A. A., Costello, P., Yoganathan, T. N., Kumagai, S., Roskelley, C. D., and Dedhar, S. (2000). The integrin linked kinase (ILK) induces

an invasive phenotype via AP-1 transcription factor-dependent upregulation of matrix metalloproteinase 9 (MMP-9). *Oncogene* 19, 5444–5452.

34. Iyer, V., Pumiglia, K., and DiPersio, C. M. (2005). α3β1 integrin regulates MMP-9 mRNA stability in immortalized keratinocytes: a novel mechanism of integrin-mediated MMP gene expression. *J. Cell Sci.* 118, 1185–1195.

35. Galvez, B. G., Matias-Roman, S., Yanez-Mo, M., Sanchez-Madrid, F., and Arroyo, A. G. (2002). ECM regulates MT1-MMP localization with beta1 or alphavbeta3 integrins at distinct cell compartments modulating its internalization and activity on human endothelial cells. *J. Cell Biol.* 159, 509–521.

36. Brooks, P. C., Stromblad, S., Sanders, L. C., von Schalscha, T. L., Aimes, R. T., Stetler-Stevenson, W. G., Quigley, J. P., and Cheresh, D. A. (1996). Localization of matrix metalloproteinase MMP-2 to the surface of invasive cells by interaction with integrin αvβ3. *Cell* 85, 683–693.

37. Ellerbroek, S. M., Fishman, D. A., Kearns, A. S., Bafetti, L. M., and Stack, M. S. (1999). Ovarian carcinoma regulation of matrix metalloproteinase-2 and membrane type 1 matrix metalloproteinase through beta1 integrin. *Cancer Res.* 59, 1635–1641.

38. Dumin, J. A., Dickeson, S. K., Stricker, T. P., Bhattacharyya-Pakrasi, M., Roby, J. D., Santoro, S. A., and Parks, W. C. (2001). Pro-collagenase-1 (matrix metalloproteinase-1) binds the alpha(2)beta(1) integrin upon release from keratinocytes migrating on type I collagen. *J. Biol. Chem.* 276, 29368–29374.

39. Munshi, H. G., Ghosh, S., Mukhopadhyay, S., Wu, Y. I., Sen, R., Green, K. J., and Stack, M. S. (2002). Proteinase suppression by E-cadherin-mediated cell-cell attachment in premalignant oral keratinocytes. *J. Biol. Chem.* 277, 38159–38167.

40. Margulis, A., Andriani, F., Fusenig, N., Hashimoto, K., Hanakawa, Y., and Garlick, J. A. (2003). Abrogation of E-cadherin-mediated adhesion induces tumor cell invasion in human skin-like organotypic culture. *J. Invest. Dermatol.* 121, 1182–1190.

41. Luo, J., Lubaroff, D. M., and Hendrix, M. J. (1999). Suppression of prostate cancer invasive potential and matrix metalloproteinase activity by E-cadherin transfection. *Cancer Res.* 59, 3552–3556.

42. Noe, V., Fingleton, B., Jacobs, K., Crawford, H. C., Vermeulen, S., Steelant, W., Bruyneel, E., Matrisian, L. M., and Mareel, M. (2001). Release of an invasion promoter E-cadherin fragment by matrilysin and stromelysin-1. *J. Cell Sci.* 114, 111–118.

43. Deryugina, E. I., Ratnikov, B. I., Postnova, T. I., Rozanov, D. V., and Strongin, A. Y. (2002). Processing of integrin alpha(v) subunit by membrane type 1 matrix metalloproteinase stimulates migration of breast carcinoma cells on vitronectin and enhances tyrosine phosphorylation of focal adhesion kinase. *J. Biol. Chem.* 277, 9749–9756.

44. Symowicz, J., Adley, B. P., Gleason, K. J., Johnson, J. J., Ghosh, S., Fishman, D. A., Hudson, L. G., and Stack, M. S. (2007). Engagement of collagen-binding integrins promotes matrix metalloproteinase-9-dependent E-cadherin ectodomain shedding in ovarian carcinoma cells. *Cancer Res.* 67, 2030–2039.

45. Conacci-Sorrell, M., Zhurinsky, J., and Ben-Ze'ev, A. (2002). The cadherin-catenin adhesion system in signaling and cancer. *J. Clin. Invest.* 109, 987–991.

46. Hazan, R. B., Phillips, G. R., Qiao, R. F., Norton, L., and Aaronson, S. A. (2000). Exogenous expression of N-cadherin in breast cancer cells induces cell migration, invasion, and metastasis. *J. Cell Biol.* 148, 779–790.

47. Suyama, K., Shapiro, I., Guttman, M., and Hazan, R. B. (2002). A signaling pathway leading to metastasis is controlled by N-cadherin and the FGF receptor. *Cancer Cell* 2, 301–314.

48. Kang, Y., and Massague, J. (2004). Epithelial-mesenchymal transitions: twist in development and metastasis. *Cell* 118, 277–279.

49 Aplin, A. E., Stewart, S. A., Assoian, R. K., and Juliano, R. L. (2001). Integrin-mediated adhesion regulates ERK nuclear translocation and phosphorylation of Elk-1. *J. Cell Biol.* 153, 273–282.

50 Yang, Z., Rayala, S., Nguyen, D., Vadlamudi, R. K., Chen, S., and Kumar, R. (2005). Pak1 phosphorylation of snail, a master regulator of epithelial-to-mesenchyme transition, modulates snail's subcellular localization and functions. *Cancer Res.* 65, 3179–3184.

51 Alexander, N. R., Tran, N. L., Rekapally, H., Summers, C. E., Glackin, C., and Heimark, R. L. (2006). N-cadherin gene expression in prostate carcinoma is modulated by integrin-dependent nuclear translocation of Twist1. *Cancer Res.* 66, 3365–3369.

52 Tan, C., Costello, P., Sanghera, J., Dominguez, D., Baulida, J., de Herreros, A. G., and Dedhar, S. (2001). Inhibition of integrin linked kinase (ILK) suppresses beta-catenin-Lef/Tcf-dependent transcription and expression of the E-cadherin repressor, snail, in APC-/- human colon carcinoma cells. *Oncogene* 20, 133–140.

53 Novak, A., and Dedhar, S. (1999). Signaling through beta-catenin and Lef/Tcf. *Cell. Mol. Life Sci.* 56, 523–537.

54 Gottardi, C. J., and Gumbiner, B. M. (2001). Adhesion signaling: how beta-catenin interacts with its partners. *Curr. Biol.* 11, R792–R794.

55 Park, J. I., Kim, S. W., Lyons, J. P., Ji, H., Nguyen, T. T., Cho, K., Barton, M. C., Deroo, T., Vleminckx, K., Moon, R. T., and McCrea, P. D. (2005). Kaiso/p120-catenin and TCF/beta-catenin complexes coordinately regulate canonical Wnt gene targets. *Dev. Cell* 8, 843–854.

56 Daniel, J. M. (2007). Dancing in and out of the nucleus: p120(ctn) and the transcription factor Kaiso. *Biochim. Biophys. Acta* 1773, 59–68.

57 Avizienyte, E., Fincham, V. J., Brunton, V. G., and Frame, M. C. (2004). Src SH3/2 domain-mediated peripheral accumulation of Src and phospho-myosin is linked to deregulation of E-cadherin and the epithelial-mesenchymal transition. *Mol. Biol. Cell* 15, 2794–2803.

58 Zhong, C., Kinch, M. S., and Burridge, K. (1997). Rho-stimulated contractility contributes to the fibroblastic phenotype of Ras-transformed epithelial cells. *Mol. Biol. Cell* 8, 2329–2344.

59 Stockton, R. A., Schaefer, E., and Schwartz, M. A. (2004). p21-activated kinase regulates endothelial permeability through modulation of contractility. *J. Biol. Chem.* 279, 46621–46630.

60 Wittchen, E. S., van Buul, J. D., Burridge, K., and Worthylake, R. A. (2005). Trading spaces: Rap, Rac, and Rho as architects of transendothelial migration. *Curr. Opin. Hematol.* 12, 14–21.

61 Nobes, C. D., and Hall, A. (1999). Rho GTPases control polarity, protrusion, and adhesion during cell movement. *J. Cell Biol.* 144, 1235–1244.

62 Yamada, S., and Nelson, W. J. (2007). Localized zones of Rho and Rac activities drive initiation and expansion of epithelial cell-cell adhesion. *J. Cell Biol.* 178, 517–527.

63 Jeon, T. J., Lee, D. J., Merlot, S., Weeks, G., and Firtel, R. A. (2007). Rap1 controls cell adhesion and cell motility through the regulation of myosin II. *J. Cell Biol.* 176, 1021–1033.

64 Carter, W. G., Wayner, E. A., Bouchard, T. S., and Kaur, P. (1990). The role of integrins α2β1 and α3β1 in cell-cell and cell-substrate adhesion of human epidermal cells. *J. Cell Biol.* 110, 1387–1404.

65 Larjava, H., Peltonen, J., Akiyama, S. K., Yamada, S. S., Gralnick, H. R., Uitto, J., and Yamada, K. M. (1990). Novel function for beta 1 integrins in keratinocyte cell-cell interactions. *J. Cell Biol.* 110, 803–815.

66 Symington, B. E., Takada, Y., and Carter, W. G. (1993). Interaction of integrins alpha 3 beta 1 and alpha 2 beta 1: potential role in keratinocyte intercellular adhesion. *J. Cell Biol.* 120, 523–535.

67 Wang, Z., Symons, J. M., Goldstein, S., McDonald, A., Miner, J. H., and Kreidberg, J. A. (1999). a3b1 integrin regulates epithelial cytoskeletal organization. *J. Cell Sci.* 112, 2925–2935.

68. Hodivala-Dilke, K. M., DiPersio, C. M., Kreidberg, J. A., and Hynes, R. O. (1998). Novel roles for α3β1 integrin as a regulator of cytoskeletal assembly and as a transdominant inhibitor of integrin receptor function in keratinocytes. *J. Cell Biol.* 142, 1357–1369.
69. Choma, D. P., Pumiglia, K., and DiPersio, C. M. (2004). Integrin α3β1 directs the stabilization of a polarized lamellipodium in epithelial cells through activation of Rac1. *J. Cell Sci.* 117, 3947–3959.
70. Chattopadhyay, N., Wang, Z., Ashman, L. K., Brady-Kalnay, S. M., and Kreidberg, J. A. (2003). alpha3beta1 integrin-CD151, a component of the cadherin-catenin complex, regulates PTPmu expression and cell-cell adhesion. *J. Cell Biol.* 163, 1351–1362.
71. Yauch, R. L., Kazarov, A. R., Desai, B., Lee, R. T., and Hemler, M. E. (2000). Direct extracellular contact between integrin α3β1 and TM4SF protein CD151. *J. Biol. Chem.* 275, 9230–9238.
72. Shigeta, M., Sanzen, N., Ozawa, M., Gu, J., Hasegawa, H., and Sekiguchi, K. (2003). CD151 regulates epithelial cell-cell adhesion through PKC- and Cdc42-dependent actin cytoskeletal reorganization. *J. Cell Biol.* 163, 165–176.
73. Yanez-Mo, M., Tejedor, R., Rousselle, P., and Sanchez-Madrid, F. (2001). Tetraspanins in intercellular adhesion of polarized epithelial cells: spatial and functional relationship to integrins and cadherins. *J. Cell Sci.* 114, 577–587.
74. Nguyen, B. P., Ryan, M. C., Gil, S. G., and Carter, W. G. (2000). Deposition of laminin 5 in epidermal wounds regulates integrin signaling and adhesion. *Curr. Opin. Cell Biol.* 12, 554–562.
75. Marinkovich, M. P. (2007). Tumour microenvironment: Laminin 332 in squamous-cell carcinoma. *Nat. Rev. Cancer* 7, 370–380.
76. Zhang, F., Tom, C. C., Kugler, M. C., Ching, T. T., Kreidberg, J. A., Wei, Y., and Chapman, H. A. (2003). Distinct ligand binding sites in integrin alpha3beta1 regulate matrix adhesion and cell-cell contact. *J. Cell Biol.* 163, 177–188.
77. Avizienyte, E., and Frame, M. C. (2005). Src and FAK signalling controls adhesion fate and the epithelial-to-mesenchymal transition. *Curr. Opin. Cell Biol.* 17, 542–547.
78. McLean, G. W., Carragher, N. O., Avizienyte, E., Evans, J., Brunton, V. G., and Frame, M. C. (2005). The role of focal-adhesion kinase in cancer–a new therapeutic opportunity. *Nat. Rev. Cancer* 5, 505–515.
79. Quadri, S. K., Bhattacharjee, M., Parthasarathi, K., Tanita, T., and Bhattacharya, J. (2003). Endothelial barrier strengthening by activation of focal adhesion kinase. *J. Biol. Chem.* 278, 13342–13349.
80. Cheresh, D. A., Leng, J., and Klemke, R. L. (1999). Regulation of cell contraction and membrane ruffling by distinct signals in migratory cells. *J. Cell Biol.* 146, 1107–1116.
81. Hsia, D. A., Mitra, S. K., Hauck, C. R., Streblow, D. N., Nelson, J. A., Ilic, D., Huang, S., Li, E., Nemerow, G. R., Leng, J., Spencer, K. S., Cheresh, D. A., and Schlaepfer, D. D. (2003). Differential regulation of cell motility and invasion by FAK. *J. Cell Biol.* 160, 753–767.
82. Choma, D. P., Milano, V., Pumiglia, K. M., and DiPersio, C. M. (2007). Integrin alpha3beta1-dependent activation of FAK/Src regulates Rac1-mediated keratinocyte polarization on laminin-5. *J. Invest. Dermatol.* 127, 31–40.
83. Xu, G., Craig, A. W., Greer, P., Miller, M., Anastasiadis, P. Z., Lilien, J., and Balsamo, J. (2004). Continuous association of cadherin with beta-catenin requires the non-receptor tyrosine-kinase Fer. *J. Cell Sci.* 117, 3207–3219.
84. Du, J., and Xu, J. (1995). [Chemical constituents of Impatiens siculi fer Hook. f.]. *Zhongguo Zhong Yao Za Zhi* 20, 232–233, 253.
85. Arregui, C., Pathre, P., Lilien, J., and Balsamo, J. (2000). The nonreceptor tyrosine kinase fer mediates cross-talk between N-cadherin and beta1-integrins. *J. Cell Biol.* 149, 1263–1274.
86. Li, H., Leung, T. C., Hoffman, S., Balsamo, J., and Lilien, J. (2000). Coordinate regulation of cadherin and integrin function by the chondroitin

sulfate proteoglycan neurocan. *J. Cell Biol.* **149**, 1275–1288.

87 Grosheva, I., Shtutman, M., Elbaum, M., and Bershadsky, A. D. (2001). p120 catenin affects cell motility via modulation of activity of Rho-family GTPases: a link between cell-cell contact formation and regulation of cell locomotion. *J. Cell Sci.* **114**, 695–707.

88 Braga, V. M., Machesky, L. M., Hall, A., and Hotchin, N. A. (1997). The small GTPases Rho and Rac are required for the establishment of cadherin-dependent cell-cell contacts. *J. Cell Biol.* **137**, 1421–1431.

89 Kitayama, H., Sugimoto, Y., Matsuzaki, T., Ikawa, Y., and Noda, M. (1989). A ras-related gene with transformation suppressor activity. *Cell* **56**, 77–84.

90 Kooistra, M. R., Dube, N., and Bos, J. L. (2007). Rap1: a key regulator in cell-cell junction formation. *J. Cell Sci.* **120**, 17–22.

91 Knox, A. L., and Brown, N. H. (2002). Rap1 GTPase regulation of adherens junction positioning and cell adhesion. *Science* **295**, 1285–1288.

92 Hogan, C., Serpente, N., Cogram, P., Hosking, C. R., Bialucha, C. U., Feller, S. M., Braga, V. M., Birchmeier, W., and Fujita, Y. (2004). Rap1 regulates the formation of E-cadherin-based cell-cell contacts. *Mol. Cell. Biol.* **24**, 6690–6700.

93 Price, L. S., Hajdo-Milasinovic, A., Zhao, J., Zwartkruis, F. J., Collard, J. G., and Bos, J. L. (2004). Rap1 regulates E-cadherin-mediated cell-cell adhesion. *J. Biol. Chem.* **279**, 35127–35132.

94 Yajnik, V., Paulding, C., Sordella, R., McClatchey, A. I., Saito, M., Wahrer, D. C., Reynolds, P., Bell, D. W., Lake, R., van den Heuvel, S., Settleman, J., and Haber, D. A. (2003). DOCK4, a GTPase activator, is disrupted during tumorigenesis. *Cell* **112**, 673–684.

95 Fukuhra, S., Sakurai, A., Yamagishi, A., Sako, K., and Mochizuki, N. (2006). Vascular endothelial cadherin-mediated cell-cell adhesion regulated by a small GTPase, Rap1. *J. Biochem. Mol. Biol.* **39**, 132–139.

96 Fukuyama, T., Ogita, H., Kawakatsu, T., Fukuhara, T., Yamada, T., Sato, T., Shimizu, K., Nakamura, T., Matsuda, M., and Takai, Y. (2005). Involvement of the c-Src-Crk-C3G-Rap1 signaling in the nectin-induced activation of Cdc42 and formation of adherens junctions. *J. Biol. Chem.* **280**, 815–825.

97 Fukuyama, T., Ogita, H., Kawakatsu, T., Inagaki, M., and Takai, Y. (2006). Activation of Rac by cadherin through the c-Src-Rap1-phosphatidylinositol 3-kinase-Vav2 pathway. *Oncogene* **25**, 8–19.

98 Kinbara, K., Goldfinger, L. E., Hansen, M., Chou, F. L., and Ginsberg, M. H. (2003). Ras GTPases: integrins' friends or foes? *Nat. Rev. Mol. Cell. Biol.* **4**, 767–776.

99 Arthur, W. T., Quilliam, L. A., and Cooper, J. A. (2004). Rap1 promotes cell spreading by localizing Rac guanine nucleotide exchange factors. *J. Cell Biol.* **167**, 111–122.

100 Maillet, M., Robert, S. J., Cacquevel, M., Gastineau, M., Vivien, D., Bertoglio, J., Zugaza, J. L., Fischmeister, R., and Lezoualc'h, F. (2003). Crosstalk between Rap1 and Rac regulates secretion of sAPPalpha. *Nat. Cell Biol.* **5**, 633–639.

101 Su, L., Hattori, M., Moriyama, M., Murata, N., Harazaki, M., Kaibuchi, K., and Minato, N. (2003). AF-6 controls integrin-mediated cell adhesion by regulating Rap1 activation through the specific recruitment of Rap1GTP and SPA-1. *J. Biol. Chem.* **278**, 15232–15238.

102 Katagiri, K., Maeda, A., Shimonaka, M., and Kinashi, T. (2003). RAPL, a Rap1-binding molecule that mediates Rap1-induced adhesion through spatial regulation of LFA-1. *Nat. Immunol.* **4**, 741–748.

103 Lafuente, E. M., van Puijenbroek, A. A., Krause, M., Carman, C. V., Freeman, G. J., Berezovskaya, A., Constantine, E., Springer, T. A., Gertler, F. B., and Boussiotis, V. A. (2004). RIAM, an Ena/VASP and Profilin ligand, interacts with Rap1-GTP and mediates Rap1-induced adhesion. *Dev. Cell* **7**, 585–595.

104 Krugmann, S., Williams, R., Stephens, L., and Hawkins, P. T. (2004). ARAP3 is a PI3K- and rap-regulated GAP for RhoA. *Curr. Biol.* **14**, 1380–1384.

105 Vasioukhin, V., Bauer, C., Yin, M., and Fuchs, E. (2000). Directed actin polymerization is the driving force for

epithelial cell-cell adhesion. *Cell 100*, 209–219.
106 Watt, F. M. (2002). Role of integrins in regulating epidermal adhesion, growth and differentiation. *EMBO J. 21*, 3919–3926.
107 Parks, W. C. (2007). What is the alpha2beta1 integrin doing in the epidermis? *J. Invest. Dermatol. 127*, 264–266.
108 Raghavan, S., Bauer, C., Mundschau, G., Li, Q., and Fuchs, E. (2000). Conditional ablation of beta1 integrin in skin. Severe defects in epidermal proliferation, basement membrane formation, and hair follicle invagination. *J. Cell Biol. 150*, 1149–1160.
109 Brakebusch, C., Grose, R., Quondamatteo, F., Ramirez, A., Jorcano, J. L., Pirro, A., Svensson, M., Herken, R., Sasaki, T., Timpl, R., Werner, S., and Fassler, R. (2000). Skin and hair follicle integrity is crucially dependent on beta 1 integrin expression on keratinocytes. *EMBO J. 19*, 3990–4003.
110 McLean, G. W., Brown, K., Arbuckle, M. I., Wyke, A. W., Pikkarainen, T., Ruoslahti, E., and Frame, M. C. (2001). Decreased focal adhesion kinase suppresses papilloma formation during experimental mouse skin carcinogenesis. *Cancer Res. 61*, 8385–8389.
111 Malliri, A., van der Kammen, R. A., Clark, K., Van Der Valk, M., Michiels, F., and Collard, J. G. (2002). Mice deficient in the Rac activator Tiam1 are resistant to Ras-induced skin tumours. *Nature 417*, 867–871.
112 Grose, R., Hutter, C., Bloch, W., Thorey, I., Watt, F. M., Fassler, R., Brakebusch, C., and Werner, S. (2002). A crucial role of beta 1 integrins for keratinocyte migration in vitro and during cutaneous wound repair. *Development 129*, 2303–2315.
113 McLean, G. W., Komiyama, N. H., Serrels, B., Asano, H., Reynolds, L., Conti, F., Hodivala-Dilke, K., Metzger, D., Chambon, P., Grant, S. G., and Frame, M. C. (2004). Specific deletion of focal adhesion kinase suppresses tumor formation and blocks malignant progression. *Genes Dev. 18*, 2998–3003.
114 Lopez-Rovira, T., Silva-Vargas, V., and Watt, F. M. (2005). Different consequences of beta1 integrin deletion in neonatal and adult mouse epidermis reveal a context-dependent role of integrins in regulating proliferation, differentiation, and intercellular communication. *J. Invest. Dermatol. 125*, 1215–1227.
115 Hamelers, I. H., Olivo, C., Mertens, A. E., Pegtel, D. M., van der Kammen, R. A., Sonnenberg, A., and Collard, J. G. (2005). The Rac activator Tiam1 is required for $\alpha3\beta1$-mediated laminin-5 deposition, cell spreading, and cell migration. *J. Cell Biol. 171*, 871–881.
116 Tscharntke, M., Pofahl, R., Chrostek-Grashoff, A., Smyth, N., Niessen, C., Niemann, C., Hartwig, B., Herzog, V., Klein, H. W., Krieg, T., Brakebusch, C., and Haase, I. (2007). Impaired epidermal wound healing in vivo upon inhibition or deletion of Rac1. *J. Cell Sci. 120*, 1480–1490.
117 DiPersio, C. M. (2007). Double duty for Rac1 in epidermal wound healing. *Sci. STKE 2007*, pe33.
118 Frank, D. E., and Carter, W. G. (2004). Laminin 5 deposition regulates keratinocyte polarization and persistent migration. *J. Cell Sci. 117*, 1351–1363.
119 Sander, E. E., van Delft, S., ten Klooster, J. P., Reid, T., Van Der Kammen, R. A., Michiels, F., and Collard, J. G. (1998). Matrix-dependent Tiam1/Rac signaling in epithelial cells promotes either cell-cell adhesion or cell migration and is regulated by phosphatidylinositol 3-kinase. *J. Cell Biol. 143*, 1385–1398.
120 Malliri, A., van Es, S., Huveneers, S., and Collard, J. G. (2004). The Rac exchange factor Tiam1 is required for the establishment and maintenance of cadherin-based adhesions. *J. Biol. Chem. 279*, 30092–30098.
121 Enserink, J. M., Price, L. S., Methi, T., Mahic, M., Sonnenberg, A., Bos, J. L., and Tasken, K. (2004). The cAMP-Epac-Rap1 pathway regulates cell spreading and cell adhesion to laminin-5 through the alpha3beta1 integrin but not the alpha6beta4 integrin. *J. Biol. Chem. 279*, 44889–44896.

122 Grinnell, F. (1992). Wound repair, keratinocyte activation and integrin modulation. *J. Cell Sci. 101(Pt 1)*, 1–5.
123 Santoro, M. M., and Gaudino, G. (2005). Cellular and molecular facets of keratinocyte reepithelization during wound healing. *Exp. Cell Res. 304*, 274–286.
124 Chrostek, A., Wu, X., Quondamatteo, F., Hu, R., Sanecka, A., Niemann, C., Langbein, L., Haase, I., and Brakebusch, C. (2006). Rac1 is crucial for hair follicle integrity but is not essential for maintenance of the epidermis. *Mol. Cell. Biol. 26*, 6957–6970.
125 Brieher, W. M., and Gumbiner, B. M. (1994). Regulation of C-cadherin function during activin induced morphogenesis of Xenopus animal caps. *J. Cell Biol. 126*, 519–527.

Index

a

actin 221
    – binding domain 111
    – cytoskeleton 73
    – regulatory protein 77ff.
actin related protein (Arp)
    – Arp 2/3 complex 77ff.
actinomyosin cytoskeleton 228
ADAM family 116
ADAP/SKAP-55 17
adaptor protein (AP)
    – AP1B 253ff.
    – AP2 94, 173
adenomatous polyposis coli (APC) 156ff., 278
adherens junction (AJ) 52ff., 153ff., 169ff., 178, 235, 251
    – cadherin 274
    – cancer 163
    – disassembly 256
    – endothelial 169ff., 178ff.
    – epithelial 153ff.
    – protein 252ff.
adhesion 109ff.
    – cell 279
    – cell-cell 273ff.
    – cell-matrix 273ff.
    – dynamics 71ff.
AF-6 284
afadin 187, 284
agonist receptor 16
Akt/protein kinase B 55ff., 124
androgen receptor (AR) 59
    – coactivator (ARA55) 51
animal model 236
AP, see adaptor protein
AP1 (activator protein) 225
aquaporin (AQP)-3 254
Arap3 285

Arf 183
Arf6 93ff., 263
ArfGTPase-activating protein (ARF-GAP) 48ff.
ArfGTPase-activating protein containing SH3, ankyrin repeat and pleckstrin homology domain-1 (ASAP1) 27ff.
arm motif 154
armadillo repeat 157
    – domain (ARM domain) 172
    – gene deleted VeloCardioFacial (ARVCF) 160, 172
armadillo repeat protein 153ff., 240
    – cancer 163
β-arrestin-2 264
arrhythmogenic right ventricular cardiomyopathy/dysplasia (ARVC/D) 239ff.
astacin family 116
atherosclerosis 206
autoimmune disease 121
    – hemidesmosome 121
axin/APC/GSK-3β complex 157ff.

b

barrier function 219
blistering skin disease 239
bone morphogenetic protein 1 (BMP1) 116
BRAG2 93ff.
bullous pemphigoid (BP) 112ff., 123
    – BP180 109ff.
    – BP230 109ff.
    – BPAG1 109ff.
    – BPAG2 109ff.

c

C3G 283
cadherin 153, 251ff., 274ff.
    – cell-cell adhesion 283
    – cytoplasmic domain 172

– desmosomal 236ff.
– E 52ff., 153ff., 170, 224ff., 241, 252ff., 275ff.
– endocytosis 258ff.
– epidermis 285
– FAK 282
– integrin crosstalk 275ff.
– internalization 260
– juxtamembrane domain (JMD) 154
– N 153, 171, 255ff., 282
– P 153
– R 153, 170, 259
– recycling 256
– trafficking 251ff.
– VE, see VE-cadherin

CagA 228
calcium
– cytosolic 202
calcium- and integrin-binding protein (CIB) 10
CAlDAG-GEFI 14
calpain 11, 76ff.
– 2 31, 73
cancer 123, 163
– adherens junction 163
– armadillo repeat protein 163
CAP1 (contraception-associated protein 1)/Prss8 (prostasin) 226
carboxy-terminal Src kinase (Csk) 31ff., 51, 179
casein kinase II (CK2) 156
catenin 172
– α 154ff., 177ff., 255, 274
– β 55, 153ff., 174ff., 225, 240f., 251ff., 274ff.
– γ 153ff., 174, 239ff.
– binding domain (CBD) 174
– C-terminal catenin-binding domain (CBD) 172
– p120 155ff., 172f., 225, 258ff., 274
– plakoglobin 153ff., 174, 239ff.
– regulation 258
caveolae 90ff.
– endocytosis 90ff.
caveosome 95
CD151 109ff., 280
cell
– behaviour 136ff.
– motile 71ff.
cell adhesion 31, 96f., 279
cell-cell adhesion 273ff.
– cadherin 283
cell-cell junction 178
– tumor 178

cell division cycle (Cdc)
– Cdc2 252
– Cdc42 35, 55, 75, 97, 161, 177ff., 221ff., 279ff.
cell invasion 32
cell matrix adhesion 135, 273ff.
– integrin-mediated 284
– three dimensional 135
cell migration 32, 53
cell proliferation 36
cell spreading 31
cell survival 36
cellular contractility 141
channel permeability 202
chondroitin sulphate proteoglycan (CSPG) 136
cicatricial pemphigoid (CP) 122
– ocular (OCP) 122
clathrin 90ff.
– adaptor protein complex (AP-2) 173, 259
– coated pit/vesicle 90ff.
claudin 220ff., 243
collagen 114, 136ff.
– IV 136
– gel 136ff.
– phagocytosis 99
colony stimulating factor (CSF) 1 78
compliancy 140
– matrix 140f.
connexin
– α subfamily 200
– β subfamily 200
– assembly 198f.
– atherosclerosis 206
– binding protein 201
– Cx32 197ff.
– Cx36 197ff.
– Cx43 197ff.
– human disease 203ff.
– interaction 199
– phosphorylation 202
– structure 198f.
– vessel inflammation 206
contractility 141
cortactin 78
Coxsackie and adenovirus receptor (CAR) 220
Crk 48ff.
Crk-associated substrate (Cas) adaptor protein 27ff.
Crumbs (Crb)/Pals/Patj 222f.
Csk, see carboxy-terminal Src kinase
Csk homologous kinase (Chk) 51

cyclin D1   37, 225
cyclin-dependent kinase (cdk)
   – cdk5   29
   – inhibitor   37
cytoskeleton   141, 156, 178ff.
   – protein   221
cytotoxic necrotizing factor (CNF-1)   184f.

d

degradation   156
density-enhanced phosphatase 1 (DEP-1)
   182, 264
desmocollin (Dsc)   236f.
desmoglein (Dsg)   236f.
desmoplakin (Dsp)   112, 176, 240
desmosome   158, 235ff.
   – cadherin   236ff.
   – development   235ff.
   – disease   235ff.
   – intermediate filament
      (IF)-anchoring   235
   – molecular composition   236
desmosomal protein
   – accessory   243
   – embryonic development   243
differentiation
   – transcription mediated   225
dishevelled (Dsh)   157
dopamine transporter (DAT)   51

e

E-cadherin, *see* cadherin
ectodermal dysplasia/skin fragility (EDSF)
   242
elastic modulus   140
EMP protein   243
endocytosis   90ff., 258ff.
   – cadherin   262
   – caveolae   90ff.
   – clathrin independent   257
   – clathrin mediated   90ff., 257
   – β1-integrin   99
   – integrin-dependent   95
endoplasmic reticulum (ER)
   – ER-Golgi intermediate compartment
      (ERGIC)   199
endothelial adherens junction   169ff., 178
   – serine/threonine phosphorylation
      182
endothelial cell selective adhesion molecule
   (ESAM)   220
Engelbreth-Holm-Swarm (EHS) sarcoma
   136
entactin   136

Epac   186
epiligrin   115
epidermal blistering disease   257
epidermal growth factor (EGF)   56, 76, 116,
   124, 257
   – receptor (EGFR)   124f., 156ff.
epidermis   285
epidermolytic palmoplantar keratoderma
   240
epithelial-mesenchymal transition (EMT)
   49ff., 256, 273
epithelium
   – barrier function   219
   – polarity   222
   – simple   217ff.
   – stratified   217ff.
ErbB2/ErbB3   125, 224
ERK   51ff., 76, 125
Erk2/MAP kinase cascade   29
exocyst   254
exocytosis   252, 265
extracellular matrix (ECM)   3f., 25, 36, 47,
   71, 89ff., 135, 273
   – remodelling   98ff.
extracellular rearrangement   7

f

FAK family interacting protein of 200 kDa
   (FIP200)   30
FAK-related non-kinase (FRNK)   27ff.
fence function   224
Fer   158, 178, 282
FERM domain   10, 26ff., 223
fibrin gel   138
fibroblast growth factor receptor (FGFR)
   263
   – FGFR1   171
fibronectin (FN)   91ff., 112
   – cell derived   138
   – matrix   138
focal adhesion   47f., 139ff.
   – assembly   74f.
   – disassembly   75f.
   – dynamics   72f.
   – formation   142
focal adhesion kinase (FAK)   25ff., 50ff.,
   73ff., 139
   – activation   28f.
   – cadherin   282
   – *in vivo* function   37f.
   – inhibition   30
   – knockout mouse   37f.
   – proline rich (PR) motif   27
   – signalling pathway   30f.

– structure 26f.
– tyrosine phosphorylation 28f.
focal adhesion targeting (FAT) domain 26ff.
FRET (fluorescens resonance energy transfer) 186
FRET (Förster resonance energy transfer) 7ff.
frizzled (Frz) 157
Fyn 158ff., 178

g
G protein coupled receptor (GPCR) 264
gap junction 197ff., 235
– channel 202
gene expression
– cross regulation 277
generalized atrophic epidermolysis bullosa (GABEB) 118f.
GIT1 50f.
Glanzmann thrombasthenia 6
glycogen synthase kinase 3β (GSK-3β) 156ff., 278
Grb and Mig (GM) domain 34
Grb2 29
Grb7 34
GTPase activating protein (GAP) 183, 221
GTPase regulator associated with FAK (GRAF) 27ff.
guanine nucleotide exchange factor (GEF) 14, 93ff., 161, 186, 221, 283
– GEF-H1 225
guanosine triphosphatase (GTPase) 35, 265, 279
– small 183
GUK domain 221

h
HDMEC-1 176
heat shock protein (Hsp) 27 51
hemidesmosome 109ff.
– assembly 116f.
– component 109ff.
– disassembly 116f.
hepatocyte growth factor (HGF) 56, 279
– receptor 164, 263
– regulated tyrosine kinase substrate (Hrs) 257
Herlitz JEB (HJEB) 120
HGF, see hepatocyte growth factor
histamine 184
human microvascular endothelial cell-1 (HMEC-1) 175

human umbilical vein endothelial cell (HUVEC) 176
hydrogen peroxide-inducible clone 5 (Hic-5) 49ff.
hypotrichiosis 239

i
Ig superfamily molecule 6
IKK1 226f.
inflammation
– vessel 206
inflammatory disease 228
innexin 198
insulin receptor substrate (IRS)
– IRS-1 124
– IRS-2 124
integrin 3ff., 75ff., 89ff., 112ff., 143
– α subunit 3ff., 89ff.
– α6β4 112ff.
– αIIbβ3 6ff.
– β1 9ff., 277ff.
– β2 9ff.
– β3 9ff.
– β4 112ff.
– β subunit 3ff., 89ff., 143
– β-tail 11ff., 28
– affinity 15, 284
– avidity 284
– cadherin crosstalk 274ff.
– cancer 123
– conformation 7
– cytoplasmic domain 9, 94f., 143
– endocytosis 99
– mediated adhesion 277
– recycling 284
– talin interaction 10
– wound healing 123
integrin activation 5ff.
– regulation 13
– structural basis 6
integrin linked kinase (ILK) 48ff., 73, 277
integrin receptor 3f.
integrin signaling 3f., 25ff., 51ff.
– EMT 53f.
– focal adhesion kinase (FAK) 25ff.
– inside-out 4ff.
– outside-in 4f.
integrin trafficking 89ff.
– cell adhesion 96f.
– regulation 92ff.
intermediate filament (IF)-anchoring desmosome 235
internalization 90ff., 260
– sorting motif 94

– tight junction 228
intracellular adhesion molecule (ICAM) 6, 91
intracellular force generation 278
intracellular rearrangement 8
invadopodia 72ff.
– architecture 77
– dynamics 76
IQGAP 177ff.

j
c-jun N-terminal kinase (JNK) 125
junction
– assembly 183
– dynamics 251ff.
– gap 197ff., 235
– maturation 225
– tight 217ff., 235
junctional adhesion molecule (JAM) 220ff.
junctional epidermolysis bullosa (JEB) 118ff.
– pyloric atresia (PA-JEB) 118
juxtamembrane domain (JMD) 154ff., 172f., 251

k
Kaiso 162
kalinin 115
kissing point 218

l
LAMA3 120
LAMB3 120
LAMC2 120
laminin 117, 136
– laminin-111/laminin-1 113
– laminin-332/laminin-5 109ff., 120ff.
leukocyte 6ff.
– adhesion deficiency (LAD) 6ff.
leupaxin 49f.
lichen planus pemphigoides (LPP) 122
lin-11, isl-1, mec-3 (LIM) domain 51
lipid raft 90ff.
lymphoid enhancer factor (Lef) 157, 176, 225, 276ff.
– LEF1 55
lysosomal degradation 98
lysosome 258

m
macrophage-stimulating protein (MSP) 117
macula adherentes 235
MAGI 221
MAGI-1 187

MAGUK (membrane-associated guanylate kinase-like homologues) 221ff.
mammalian tolloid (mTLD) 116
– like 1 (mTLL1) 116
– like 2 (mTLL2) 116
matrigel®/reconstituted basement membrane (rBM) 136ff.
matrix 140
– adhesion formation 145
– compliance 140
– engineered three dimensional 138
– fibronectin 138
matrix metalloproteinase (MMP) 35, 77, 98ff., 139, 275
– MMP-2 116
– MMP-14/MT1-MMP 100, 116
– MMP-19 116
membrane element 112
c-met 164, 263
microbial pathogen 95
microfluidic chamber 138
mitogen-activated protein (MAP) kinase (MAPK) 51ff.
– cascade 29, 202, 239, 263
– ERK 29, 51ff., 76, 125
– p38 54, 125, 239
model three dimensional matrix 136ff.
– cellular response 144
MUPP1 221
muscular dystrophy (MD)
– epidermolysis bullosa simplex (MD-EBS) 118f.
MT1-MMP 100
myosin-light chain (MLC) 141
– kinase (MLCK) 76

n
N-cadherin, see cadherin
neonatal ichthyosis-sclerosing cholangitis (NISCH) syndrome 228
nicein 115
non-Herlitz JEB (non-HJEB) 118ff.

o
occludin 219ff.
oculodentodigital dysplasia (ODDD) 205
oral pemphigoid (OP) 122

p
p21-activated kinase (PAK) 73ff., 183
$p21^{waf1}$ 37
$p27^{kip1}$ 37
p38MAPK 54, 125, 239
p53 124

p120 catenin 155ff., 172f., 225, 258ff., 274
p120RasGAP 56f.
p190RhoGAP 55ff., 161, 262
P-cadherin, see cadherin
Pals/Patj/Crbs 222ff.
pannexin-1 198
Par3/Par6/aPKC complex 222f.
paxillin 76ff., 282
– family 47ff.
– kinase linker (PKL) 48ff.
– leucine rich (LR) motif 50
– structure 50f.
– superfamily 47ff.
PDZ domain 221
– multi 224
PECAM (cell-cell adhesion molecule) 14
pemphigus
– foliaceus (PF) 238
– vulgaris (PV) 238, 257
peroxisome proliferators-activated receptor (PPAR) γ 51
Perp 243
phagocytosis 99
phosphatase and tensin homologue deleted from chromosome 10 (PTEN) 30ff.
phosphatidylinositol kinase-3 (PI3K) 34ff., 124
phosphatidylinositol 4-phosphate 5-kinase (PIP5K) 94
phosphatidylinositol phosphate kinase type Iγ(PIPKIγ) 253f.
phosphatidylinositol phosphate kinase type Iγ-90 (PIPKIγ-90) 11
phosphatidylinositol phosphate kinase type Iγ661 (PIPKIγ661) 75
phosphoinositide 75
phosphotyrosine-binding (PTB) domain 10ff.
photoreceptor polarization 223
PKA 186
PKD1 93
PKL/GIT2 48ff.
plakin protein 109ff.
plakoglobin, see catenin
plakophilin (Pkp) 242
– Pkp2 242ff.
– Pkp3 242ff.
plaque protein 110
platelet 6ff.
– derived growth factor (PDGF) 76, 92
plectin 110ff.
– MD-EBS 119
PMP-22 243
podosome 72ff.
– PTA assembly 78
– PTA disassembly 80
– type adhesion (PTA) 77f.
polydimethyl siloxand (PDMS) channel 139
profilin 142
prostacyclin (PGI2) 186
prostaglandin E_2 (PGE2) 186
protein kinase B 55ff.
protein kinase C 73ff., 93, 182, 202, 280
– atypical (aPKC) 55, 223
– aPKCζ 223
– aPKCι/λ 223
– Par3/Par6/aPKC complex 222f.
– PKCα 16, 93f., 117
– PKCδ 117
– PKCε 93
protein tyrosine kinase 2 (PYK2) 51ff., 78f.
protein tyrosine phosphatase (PTP) 30
– PTPμ 179
– PTP-PEST 31f., 51
proteolysis
– extracellular 275
pyloric atresia junctional epidermolysis bullosa (PA-JEB) 118

r

R-cadherin, see cadherin
Rab 183
– GTPase family 92
– Rab4 92ff.
– Rab5 93, 257
– Rab7 257
– Rab8 100
– Rab11 92ff.
– Rab21 93f.
Rac 55, 75ff., 143, 183, 262, 279ff.
– 1 161, 252, 282
– 2 79
– GEF 183, 284
Ran 183
Rap1 14ff., 186, 284
– GTPase 186
Rap1-GTP interacting adaptor molecule (RIAM) 14ff.
RapL/NORE1B 14
Ras 125, 178ff., 283
receptor-interacting protein (RIP) 36
receptor tyrosine kinase (RTK) 262ff.
reconstituted basement membrane (rBM) 136ff.
Rho 141ff., 183, 262, 279ff.
– GEF-H1 225
– kinase (ROCK) 31ff., 55ff., 141ff.
– p190RhoGAP 161
Rho GTPase 48, 73ff., 161, 183, 265
– regulation 261

RhoA 55ff., 161
– GTPase activating protein (GAP) 57
Ron 118
rosette 77

s

S-phase kinase-associated protein 2 (skp2) 37
scaffolding protein 221
Sec 6/8 complex 254
serine/threonine phosphorylation 182
SH2 domain containing protein 28ff.
SH3 domain 221
SHP-2 73ff., 125, 179
signal transduction pathway
– intracellular 281
signalling 202
– endothelial adherens junction 169ff.
signalling molecule 79
skin disease 118
Smad2 125
Smad3 51, 125
SNARE 254
sorting
– nexin 1 (SNX1) 263
– polarized 252
Spa1 284
specificity-determining loop (SDL) 113
Src 31ff., 56, 73ff., 202, 257
– C-terminal Src kinase (Csk) 31ff., 51, 179
stratifying epithelia 225
striate palmoplantar keratoderma (SPPK) 237ff.
syndesmos 51

t

T cell factor (Tcf)/lymphoid enhancer factor (Lef) 157, 176, 225, 276ff.
TAC 252
talin 9ff., 75, 143
targeting patch 265
tetraspanin 109
TIAM1 223, 284f.
tight junction 217ff., 235
– barrier function 222
– disease 227
– epithelial barrier function 219
– epithelial polarity 222
– formation 223
– internalization 228
– signalling 225
– stratifying epithelia 225
– ultrastructure 218
tissue remodelling 47ff.

trafficking
– endocytic 265
– exocytic 265
– post-Golgi 265
trans Golgi network (TGN) 199, 252
transferring receptor 94
transforming growth factor (TGF) β 49ff., 125, 184, 224
transmembrane component 219
transmembrane propagation 7
tricellulin 220
tuba 221
tumor
– cell-cell junction 178
tumor necrosis factor (TNF)
– TNFα 80, 178ff.
– receptor-associated factor (TRAF) 51
tyrosine activation motif (TAM) 117
tyrosine kinase 282
tyrosine phosphorylation 76, 178f.

u

ubiquitination 156

v

vascular endothelial growth factor (VEGF) 125, 202
– receptor 2 (VEGFR2) 171, 264
Vav2 183, 284
VE-cadherin, see cadherin
VE-PTP 182
vessel inflammation 206
vitronectin 98

w

Wiskott-Aldrich syndrome protein (WASP) 77ff.
– interacting protein (WIP) 78
– neural (N-WASP) 78
Wnt 156
– signalling 156ff., 241
wound healing 123

y

Yes 161

z

Zeste-white3 157
ZONAB (Zo-1 associated nucleic acid binding protein) 225
zonula occludens (ZO) 218ff., 235
– ZO-1 52, 221ff.
– ZO-2 201, 221ff.
– ZO-3 224